Legal AI: Optimizing Firm Workflows

Vas Edelen

Contents

Introduction

As we stand at the intersection of technology and legal practice, the landscape before us is one of profound transformation. The rise of artificial intelligence (AI) presents law firms and legal professionals with a remarkable opportunity to redefine their operations, enhance client satisfaction, and influence case outcomes more positively than ever before. "Legal AI: Optimizing Firm Workflows" is meticulously crafted to act as a pivotal guide for this transformation, enabling legal experts, from lawyers to paralegals, to traverse the new terrain of legal AI. This book delves deep into the vast potential of AI to revamp the essence of legal work—spanning research, document review, litigation support, and contract management. The intent is unequivocal: to empower legal practitioners to not just survive but flourish in the changing market, using AI as a catalyst for enhanced efficiency, precision, and extraordinary client service.

AI as a Keystone in Modern Legal Frameworks

In an era marked by the swift and complex evolution of AI, the legal profession is poised to undertake a monumental shift in practice. AI stands as a transformative force, propelling legal professionals towards heightened operational efficiency and improved case outcomes. It signifies not just the automation of tasks but the reinvention of the legal workflow, demanding acute comprehension of AI's capabilities and its judicious application.

The initial trepidation within the legal sector towards AI is being eclipsed by the acknowledgment of its considerable advantages. Embracing AI, however, comes with its own set of complexities, from data privacy to ethical considerations and job security. These challenges call for thoughtful navigation to ensure that AI's integration upholds the caliber and integrity of legal work.

AI's influence in law transcends productivity—it heralds a new epoch of insight, offering informed decisions and predictive foresight that can reshape legal strategies. Yet, integrating AI successfully necessitates more than technology—it requires a paradigm shift in the culture of legal organizations towards perpetual learning and evolution. Legal professionals are tasked with acquiring new competencies and ethical paradigms to leverage AI responsibly.

Furthermore, AI has the potential to democratize legal services, making advice more accessible and promoting equity within the legal system. Nevertheless, the profession is at a crucial juncture, and the path ahead is strewn with both hurdles and prospects. Legal practitioners must recognize that AI, while not a cure-all, is a potent tool that can substantially elevate legal practice.

This book unfolds as a comprehensive guide, steering legal professionals to engage with AI proactively. It endows readers with the necessary knowledge to exploit AI's full potential, ensuring their role as drivers, not mere observers, in this technological shift.

Embracing AI in legal practice is more than about operational efficiency or cost reduction; it is about fundamentally enriching the practice of law, ensuring justice is dispensed most effectively and equitably. As we pave the way for AI's role in law, we must proceed with both caution and hope, aiming to use technology to reinforce the legal profession's highest ideals. We envision a future where law practice is not just technologically advanced but also more streamlined and efficient, with this book serving as a blueprint to achieve that reality.

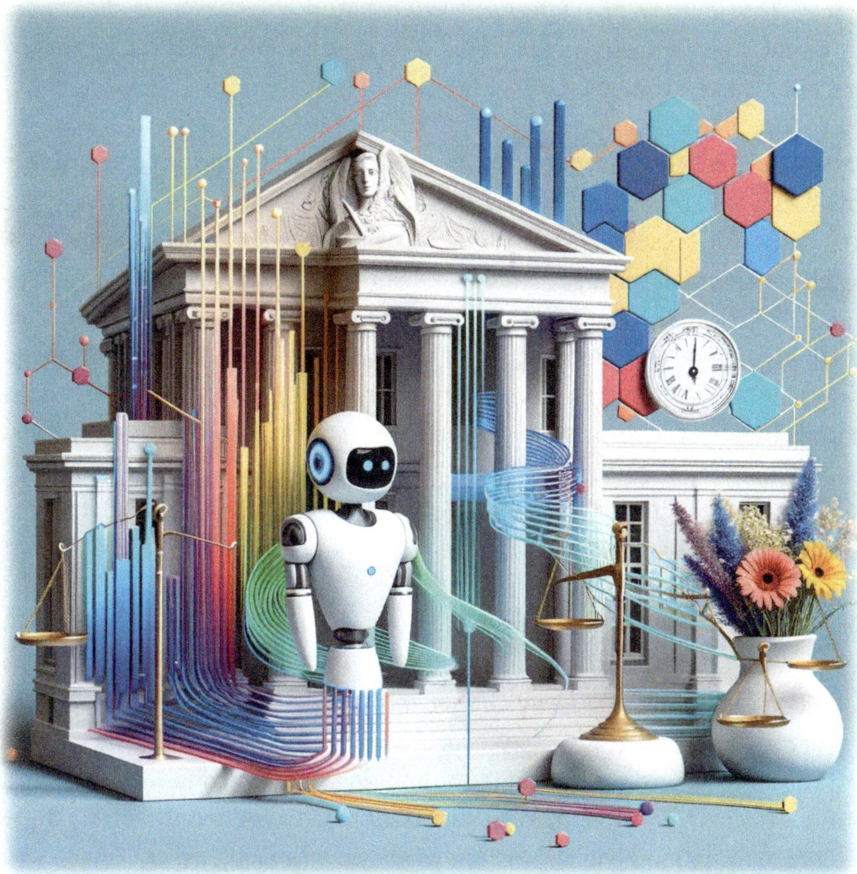

Chapter 1:

AI: Transforming the Legal Landscape

The onset of artificial intelligence (AI) in the legal domain signals a pivotal reimagining of traditional legal processes and client relations. The dawn of AI in legal practices heralds a significant transformation, one that promises to redefine the traditional paradigms of legal work, client engagement, and the very fabric of legal reasoning and processes. As legal professionals, our venture into this new era is not merely about adopting technology; it's about embracing an ally that augments our capabilities, refines our efficiency, and elevates the quality of service we offer to our clients. The journey of AI from a conceptual novelty to a practical asset within the legal industry underscores the rapidly evolving landscape where technology and law converge.

Understanding the concepts of AI, machine learning, and deep learning is paramount for legal professionals aiming to harness these technologies effectively. At its core, AI in legal practices is not about replacing the lawyer but enhancing their expertise. Machine learning algorithms, powered by vast legal datasets, can uncover insights, predict outcomes, and streamline tasks that traditionally consumed hours of manual labor. This ability to process and analyze data at unprecedented scales and speeds enables legal practitioners to focus on strategy and client counseling, tasks that demand human judgment and creativity. The advent of AI in legal practices represents a paradigm shift, moving from mere data management to deep, analytical, and predictive capabilities.

The disruption brought about by AI in conventional legal workflows cannot be overstated. From document review to legal research, contract analysis, and litigation support, AI technologies offer a new lens through which legal work can be viewed and executed. The transition towards AI-enhanced workflows demands a reevaluation of traditional roles and tasks within legal firms. Embracing AI requires not just technological adoption but a cultural shift within organizations, fostering an environment where innovation thrives. As we stand at the dawn of AI in legal practices, the potential for AI to improve operational efficiency, client satisfaction, and case outcomes is immense. The journey of integrating AI into legal practices is challenging yet rewarding, paving the way for a more dynamic, efficient, and insightful legal profession.

The Emergence of AI in Legal Operations

The integration of Artificial Intelligence (AI) into the legal profession marks a pivotal transformation in both the execution and perception of legal work. This chapter invites the reader on an exploratory journey through AI's incipient entry into legal services, its present landscape, and its potential future as a transformative force. For legal professionals, comprehending this evolution is imperative—not only to maintain relevance but to leverage the manifold opportunities AI presents.

AI's narrative within legal services is a chronicle of incremental triumph, steeped in computational experiments dating back to the mid-20th century. Yet, it was at the dawn of the 21st century that AI technologies started making indelible impacts. These developments, in conjunction with the proliferation of data and advances in computational abilities, propelled AI into a practicable instrument for legal utilities.

Initially, law firms and legal departments greeted AI with a blend of curiosity and caution, perceiving it as a harbinger of disruption to traditional legal roles. This sentiment evolved as the tangible benefits of AI crystallized. AI's proficiency in dissecting and analyzing copious legal documents in mere fractions of the time required by humans has become a linchpin in legal operations.

Pioneering the adoption of AI in legal research, Luminance Technologies has played a pivotal role in supplanting conventional research methodologies with AI-enhanced efficiency and precision. Legal professionals have embraced these AI algorithms, which have revolutionized the search process through case law and statutes, resulting in more thorough and expedited research outcomes. The ripple effects are considerable: heightened procedural efficiency and the bolstering of legal advice and decisions.

Document review and management, given the document-intensive nature of the legal field, stands as another realm revolutionized by AI. Automating the organization, identification, and categorization of documents, AI has honed critical processes, mitigated error margins, and ensured that vital information is neither overlooked nor misjudged.

In litigation support, predictive analytics—a facet of AI—has garnered a foothold by drawing on historical data to forecast case outcomes with remarkable precision. This foresight not only informs strategic planning but also equips legal professionals with a vantage point, enhancing their ability to advise clients more effectively.

AI's expansion within legal services continues unabated, with contract analysis, compliance monitoring, and risk management emerging as the latest frontiers being conquered by AI. These tools deftly pinpoint risks and compliance matters much faster than human oversight could, advocating for proactive legal strategies over-reactive responses.

The surge of AI-powered chatbots and virtual assistants has also been transformative in client service domains within the legal sector. These tools offer instantaneous responses to client inquiries, schedule appointments, and can draft rudimentary legal documents, elevating client satisfaction while liberating legal experts to focus on more intricate matters.

However, the amalgamation of AI in legal services has faced its share of hurdles. The concerns of data privacy, ethical considerations, and the propensity for bias within AI systems are pivotal issues confronting the legal community. Moreover, a discernible divide in AI literacy among legal practitioners punctuates the need for exhaustive education and training initiatives.

Financial considerations are pivotal in the adoption of AI within legal services. The infusion of AI necessitates significant investments in technological infrastructure and specialized talent. This financial barrier is particularly formidable for small firms and solo practitioners, potentially exacerbating the disparities between different echelons of legal service providers.

Looking forward, the prospects of AI in legal services are undeniably optimistic. As AI technology grows more sophisticated and its applications more pervasive, its integration into legal practice is expected to escalate. The legal profession is concurrently undergoing a cultural metamorphosis, with AI literacy becoming a fundamental competency. Academic legal institutions are starting to integrate AI into their curricula, equipping the forthcoming cadre of lawyers for a future where AI is entwined with legal practice. This paradigm shift transcends the mere adoption of new technologies; it signifies an embracement of innovation and efficiency.

In summation, AI's ascendance in legal services is a hallmark of one of the legal profession's most profound evolutions. Poised on the brink of this new epoch, it is incumbent upon legal professionals to not only grasp the capabilities of AI but also to engage with it actively. The ultimate objective is to deploy AI to amplify operational efficiency, enrich client experiences, and positively influence case outcomes. For those prepared to embrace this change, the prospects are boundless.

As we chart the course of AI's integration into legal practices, an intricate understanding of both the prospects and the challenges is indispensable. The odyssey of AI in the legal realm is an ongoing saga, unfolding with each technological advancement and each success story. Looking ahead, AI's role in legal services is set for further expansion, reshaping the landscape of legal practice in ways that are only beginning to unfold.

Decoding AI, Machine Learning, and Deep Learning in Law

In the legal sector's evolution, AI, machine learning, and deep learning have become key terms, each representing a specific aspect of computational intelligence. For legal professionals, understanding these concepts is crucial for recognizing the transformative impact of technology in their field. AI, or Artificial Intelligence, is the broad endeavor to replicate human intelligence in machines, enabling them to undertake tasks that usually require human intellect and reasoning. Companies like Casetext are at the forefront of integrating AI in legal research, showcasing how AI can enhance the efficiency and accuracy of legal document analysis.

Machine learning, a subset of AI, emphasizes machines' ability to learn from experience and improve over time without being explicitly programmed for each task. This is similar to how legal professionals enhance their skills - through repeated exposure and learning. Casetext leverages machine learning to analyze legal documents, enabling the system to learn from vast legal databases and improve its ability to identify relevant legal precedents and insights.

Deep learning, a more complex subset of machine learning, mimics the human brain's neural networks, allowing for the analysis of large and intricate datasets. In the legal field, Casetext uses deep learning for natural language processing, enhancing the system's ability to understand and analyze legal language, making it possible to sift through extensive legal documents efficiently.

The legal profession's cautious approach to adopting new technologies is understandable, given its foundation in precedent and tradition. However, the parallels between legal analysis and AI's capabilities are striking. Both processes require the analysis of large volumes of information to identify relevant patterns or precedents, and AI can perform these tasks with unparalleled speed and scale.

Integrating AI into legal practices, as demonstrated by Casetext, does not replace human lawyers but instead supports them. AI excels in processing large amounts of data, which complements the nuanced understanding, ethical considerations, and persuasive argumentation that human lawyers provide. This synergy allows legal professionals to focus on more strategic and intellectually rewarding aspects of their work.

Automation through AI in legal workflows, exemplified by Casetext's document sorting and precedent retrieval capabilities, enhances efficiency, and reduces the chance of human error, resulting in a more agile and reliable practice.

The misconception that AI might replace legal professionals overlooks the fact that AI is a tool designed to augment, not replace, the human aspect of legal work. It facilitates faster and more accurate execution of tasks, enabling lawyers to delve deeper into the analytical and creative aspects of their jobs.

The training of machine learning models, crucial for their effectiveness, depends on the quality of data they receive. Legal professionals, therefore, play a vital role in guiding these AI systems, as seen with Casetext, ensuring they operate within legal and ethical boundaries and that the 'human touch' in interpreting the law is maintained.

However, integrating AI like Casetext into legal practices comes with challenges, including concerns about data privacy, security, and the need for ethical decision-making to prevent biases in AI-assisted outcomes.

The deep learning capabilities of Casetext, especially in natural language processing, have revolutionized client interactions and legal document analysis by automating the extraction of relevant information and streamlining communication, thereby improving service quality and client satisfaction.

The shift towards AI in legal research, led by platforms like Casetext, represents a significant change from traditional methods. AI-powered tools can rapidly identify relevant case laws and statutes, marking a leap in efficiency and resource management.

Furthermore, the predictive analytics of AI, used by Casetext, offer insights into potential legal outcomes based on historical data analysis, aiding in litigation and negotiation strategy formulation.

The adoption of AI in legal practices signifies a shift towards continuous learning and adaptation, with Casetext being a prime example of how legal professionals must evolve with the technology. The legal industry is on the verge of a technological revolution that promises not just automation but a transformation of legal service delivery.

In conclusion, the integration of AI, machine learning, and deep learning in the legal profession, as exemplified by Casetext, illustrates the potential for these technologies to empower legal professionals. With careful implementation and a commitment to ethical standards, AI can complement and elevate traditional legal practices, ushering in a new era of efficiency and innovation in the legal domain.

How AI is Reshaping Legal Operations

The incursion of AI into the legal profession signifies more than just a technological upgrade; it beckons a profound transformation in how legal operations have traditionally been conducted. For decades, lawyers, paralegals, and legal experts have adhered to workflows deeply rooted in precedent, relying heavily on manual research, document review, and one-on-one consultations. The landscape, however, is shifting inexorably towards an era where these conventional practices are being challenged and redefined by artificial intelligence, notably by platforms like Blue J Legal and LegalSifter.

Understanding AI's disruption requires a granular examination of its impact across various facets of legal work. Blue J Legal's predictive analytics, for instance, is revolutionizing legal research, the backbone of legal practice, by enabling lawyers to forecast case outcomes in tax and employment law. This AI-powered capability sifts through vast legal databases in moments, enhancing the accuracy and relevance of the findings and enriching the legal strategies devised by practitioners.

Document review and analysis, another cornerstone of legal practice, similarly benefits from AI's intervention. Technologies like predictive coding and machine learning, as applied by LegalSifter, rapidly identify patterns and significant information within documents. This not only accelerates the review process but also increases its thoroughness and precision, enabling legal professionals to extract vital insights from data that might have been overlooked manually.

AI is also transforming contract analysis and management. Tools provided by LegalSifter automate the review and drafting of contracts, identify clauses that deviate from standards, and suggest modifications. This streamlines workflow, mitigates risks associated with human error, and stands as a testament to AI's practical utility in day-to-day legal operations.

Moreover, the function of legal advice itself is being transformed with the advent of AI chatbots and virtual assistants. Blue J Legal and LegalSifter contribute to this field by offering preliminary legal guidance and drafting capabilities. While not a substitute for in-depth legal counsel, they offer a cost-effective and efficient first line of assistance, enhancing client service and satisfaction.

In litigation support and prediction, Blue J Legal's tools offer probabilistic assessments of case outcomes, helping legal teams to strategize more effectively. Though this application is in its nascent stages, the implications are profound, challenging traditional approaches to case preparation and strategy formulation.

On the operational side, AI's potential for revolutionizing legal billing, timekeeping, and practice management is becoming more evident. Automated systems, like those possibly integrated into platforms such as Blue J Legal and LegalSifter in the future, can track billable hours with greater accuracy, predict case costs, and manage client accounts, thereby improving transparency and trust.

Compliance and risk management, areas ripe for AI-induced innovation, are also benefiting from the predictive analytics offered by Blue J Legal. Law firms can better anticipate potential legal risks and compliance issues, allowing for proactive management strategies.

While assessing AI's disruption, the challenges and concerns it raises, such as ethical implications, the potential for job displacement, and the need for robust data privacy and security measures, must be acknowledged. Moreover, AI's integration demands a re-skilling of the workforce, with lawyers and legal professionals needing to acquaint themselves with these new technologies.

Resistance to change is another significant barrier. The legal profession, known for its conservatism, may view AI's ascent with skepticism. Overcoming this requires demonstrating AI's value through tangible improvements in efficiency, client satisfaction, and case outcomes—benefits that Blue J Legal and LegalSifter exemplify.

Despite these challenges, AI offers the promise of enhanced efficiency, deeper insights, and the democratization of legal services. It allows lawyers to focus on higher-value activities, such as strategizing and client counseling, elevating their role in their client's lives.

The transformation AI promises in legal workflows is not merely incremental but revolutionary. Its successful integration demands a strategic approach that includes the adoption of AI technologies like Blue J Legal and LegalSifter and cultural and organizational adjustments within legal practices.

In conclusion, AI's disruption of conventional legal workflows heralds a new era in legal practice. Navigating this new landscape requires careful consideration of the ethical, operational, and cultural challenges it presents. For forward-thinking legal practitioners, the journey towards AI integration offers an unparalleled opportunity to redefine the practice of law in the 21st century.

As we continue to delve deeper into the implications of AI for legal practices in subsequent chapters, the role of innovators like Blue J Legal and LegalSifter becomes increasingly significant. The true extent of AI's impact remains to be seen, but its capacity to transform the legal profession is undeniable and already underway.

Embracing the AI Renaissance in Legal Practices

As we stand at the cusp of a transformative era in legal services, the integration of Artificial Intelligence (AI) represents one of the most significant shifts in the profession's history. This concluding section reflects on the broader implications of AI's integration, as exemplified by pioneering entities like Luminance Technologies, Blue J Legal, Casetext, and LegalSifter. These companies not only exemplify the practical utility of AI in law but also herald a new paradigm for legal practices.

Luminance Technologies, by revolutionizing legal research through AI, has set a precedent for efficiency and accuracy, enabling lawyers to navigate the vast sea of legal precedents and legislation with unprecedented ease. Blue J Legal's predictive analytics have redefined the landscape of tax and employment law, offering practitioners foresight and strategic depth in their legal proceedings. Similarly, Casetext has demonstrated the profound impact of AI in legal document analysis, providing legal professionals with tools that enhance their research and case preparation capabilities. LegalSifter's role in streamlining contract analysis and management further illustrates AI's capacity to refine and expedite legal workflows.

The journey of AI in legal services, from its nascent stages to its current prominence, underscores a broader narrative of technological evolution and cultural adaptation within the legal profession. The adoption of AI technologies has shifted from a cautious exploration to an essential strategy for enhancing operational efficiency, improving client satisfaction, and ensuring more accurate and effective legal outcomes.

However, this journey is not without its challenges. Issues of data privacy, ethical considerations, and the potential for algorithmic bias remain at the forefront of discussions about AI in law. Moreover, the financial investment required to integrate AI technologies poses a significant barrier, particularly for smaller firms and solo practitioners, highlighting the need for scalable and accessible solutions.

The future of AI in legal services is bright and filled with potential. As AI technologies continue to evolve and become more sophisticated, their integration into legal practice is expected to deepen, reshaping the profession in ways we are only beginning to comprehend. This evolution is not just about adopting new technologies but also about embracing a cultural shift towards innovation, efficiency, and a client-centered approach.

In conclusion, the AI renaissance in legal practices is more than a technological revolution; it is a fundamental reimagining of how legal services are delivered and consumed. The insights and innovations brought forth by companies like Luminance Technologies, Blue J Legal, Casetext, and LegalSifter serve as beacons of this transformative journey. As legal professionals, our task is to navigate this new paradigm with foresight, adaptability, and a commitment to leveraging AI's potential to enhance the practice of law, thereby ensuring justice and legal services are more accessible, efficient, and equitable in the 21st century.

Chapter 2:

Innovating Legal Practice: The AI Research Transformation

The introduction of artificial intelligence (AI) into legal practices has earmarked a new era, most notably in the domain of legal research. Traditionally, legal research was a labor-intensive process, often daunting even for the most seasoned of professionals. It involved sifting through countless volumes of case law, statutes, and legal journals, which not only demanded significant time but also a high level of attention to detail to ensure no critical piece of information was overlooked. However, the advent of AI in legal research has profoundly transformed this scenario by streamlining research processes, thus, amplifying efficiency while minimizing human error. AI's ability to process and analyze large datasets rapidly, coupled with its capability to learn and improve over time, has ushered in a more predictive and efficient method of conducting legal research.

One of the most remarkable advancements brought about by AI in legal research is its ability to provide legal professionals with highly relevant information at an unprecedented speed. This is achieved by using sophisticated algorithms that can quickly sift through data, recognize patterns, and render predictions or suggestions based on the input it receives. This high level of precision and efficiency in retrieving and analyzing legal documents has not only saved valuable time but has also elevated the quality of legal research output. For legal teams, this revolution means that they can now tackle more complex cases with a higher degree of confidence, knowing that the insights derived from AI-powered legal research tools are both comprehensive and reliable.

Moreover, the integration of AI into legal research teams has not only optimized the research process but also facilitated a more collaborative and strategic approach to legal inquiries. AI legal research tools are designed to seamlessly integrate with existing legal databases and software, ensuring that legal teams can easily adapt to and adopt these sophisticated technologies. This smooth integration empowers legal professionals to focus more on analytical and strategic aspects of cases rather than spending excessive time on preliminary research. As a result, AI in legal research is setting a new standard, fostering a culture of innovation and efficiency that is paramount for staying ahead in the fast-paced legal industry. In embracing AI, legal practices are not merely adopting a new tool but are revolutionizing the very fabric of legal research processes for the better.

Harnessing AI: A Guide to Enhanced Legal Research

In the dynamically shifting terrain of legal technology, artificial intelligence (AI) stands as a formidable agent of change, reshaping the domain of legal research. The erstwhile era, characterized by protracted manual searches and analyses, is yielding to an era marked by AI-facilitated processes. This discourse aims to elucidate the transformative influence of AI on legal research, offering legal practitioners — lawyers, paralegals, and legal scholars — a detailed framework to utilize this avant-garde technology with adeptness.

The advent of AI in legal research heralds an epochal transition from age-old practices. Legal practitioners employing AI instruments benefit from enhanced search capabilities, foresight through predictive analytics, and semantic comprehension, all of which cumulatively amplify research efficacy and precision. This quantum leap in technological prowess transcends mere convenience to become a strategic instrument with substantial repercussions on case results and client contentment.

Proficiency in AI's functionalities is crucial for its fruitful assimilation into legal research. AI's machine learning algorithms are capable of parsing through copious amounts of legal data with alacrity. Crucially, AI is designed to refine its search criteria and results continuously, using feedback and outcomes to guide its evolution. This adaptive quality renders AI indispensable for keeping pace with the ever-changing legal landscape and jurisprudence.

Nevertheless, mastering AI in legal research demands a considered strategy. Law firms must evaluate their unique requirements and aim to pinpoint AI applications that resonate with their research endeavors. Such a bespoke approach guarantees that the AI solutions chosen are integrated into current workflows with minimal disruption, thus enhancing rather than interrupting the research continuum.

A linchpin of AI in legal research is its prescient capacity. By scrutinizing historical data and patterns, AI can conjecture likely judicial verdicts, endowing legal professionals with foresight that is crucial during case preparation. This predictive prowess, when judiciously employed, can underpin more tactically sound decisions and fortify the arguments presented in court.

Additionally, AI propels the accessibility of legal information. Leveraging natural language processing (NLP), AI systems can decode and construe legal vernacular, enabling legal practitioners to execute comprehensive research unencumbered by the intricacies of legal terminology. This facet not only streamlines research but also democratizes it, equalizing opportunities for smaller entities and individual advocates.

The deployment of AI in legal research also translates into substantial time and cost economies. Automating the most labor-intensive segments of research allows legal professionals to divert their resources towards more strategic pursuits, like client engagement and case strategy. Such enhanced efficiency can lead to more competitive pricing structures, which can improve client satisfaction and loyalty.

Education and skill development are paramount to the successful incorporation of AI in legal research. Legal teams must acquire the skills necessary to maneuver AI tools adroitly. Continuous training initiatives ensure that legal professionals are not only proficient in utilizing AI but are also conversant with the most current developments and practices in AI-assisted legal research.

Moreover, the ethical implications of deploying AI in legal research cannot be marginalized. Legal practitioners must maintain a fine equilibrium between harnessing AI's extensive benefits and ensuring adherence to the loftiest standards of legal ethics, including data privacy, transparency of AI methodologies, and the prevention of bias in AI-derived conclusions.

The prospect of cooperative AI in legal research introduces novel opportunities. By promoting collaborative platforms, AI can facilitate the secure exchange of insights and data amongst legal professionals, spurring collective enhancements in legal research methods. This collaborative modality accelerates and enriches the research process with varied perspectives and expertise.

Despite the manifest advantages, transitioning to AI-centered legal research is accompanied by challenges. Hesitance towards change, apprehensions about technological encroachment on employment, and initial investments in AI tools constitute formidable barriers. However, with a focused stratagem that accentuates AI's role in supplementing human acumen, legal establishments can surmount these obstacles efficaciously.

Peering into the future, the prospects for legal research augmented by AI are vibrant. As AI technologies advance, we can envisage tools of even greater sophistication, offering more profound insights, enhanced predictive precision, and fluid incorporation into legal procedures. Legal professionals who excel in applying AI to research will indubitably spearhead this new vanguard in legal practice.

For those commencing this technological odyssey, mastering AI for streamlined legal research is a journey marked by perpetual learning, adaptation, and strategic application. It requires embracement of innovation, dedication to ethical standards, and foresight for a future where AI and human expertise synergize to propel the legal field forward.

LexisNexis Context into legal research workflows exemplifies the next phase in this evolution. LexisNexis Context leverages advanced machine learning and natural language processing to mine tens of millions of case law documents, delivering critical insights into litigation. This powerful tool is emblematic of the AI prowess discussed earlier — analyzing language patterns and judicial decisions to aid in the formidable task of legal research.

This integration of LexisNexis Context aligns seamlessly with the overarching theme of harnessing AI for improved legal research efficiency. By embracing this sophisticated application, legal practitioners can elevate their research capabilities, discerning pertinent information with greater acuity. LexisNexis Context stands as a testament to the potential AI holds in transforming data into actionable intelligence.

In sum, the journey towards mastering AI for streamlined legal research is both an individual and collective endeavor. It involves engaging with pioneering tools like LexisNexis Context, which serves as both a beacon and a benchmark for the industry. Through the judicious application of such advanced AI tools, the legal profession is poised to achieve a new zenith of research excellence, redefining the practice of law in the AI age.

Leading the Pack: In-Depth Look at AI-Assisted Legal Research Platform

The incorporation of artificial intelligence (AI) into legal research tools marks a seminal evolution in the legal industry, revolutionizing the way legal research is conducted. This section is devoted to offering legal professionals, from the most seasoned attorneys to entry-level paralegals, an exhaustive evaluation of the leading AI legal research tools reshaping the industry. Our aim is to guide you through a comprehensive analysis of these advanced platforms, assessing their impacts on enhancing the efficiency and efficacy of legal research processes.

The impetus for adopting AI within legal research tools stems from the capabilities of AI to deliver rapid, accurate, and insightful analysis of extensive legal databases, which could be otherwise overwhelming or time-consuming for human researchers. A pivotal aspect of our scrutiny of these AI tools is the recognition that the optimal legal research tool for practice isn't necessarily the most ubiquitous one but the one that most closely aligns with the particular research needs and types of cases that the firm typically handles.

Among the vanguard of AI legal research instruments is Westlaw Edge, which exemplifies the transformative power of AI within this domain. Westlaw Edge, with its robust AI engine, is meticulously engineered to refine the legal research process while ensuring accuracy and delivering actionable insights. This platform enables legal professionals to swiftly provide clients with precise and substantiated answers. As an AI-driven legal research tool that employs predictive analytics, Westlaw Edge stands as a testament to how AI can significantly enhance the capabilities of legal researchers.

Continuing this narrative, Luminance Technologies emerges as a pioneering force, leveraging the cognitive computing power of IBM's Watson. This innovative platform interprets natural language queries with a remarkable depth of understanding, returning responses that not only highlight relevant case law but also provide a nuanced understanding of how these cases align contextually with the queries posed. Luminance Technologies epitomizes how machine learning and natural language processing can penetrate the complex linguistic fabric and intricacies inherent in legal texts.

In the same vein, LexisNexis has introduced Lex Machina to the arena, a platform that represents a quantum leap towards a more data-centric approach in legal research. Lex Machina empowers users to anticipate litigation outcomes, comprehend judicial rulings, and gain insights into the litigation history of the parties involved. Through legal analytics, AI provides a strategic edge, enhancing legal strategies and increasing the likelihood of securing favorable outcomes.

Moreover, Casetext's CARA AI has caused a paradigm shift in traditional legal research methods. Attorneys can now upload briefs, memoranda, or other legal documents to CARA AI, which in turn identifies and presents the most relevant case laws, statutes, and regulatory materials that bear directly upon the issues addressed in the documents. This feature signifies AI's profound ability to tailor research outputs to the precise context of the document, thereby augmenting the relevance and practical utility of research findings.

The ability of these diverse AI legal research tools to learn from user interactions is an attribute that cannot be overstated. This adaptive learning capability ensures that these platforms become more adept and insightful with each use, aligning themselves with the nuanced requirements of legal research and the specific preferences of their users. For law firms, this means that the investment in AI legal research tools is not just about yielding immediate enhancements in research productivity but is also about securing sustained improvements in the firm's research competencies over the long term.

While we extol the virtues of AI in legal research, it is equally imperative to confront the challenges and ethical concerns that accompany the employment of AI. Issues such as data privacy, the transparency of algorithms, and the potential biases ingrained within AI systems are at the center of contemporary debates. Legal professionals must engage in a thorough examination of the strategies AI tool providers employ to mitigate these issues. Solid ethical guidelines and rigorous data protection policies are hallmarks of AI legal research tools that are both dependable and trustworthy.

Further to these considerations is the financial aspect of integrating AI into legal research. A comprehensive cost-benefit analysis should extend beyond mere subscription fees to factor in AI's potential to diminish billable hours spent on research, which should ultimately result in cost reductions for clients and competitive pricing models for legal services, thus conferring a market advantage to the law firm.

The process of integrating AI legal research tools into existing workflows, with an emphasis on ease of adoption and requiring minimal disruption, is another vital consideration. The most efficacious AI legal research tools are those that provide user-friendly interfaces, extensive support, and ample training resources to ensure a smooth transition for legal teams, regardless of their existing technological proficiency. The seamless integration of AI tools with current legal research methodologies and databases maximizes their utility and facilitates the widespread adoption of AI technologies in the routine operations of a law firm.

Shifting client expectations also play a crucial role in this narrative. Today's clients are increasingly demanding faster turnaround times, more accurate case analyses, and predictive insights regarding case outcomes. Law firms that are equipped with advanced AI research tools are better positioned to satisfy these elevated expectations, which in turn enhances client satisfaction and promotes client loyalty. This dynamic emphasizes the importance of selecting an AI tool that not only meets the firm's research needs but also contributes to elevating the client service experience.

The process of evaluating AI legal research tools must also include a consideration of the software's ability to integrate with other legal tech tools and platforms. This interoperability enhances the value of AI research tools by enabling a more streamlined and efficient legal tech ecosystem within a firm. The capacity for these tools to work in concert with e-discovery platforms, document management systems, and legal practice management software can significantly impact a firm's overall operational efficiency.

As we delve deeper into the evaluation, it is crucial to seek feedback from current users of these AI tools. Testimonials and case studies highlighting real-world applications and benefits can provide invaluable insights into how these tools perform under various legal research scenarios. This qualitative feedback complements the quantitative analysis of features, costs, and benefits, providing a well-rounded perspective of the tool's practicality and utility.

Additionally, the environmental impact of deploying AI in legal research is an often-overlooked consideration. The computational resources required to power advanced AI algorithms can be substantial, and legal professionals who prioritize sustainability should consider the energy efficiency of AI tools as well as the environmental policies of their providers. This aligns with the broader movement towards eco-conscious business practices in various industries, including the legal sector.

Westlaw Edge, in particular, exemplifies how AI tools can blend seamlessly into a law firm's environment. Its advanced features not only streamline research but also infuse the research process with depth and precision that was previously unattainable. Westlaw Edge's sophisticated analytics, combined with a comprehensive database of legal resources, affords users a cutting-edge advantage in legal research.

In our quest for the preeminent AI legal research tool, we must also reflect on the support and development provided by the companies behind these tools. Continuous updates, improvement in functionalities, and responsive customer support are essential elements that ensure these tools remain relevant and effective in a rapidly changing legal landscape. The development teams behind platforms like Westlaw Edge, Luminance Technologies, Lex Machina, and CARA AI are dedicated to refining their products, which translates to better performance and more robust features for the end-users.

Moreover, the integration of AI tools must also be assessed in terms of data security. With law firms being prime targets for cyber threats, the security features of AI legal research tools become a paramount consideration. Encryption, secure access, and regular security audits are features that law firms should look for when choosing an AI legal research tool.

Considering the competitive nature of the legal industry, AI legal research tools also play a critical role in firm branding and differentiation. A law firm that utilizes sophisticated AI tools effectively communicates a message of innovation and forward-thinking to potential clients and partners. This not only helps attract business but also recruits top talent who are eager to work at the forefront of legal technology.

Training is another critical aspect of integrating AI into a law firm's operations. The best AI legal research tools provide comprehensive training modules that are tailored to different learning styles and competencies within a firm. This ensures that all members of the firm, irrespective of their role, can leverage the benefits of AI in their day-to-day work.

Client confidentiality is another dimension that cannot be neglected. AI legal research tools must be designed with strict protocols to ensure that sensitive client information is not compromised. This involves implementing robust access controls and regular monitoring to detect any unauthorized access or data breaches.

Finally, the true value of AI legal research tools is realized when they are embraced as part of a larger strategic vision. Firms must approach AI not as a mere addition to their existing tools but as a key component of their strategic planning. This involves aligning AI tools with the firm's goals for growth, service excellence, and innovation. The integration of AI into legal research is not an endpoint but a starting point for ongoing development and refinement.

In conclusion, identifying the leading AI legal research tool for a law firm is a multifaceted endeavor, entailing a balanced appraisal of technological features, ethical frameworks, financial considerations, integration simplicity, and strategic alignment with the firm's long-term objectives. It demands rigorous due diligence, a forward-looking approach, and a recognition of the evolving nature of legal research. The AI revolution in legal research is not merely about adopting new tools; it signifies a commitment to a new paradigm—one that recognizes AI's potential to elevate legal research from a labor-intensive task to a strategic advantage. With thoughtful evaluation and strategic deployment, AI legal research tools hold the promise of inaugurating a new era of efficiency, precision, and innovation in the legal field. The revolution is well underway, inviting legal professionals to seize the transformative power of AI to reshape the landscape of legal research for years to come.

Optimizing Legal Teams with Artificial Intelligence Integration

The advent of Artificial Intelligence (AI) within legal research teams heralds a transformative leap towards enhanced efficiency, accuracy, and depth in legal research methodologies. This pivotal shift transcends the mere adoption of innovative technology; it is a reinvention of the modus operandi of legal research. For law firms' intent on maintaining their edge in a competitive arena, the seamless integration of AI into their research apparatus is essential.

A considered, strategic assessment of the firm's existing resources and methodologies is the precursor to AI adoption. Identifying key areas ripe for AI's value addition is paramount. This could range from the automation of mundane tasks, and refinement of document analysis, to the application of predictive analytics for prognosticating legal verdicts. By pinpointing such potential, firms can customize AI integration to align with their unique objectives.

At the core of AI integration lies the necessity for education and training. All legal practitioners, including experienced lawyers and paralegals, must be versed in AI's capabilities and practicality within their professional sphere. This knowledge-sharing not only unravels the complexities of AI but also empowers the team to incorporate AI utilities naturally, thus enhancing their proficiency and the caliber of their research output.

Selecting the appropriate AI technology is pivotal. The market abounds with various legal AI instruments, each with distinct advantages and specializations. It is imperative for firms to undertake comprehensive evaluations to choose tools that resonate with their specific research exigencies and technological framework. During this vetting process, the aspects of compatibility, scalability, and technical support should be given high priority.

Upon tool selection, integration into pre-existing workflows must be undertaken with precision and care, aiming to reduce potential disruption. Often, this phase includes tailoring the AI solution to meet the bespoke needs of the firm, ensuring it augments rather than disrupts the established practices.

Data is the cornerstone of any AI operation. To harness the full potential of AI, there must be a ready supply of well-organized, secure, and AI-accessible legal databases and documents. This could necessitate a significant initial effort to digitize and classify legacy documents, but the ensuing efficiency in research processes justifies the investment.

The symbiosis between human researchers and AI instruments is a cornerstone of maximizing AI's utility. AI shines in sifting through extensive datasets rapidly, detecting patterns, and pinpointing pertinent facts. Conversely, humans offer critical analysis, contextual understanding, and strategic foresight. This collaboration unveils an unprecedented capacity to tackle intricate research tasks with elevated efficacy.

An ongoing cycle of monitoring and assessment is integral to AI integration. AI systems are inherently dynamic, adapting and evolving with new data and user interactions. Firms must institute mechanisms for continuous review of their AI tools' effectiveness, ensuring consistent value delivery and making requisite adjustments.

Client privacy and data protection are of utmost importance in the legal domain. Consequently, any AI technologies employed must adhere to stringent security measures, ensuring the integrity of sensitive information through robust encryption, stringent access controls, and comprehensive audit trails.

The human dimension of AI adoption is not to be underestimated. Effective change management is necessary to alleviate apprehensions, shape expectations, and cultivate an innovative mindset within the firm. Leadership must proactively elucidate AI's advantages, mitigate fears regarding job redundancy, and highlight AI's role in augmenting rather than supplanting the human element in legal research.

Conclusively, the successful integration of AI into legal research teams begins with a strategic blueprint, followed by thorough education, judicious technology selection, and meticulous deployment. Data stewardship, human-AI collaboration, perpetual evaluation, steadfast data security, and adept change management are all vital elements of this paradigm. Law firms that adopt a calculated and strategic approach to AI integration can not only enhance their legal research prowess but also redefine industry benchmarks, offering their clients an unrivaled level of service.

As AI's presence in legal research burgeons, we anticipate the continued maturation of technologies such as advanced natural language processing, predictive analytics, and machine learning algorithms, further enriching the insights and strategy behind legal research. The possibilities for AI to elevate the legal sector are immense, with this journey only at its inception. Law firms that embrace AI now are setting themselves up not just for immediate success but for enduring preeminence in a future driven by data.

Thus, the mandate for legal professionals is unequivocal: adopt AI with an open mindset, a methodical framework, and a commitment to perpetual learning and adaptation. By doing so, legal research teams will not merely prosper in the AI epoch but will establish new exemplars of legal research distinction.

In the context of global legal inquiries, Jus Mundi exemplifies the worldwide scope of AI's impact on legal research. As a preeminent legal and arbitration intelligence engine, Jus Mundi offers an extensive database encompassing commercial arbitration, investment arbitration, international trade law, and public international law, including the law of the sea. With exclusive partnerships and an advanced AI that adeptly navigates complex global legal datasets, Jus Mundi emerges as an indispensable asset for firms engaged in international legal matters. The platform's sophisticated AI capabilities enable legal teams to dissect and understand the nuances of global legal frameworks, making it a critical component of any internationally oriented legal research team.

Embrace the revolution of AI in legal research with foresight and resolve. Harness tools like Jus Mundi to transcend traditional research confines, positioning your firm not just for current relevance, but as a pioneer in the legal landscape of tomorrow. The integration of AI into legal research is not a mere progression; it is a transformative force, poised to redefine the contours of the legal profession for the foreseeable future.

Legal professionals worldwide are called to action: integrate AI into your practices, embracing tools like Jus Mundi to ensure that your firm not only meets the challenges of the present but shapes the future of international legal research. By incorporating AI into legal workflows, research teams enhance their capabilities to analyze and interpret complex legal issues across jurisdictions and legal systems.

Jus Mundi, with its comprehensive coverage of international law and arbitration, including specialized areas such as commercial arbitration, investment arbitration, and the law of the sea, serves as a beacon of AI's potential in global legal research. Its AI-driven engine facilitates the efficient navigation of expansive legal datasets, empowering legal teams with rapid, accurate insights and a thorough understanding of international legal landscapes. This global perspective is essential for firms that operate across borders or handle cases that involve multiple legal frameworks.

Moreover, the platform's exclusive partnerships provide access to a wealth of knowledge and expertise, offering legal researchers a robust array of resources to support their work. The integration of such a powerful tool underscores the importance of choosing AI solutions that not only improve operational efficiency but also expand the horizons of legal research.

In the evolving world of legal services, where international matters are increasingly prevalent, the ability to quickly adapt to and adopt advanced technologies like Jus Mundi's AI becomes a key differentiator. This is the future of legal research—intuitive, interconnected, and intelligent. It demands a proactive approach, a willingness to embrace change, and a dedication to excellence.

Legal research teams that combine human intellect with AI's computational power will lead the vanguard of the profession. They will redefine excellence in legal research, setting new standards for the industry and achieving outcomes that were once thought unattainable. This is not just an evolution but a revolution in the truest sense, as AI and platforms like Jus Mundi redefine what's possible in legal research, today and in the future.

Chapter 3:

Machine Learning: Reshaping Legal Document Examination

The realm of legal document review, long considered a cornerstone of litigation and due diligence processes, has been undergoing a remarkable transformation, primarily driven by advancements in artificial intelligence and machine learning technologies. The traditional method of manually sifting through mountains of documents is not only time-consuming but also prone to human error, a realization that has paved the way for AI's integration into this critical legal function. AI's ability to rapidly analyze, filter, and identify relevant documents within vast datasets has not only expedited the document review process but has also significantly increased accuracy and efficiency, thereby reshaping the landscape of legal practices worldwide.

At this juncture, it is pertinent to introduce LegalSifter, a pioneering solution that epitomizes this transformative shift. LegalSifter's AI software and expert services seamlessly blend machine learning capabilities with legal expertise to ensure contracts no longer hinder business momentum, effectively mitigating revenue leaks and reducing the risk of signing adverse agreements. The software's prowess in enhancing precision and aligning with the objectives of legal practice represents a significant leap forward in document review processes.

This evolution is markedly evident in the domain of e-discovery, where the application of AI, particularly through predictive coding, has revolutionized how legal professionals approach document analysis. Predictive coding employs algorithms to distinguish relevant from irrelevant documents based on a set of training data, effectively "learning" from attorney responses. This shift towards an AI-driven approach not only streamlines the review process but also reduces costs and mitigates the risk of overlooking critical information. As such, the legal profession stands at the cusp of a new era, where AI's role in document review signifies a departure from traditional, labor-intensive methods towards more strategic, intelligent operations.

Furthermore, the integration of AI into document review has broader implications for the practice of law, impacting everything from the speed and accuracy of legal research to the delivery of legal services. Legal teams equipped with AI capabilities are better positioned to serve their clients' needs, adapt to the rapidly changing legal landscape, and maintain a competitive edge. This chapter delves into how AI and machine learning technologies are transforming document review, highlighting not only the operational efficiencies gained but also the strategic advantages that these technologies confer upon forward-thinking legal professionals. As we explore the nuts and bolts of employing AI in document review, it becomes clear that the legal profession is on the brink of a paradigm shift, with AI acting as both a catalyst and an indispensable ally.

E-Discovery Reimagined: AI's Strategic Advancement

In the dynamic sphere of legal practices, the ascension of electronic discovery (e-discovery) heralds a paradigm shift, critically influenced by the innovative march of artificial intelligence (AI). This exploratory section sheds light on the envisioned progression of e-discovery, bolstered by AI mechanisms, and scrutinizes its influence on document review proficiency, precision, and the extensive legal industry.

AI's integration into e-discovery transcends a transient vogue, revolutionizing legal practitioners' modus operandi in document analysis and management. This discourse elucidates that AI-catalyzed e-discovery begets unparalleled prospects for legal cohorts, endeavoring to sharpen search accuracy, accelerate the review timeline, and substantially alleviate the traditional manual encumbrance.

In foresight, the linchpin of e-discovery within legal vocations is increasingly AI's prowess. Technologies like predictive coding become the vanguard of this transformation, empowering the assessment of a modest document ensemble to guide the examination of more substantial corpuses. Predictive coding's astuteness in pinpointing pertinent documents epitomizes this transition, thereby optimizing the review trajectory.

Moreover, the emergence of AI-enriched instruments portends a future wherein identifying trends, irregularities, and interrelations amid colossal data troves becomes more intricate. This acumen not only bolsters document review precision but also reveals insights erstwhile concealed by the vastness of data.

Envisioning e-discovery's prospective panorama, the significance of continuous learning algorithms in AI applications takes center stage. These self-evolving algorithms, which mature with user interactions and progressive iterations, pledge a perpetually refining e-discovery course. Thus, document review efficiency and efficacy are foreseen to enhance perpetually, adapting to the mutable nuances of legal dossiers and data aggregates.

Yet, integrating AI within e-discovery presents its own set of quandaries, such as data privacy, security, and algorithmic bias, which mandate vigilant consideration. Tackling these issues is essential to align e-discovery's future with ethical codes and legal frameworks.

Also indispensable is the element of human oversight in an AI-driven e-discovery framework. Even as AI amplifies efficiency and precision, the discerning acumen and dexterity of legal professionals remain irreplaceable. Fostering a synergistic alliance between AI technologies and legal savants is crucial to unlock the full potential of AI in e-discovery.

Preparatory training and scholarly edification are also pivotal in equipping legal practitioners for an AI-propelled e-discovery era. Comprehending the technologies at the core, as well as their strengths and limitations, is vital for employing AI with efficacy and integrity within legal precincts.

From an economic standpoint, the AI-influenced e-discovery pivot heralds substantial cost efficiencies and resource optimization for legal practices. Diminishing the duration and human resources requisites for document analysis enables legal entities to reallocate their capabilities more judiciously, potentially leading to enhanced case resolutions and client contentment.

Looking even further ahead, the assimilation of AI in e-discovery is set to engender novel methods for tackling legal conundrums. Legal analytics, an emerging domain, employs AI to prognosticate verdicts based on historical data, providing strategic acumen that can guide legal verdicts and tactical case planning.

In this AI-augmented future, e-discovery's boundaries are envisaged to extend beyond conventional legal documents. AI is likely to become instrumental in evaluating an expanded array of data types, like emails, social media interactions, and other digital communications. This progression mirrors the evolving essence of proof in the digital era and accentuates the requisite for legal practices to adapt in kind.

The influence of AI on e-discovery further encourages the collaboration between legal brigades and technologists. This cross-disciplinary synergy is critical for formulating and effectuating AI responses that resonate with the distinct requisites and impediments of legal practices.

Additionally, the global outlook on AI-determined e-discovery accentuates the need for harmonization of international norms and practices. As legal territories confront AI's implications for privacy, data protection, and trans-border data transactions, fostering global collaboration is essential for surmounting these intricate challenges.

In sum, e-discovery's destiny in legal practices is inextricably linked with AI's advancements. Although this prospect heralds enhanced efficiency, exactitude, and acumen, it also introduces challenges that necessitate deliberate contemplation and intervention. By adopting AI-driven e-discovery with an educated, principled, and cooperative stance, legal professionals can stride into this transformative epoch with assurance, reshaping the legal practice landscape for the superior.

As we venture forth, the amalgamation of AI into e-discovery will not only reconceptualize the conventional document scrutiny processes but also signal a novel epoch in legal praxis. This progression towards an AI-infused future is as exhilarating as it is formidable, demanding comprehensive cognition, nimble adaptation, and steadfast dedication to principled practices. The potentialities are extensive, and the prospects for a positive impact on the legal profession and the justice system at large are substantial.

In the vanguard of this AI-integration stands Everlaw, a cloud-based platform that empowers legal teams to collaborate effectively in the discovery, comprehension, and actioning of information. Utilized by corporate counsel, litigators, and government attorneys for internal investigations and litigation, Everlaw's e-discovery solution harnesses AI and data visualization to illuminate the hidden facets of evidence and documents.

Positioning itself as a harbinger of predictive analytics, Everlaw embodies a forward-looking instrument, priming firms for the imminent AI-enriched e-discovery landscape. Its platform offers a vantage point from where legal professionals can anticipate the trajectory of legal cases and refine their strategies accordingly.

The implementation of Everlaw's tools in the e-discovery process exemplifies the merger of foresight and technology, preparing legal entities to thrive in an environment where data is not only voluminous but also complex and multifaceted. With its cutting-edge features, Everlaw represents a microcosm of the broader shift toward AI-driven legal practices, showcasing what is possible when innovation meets the meticulous demands of the legal profession.

Incorporating Everlaw into this discussion of e-discovery's future underscores the practical applications of AI in law. It's a testament to how platforms can not only keep pace with but also drive the evolution of legal technologies, equipping practitioners with the means to exceed the current benchmarks of document review and legal inquiry.

Ultimately, as legal practices embrace platforms like Everlaw and integrate AI into their workflows, they are setting the stage for a revolutionized e-discovery process that is as transformative as it is inevitable. This trajectory is marked by a commitment to excellence, continuous improvement, and an unwavering focus on delivering justice in the digital age.

Through AI's lens, we foresee an e-discovery realm replete with sophistication, where the intricacies of legal data are navigated with precision and foresight. In this realm, Everlaw emerges as a pivotal force, driving the evolution of e-discovery with innovation and insight at the forefront of legal technology. The fusion of AI into e-discovery, therefore, is not merely a distant promise but a present reality, sculpting the future of legal practices day by day.

The Efficiency Revolution: AI in Document Review

In the dynamic realm of legal practices, the quest for enhanced efficiency while upholding accuracy has been a perpetual priority. The emergence of Artificial Intelligence (AI) has catalyzed a revolutionary shift in document review methodologies, facilitating speed and precision on an unprecedented scale. Legal professionals find themselves at the forefront of a transformative movement, poised to reshape the very framework of legal document analysis.

Traditionally, document review has been a laborious endeavor, demanding considerable time and resources. The intricate demands of legal work necessitate thorough examination, rendering the review process both pivotal and strenuous. Yet, the advent of AI in the legal sector signifies a rapid paradigm shift.

AI epitomizes a potent suite of tools, adept at navigating through extensive data effortlessly. Machine learning algorithms empower AI systems to identify patterns, categorize documents, and predict relevancy with impressive accuracy. This evolution not only expedites the document review process but also curtails the likelihood of errors.

Initial forays into AI-driven document review faced skepticism, centered around concerns of accuracy, dependability, and the risk of overlooking vital information. However, empirical analyses and case studies have consistently affirmed AI's capacity to equal, and often exceed, human proficiency in document review tasks.

A pivotal innovation in this sphere is predictive coding. This approach utilizes machine learning to extrapolate a legal expert's decision-making on a subset of documents, applying these judgments across a broader dataset. This breakthrough enables legal teams to concentrate their review efforts on the most pertinent documents, thereby optimizing their case strategies.

Moreover, AI-driven document review heralds substantial cost efficiencies. The conventional billable hours linked with manual review can be significantly minimized, facilitating optimal resource allocation within law firms. This shift not only bolsters operational efficiency but also sharpens competitive edges in legal service pricing.

Despite its benefits, AI integration in legal practices faces hurdles. The effectiveness of AI systems largely hinges on the quality of their training data. Given the diversity in legal document structure and complexity, compiling a comprehensive training dataset is crucial. Additionally, continuous oversight by legal experts is essential to ensure AI's accurate interpretation of legal nuances.

To fully leverage AI in document review, law firms must commit to extensive training and foster an innovative culture. Legal practitioners need to comprehend the strengths and limitations of AI technologies for successful implementation. Nurturing an environment conducive to innovation is vital, encouraging legal professionals to explore and refine AI integration methods.

Ethical considerations also play a critical role in AI deployment. Given AI's interaction with sensitive and confidential data, securing client information is paramount. Legal entities must navigate these ethical landscapes diligently, adhering to stringent confidentiality and data protection norms.

The evolution of legal document review is unmistakably influenced by AI advancements. As AI technology evolves, its applications in legal practices are set to broaden, encompassing contract analysis, litigation support, and beyond.

AI's potential to revolutionize document review is vast, offering pathways to enhanced efficiency, accuracy, and cost-effectiveness. Law firms embracing this technology stand to gain significant operational and client service enhancements.

However, integrating AI in legal document review demands meticulous planning and execution. Law firms must address technical, ethical, and operational challenges to fully harness AI's capabilities. A strategic approach, emphasizing training, process optimization, and ethical considerations is imperative.

Conclusively, AI's role in efficient document review marks a significant advancement in legal practice. While challenges persist, the opportunities it presents are immensely promising. By adopting AI, legal professionals can initiate a new era of efficiency, precision, and innovation in document review, thereby redefining industry standards and enriching client services.

The future of legal document review, increasingly intertwined with AI, beckons a rapid and effective adaptation by legal practitioners. With careful implementation and a commitment to continual learning and adaptation, the legal sector can tap into AI's power to achieve unprecedented efficiency and effectiveness in document review and beyond.

The journey toward embracing AI in legal practices is just beginning, and the scope for growth and improvement is boundless. Through AI, the legal profession can not only refine its current practices but also redefine potential, paving the way for a future where legal services are more accessible, efficient, and effective.

Within this evolving landscape, Relativity emerges as a pivotal player, reshaping the delivery of legal services with its cutting-edge AI capabilities. Renowned for its comprehensive e-discovery and compliance solutions, Relativity's flagship platform, RelativityOne, is instrumental in managing vast data volumes and streamlining the discovery process in litigation, investigations, and compliance endeavors.

Relativity's prowess lies in its advanced search functionalities, machine learning technology for data analysis and categorization, and its adeptness in managing large datasets. These features collectively empower legal teams to efficiently organize and review documents, thereby adapting to the ever-changing legal environment and maintaining a competitive stance.

The integration of Relativity's AI in document review processes revolutionizes how legal professionals approach and handle data. Its machine learning algorithms facilitate the swift identification, classification, and analysis of documents, streamlining the review process and enhancing accuracy and productivity.

Moreover, Relativity's AI-driven tools significantly reduce the time and resources traditionally required for document review. By automating routine tasks and enabling more focused and strategic review efforts, legal teams can allocate their expertise where it's most impactful, thus enhancing overall case management and outcomes.

The software's predictive coding feature is particularly noteworthy, allowing legal professionals to apply their judgment on a sample of documents and then extend those decisions across entire datasets. This not only speeds up the review process but also ensures consistency and accuracy in document categorization and relevancy assessment.

Relativity's platform also fosters collaboration among legal teams, enabling seamless sharing and analysis of documents. This collaborative environment, enhanced by AI, facilitates a more integrated approach to case preparation and strategy development, ensuring all team members are aligned and informed.

Additionally, Relativity's community-driven marketplace offers a wealth of applications and integrations, further extending the platform's capabilities. This ecosystem allows legal teams to customize their use of RelativityOne, integrating additional tools and services that complement and enhance the AI-driven document review process.

The impact of Relativity's AI on the legal industry extends beyond operational efficiencies and cost savings. It also plays a crucial role in upholding legal standards and ensuring thorough and unbiased document review. By automating processes susceptible to human error and fatigue, AI helps maintain the integrity and thoroughness of legal investigations and discovery processes.

Relativity's commitment to security and privacy is another critical aspect of its AI integration. In an era where data breaches and privacy concerns are rampant, Relativity ensures that its AI tools adhere to stringent security protocols, safeguarding sensitive and confidential information throughout the document review process.

As legal practices continue to evolve, the need for robust, AI-powered solutions like Relativity becomes increasingly apparent. Its ability to handle complex data analysis and review tasks efficiently positions legal firms to navigate the complexities of modern legal challenges effectively.

In conclusion, Relativity's AI not only transforms the landscape of document review but also sets a new benchmark in the delivery of legal services. Its advanced AI features, coupled with a comprehensive suite of tools and a collaborative ecosystem, enable legal teams to adapt swiftly to the changing landscape and stay ahead in the competitive legal market. By embracing Relativity's AI-driven solutions, law firms can achieve greater efficiency, accuracy, and client satisfaction, thereby redefining the standards of excellence in the legal profession.

As we venture further into the AI era in legal document review, the integration of platforms like Relativity will undoubtedly become a cornerstone of successful legal practice. The journey towards more efficient, accurate, and innovative legal services is ongoing, and with AI and Relativity at the helm, the possibilities for improvement and advancement in legal document review are boundless.

The Rise of Predictive Coding in Legal Tech

Within the legal profession, the rise of artificial intelligence heralds an era of transformative change, especially in the domain of document review. At the forefront of AI-driven innovations in this field is predictive coding, a sophisticated method that synergizes human legal expertise with advanced machine learning algorithms. This technique has rapidly become a formidable tool for navigating the ever-expanding sea of electronic documents. This chapter delves into the profound impact of predictive coding on document analysis, showcasing how it revolutionizes efficiency and accuracy in legal practices.

Predictive coding is fundamentally engineered to amplify the capabilities of legal teams, equipping them to tackle the burgeoning volumes of electronic documentation efficiently. This method learns from a subset of documents reviewed by legal specialists, assimilating the criteria deemed relevant. It then extends this learned logic across expansive datasets, identifying documents that are likely pertinent to the case. The repercussions of this technology are game-changing, transforming document review from a laborious, manual process into a strategic operation that conservatively reserves human intervention for the most pivotal content.

The time-saving aspect of predictive coding is indisputable. Legal practitioners can now perform in hours which previously took days or even weeks, translating into substantial cost reductions for clients. Additionally, the minimized workload for legal experts allows law firms to reallocate their talents more strategically, focusing on case strategy and client relationships, rather than the monotonous chore of sifting through documents.

Despite its advantages, the reception of predictive coding is not without its detractors. Critics voice concerns about the accuracy of such technologies, fearing that an excessive dependence on algorithmic processing might lead to overlooking vital documents. Addressing these apprehensions requires an in-depth exploration of machine learning mechanics and tangible proof of the methodology's effectiveness, which this chapter approaches with thorough examination.

Empirical studies and practical deployments of predictive coding in document analysis provide convincing demonstrations of its reliability and precision. These examples, hailing from various legal scenarios, reveal not only the technological prowess but also the strategic insight essential to utilize predictive coding to its full potential.

Predictive coding's integration into document review processes provokes pivotal discussions regarding the role of legal professionals. Rather than rendering attorneys redundant, it bolsters their interpretive and strategic capacities, liberating them from the drudgery of conventional document review. This technology acts as an auxiliary guide, supporting but never supplanting the irreplaceable human factor in legal analysis.

Anticipating the future trajectory of predictive coding entails its convergence with other AI technologies. Natural language processing, for instance, can refine the system's contextual and semantic text comprehension, thereby elevating sorting precision. These advancements foretell a further maturation of predictive coding, positioning it as an even more essential element in the legal arsenal.

From the clientele's vantage point, the merits of predictive coding surpass mere cost-effectiveness. Modern clients demand transparency and accountability within legal proceedings. Predictive coding's analytical capabilities deliver nuanced insights into document review, enabling law firms to provide clients with data-driven justifications for their strategies and decisions.

Inculcating a robust understanding of predictive coding among legal professionals is paramount for harnessing its full benefits. Practitioners must familiarize themselves not solely with the operational aspects but also with the strategic implications of integrating AI into their workflows. Empowering teams with the acumen to adeptly employ predictive coding is a pivotal factor in the success of law firms transitioning into AI-augmented document analysis.

Ethical issues also assume significance in discussions about predictive coding. The potential of this technology to sway case outcomes imparts it with considerable influence. To ensure responsible use of predictive coding, the legal field must observe ethical standards that promote transparency, accountability, and fairness in AI applications.

The legal precedents regarding predictive coding are still emerging. Judicial systems are starting to acknowledge its utility, yet a comprehensive legal framework to regulate its usage is in the developmental phase. This evolving judicial environment presents both prospects and challenges for legal entities as they steer through the judicial reception of predictive coding in document review.

The economic ramifications of predictive coding are profound. By drastically curtailing the time and effort requisite for document analysis, predictive coding has the potential to revolutionize the economic structure of legal services. Such transformation could democratize legal representation, making it more accessible and affordable for a wider spectrum of clients.

Envisioning the path ahead, predictive coding indicates a trend towards a more automated future in document review. Nevertheless, this trend does not undermine the necessity for a human touch that remains paramount in legal analysis. Instead, it redefines the lawyer's role, underscoring the importance of cognitive prowess, strategic planning, and profound legal acumen. In this new era, predictive coding emerges not merely as a tool but as a facilitator for the evolution of the legal vocation.

Concludingly, predictive coding stands at the vanguard of AI's integration into legal practice. It embodies AI's promise to elevate the effectiveness, precision, and strategic complexity of legal services. As law firms chart their course through the complexities of predictive coding implementation, they must remain cognizant of its challenges, ethical implications, and the imperative for continual human oversight. The voyage has commenced, and although the path may be intricate, the destination—a transformed legal industry distinguished by innovation and enhanced access to justice—is compellingly worthwhile.

This section has elucidated the transformative impact of predictive coding on document analysis, affirming its status as a pivotal change agent in the legal realm. As legal entities and practitioners acclimatize to this altered terrain, the insights and principles delineated herein will be pivotal in steering their path toward the effective adoption and optimization of AI technology within their practices.

Logikcull Integration:

The narrative of predictive coding's ascendancy in legal technology would be incomplete without recognizing platforms that are leading the charge in this transformation. Enter Logikcull, a vanguard cloud-based eDiscovery software that empowers legal and IT teams to unearth and safeguard crucial information. This platform exemplifies innovation in legal tech, delivering a suite of features that synergize perfectly with the tenets of predictive coding.

Logikcull's culling intelligence is a testament to the power of automation, categorizing data into intelligible search filters that simplify discovery. It takes the laborious process of data sorting and transforms it into a sleek, automated affair by indexing, deduplicating, and organizing data with intelligent filters. This system works in tandem with the principles of predictive coding, aligning the objectives of swift, and precise document retrieval with the technological means to achieve them.

The platform's ability to facilitate tagging, redacting, annotating, and sharing documents mere minutes after upload represents a quantum leap in efficiency. Coupled with the predictive coding process, this rapid functionality ensures that crucial information is highlighted and secured with unprecedented speed, enhancing the pace and accuracy of case preparations.

Logikcull's drag-and-drop data ingestion, processing, and organization features herald a new age of user-friendly eDiscovery. This intuitive approach dovetails with predictive coding's objective to make document analysis more accessible and less time-consuming for legal professionals.

Automated, self-serve document productions within Logikcull stand out as emblematic of the software's capacity to streamline complex legal processes. By integrating predictive coding, the platform can autonomously generate relevant documents, reducing the need for exhaustive manual review and fostering a more efficient discovery phase.

With permission-based access and detailed activity reporting, Logikcull underscores the collaborative aspect of modern legal work. These features enhance the synergy between team members, expediting the document review process and ensuring that each action taken within the system is recorded and accountable. The alignment with predictive coding's methodology here is clear, as both advocate for an approach that maximizes efficiency while maintaining stringent oversight and control.

Security is another cornerstone of Logikcull's offering, safeguarding against sanctions and data breaches—a concern that resonates with the ethical use of predictive coding. By providing a secure platform, Logikcull assures that the efficiencies gained through predictive coding do not come at the expense of client confidentiality and data integrity.

As we explore the nexus of predictive coding and Logikcull, it becomes evident that the latter is not just an eDiscovery tool but a beacon for the entire legal profession's future direction. Logikcull stands out for its automated, cloud-based prowess, offering predictive coding capabilities that manage e-discovery with remarkable efficiency. In doing so, it not only streamlines the process but also reinforces the strategic, human-centric approach that is the hallmark of effective legal practice.

By weaving Logikcull into the fabric of document analysis, law firms can harness the full potential of predictive coding. The platform's sophisticated automation and user-centric design bolster the capability to manage vast volumes of data with precision. Consequently, Logikcull becomes an indispensable ally in the quest to transform legal document analysis, embodying the principles of predictive coding and propelling legal practices towards a future of heightened efficiency, transparency, and informed decision-making.

As we stand at this juncture, predictive coding, bolstered by platforms like Logikcull, is a potent testament to the harmonious potential of human expertise and AI. It assures us that in this rapidly evolving legal landscape, technology is an ally—not an adversary—to the legal mind. This partnership promises not just a redefined legal profession but also a justice system that is more accessible, equitable, and efficient for all.

Chapter 4:

Enhancing Contract Management with AI

In the contemporary legal sector, the implementation of Artificial Intelligence (AI) has been a game-changer, particularly in the realm of contract analysis and management. The age-old practice of meticulous manual review is being phased out, giving way to the age of AI, where efficiency and precision are paramount. This innovative technology has revolutionized traditional contract management, which was once a cumbersome and error-prone process. AI now offers a streamlined approach that covers the entire contract lifecycle management, from inception to execution, allowing legal professionals to channel their expertise into strategic decision-making rather than getting bogged down with administrative tasks. Such transformation promotes heightened productivity and a marked reduction in oversight, ensuring that contracts are not only compliant but also robust against potential legal challenges.

ThoughtRiver, a leading force in legal tech, exemplifies the cutting-edge application of AI within this sector. Its sophisticated platform is specially tailored to the nuanced requirements of contract analysis. By employing advanced natural language processing (NLP), ThoughtRiver automates contract review, identifies risks, and suggests amendments, effectively bridging the gap between legal foresight and operational efficiency. As a result, law firms and in-house legal teams can focus on strategic initiatives with the assurance that the day-to-day contract management is both proactive and precise.

The capabilities of AI, particularly through platforms like ThoughtRiver, extend far beyond routine tasks. They envelop sophisticated contract creation, review, compliance monitoring, and drafting, benefiting those who manage the increasing volume and complexity of contracts faced by legal departments and law firms on a daily basis. Moreover, the application of machine learning algorithms means these systems improve with each use. As they evolve, they become adept at detecting patterns and foreseeing issues that might escape even the most experienced legal professionals.

In the legal landscape, where precision is paramount, Artificial Intelligence (AI) has revolutionized contract analysis and management, heralding a new age of efficiency and accuracy. Gone are the days when manual contract review was the standard, a process fraught with labor intensity and vulnerability to error. Now, AI-driven technologies like those developed by ThoughtRiver have reshaped this arena, streamlining the entire contract lifecycle management from creation to compliance. This pivotal shift has empowered legal professionals to direct their focus toward strategic analysis and decision-making.

The implementation of ThoughtRiver's AI solutions in the contract management process signifies a remarkable leap forward. The platform's adept use of natural language processing (NLP) to review contracts ensures that legal teams can swiftly identify potential risks and navigate complex agreements with confidence. With such technology, the intricate process of contract vetting becomes a far less daunting task. This allows for an unprecedented level of oversight, ensuring every contract is not only compliant with current regulations but also fortified against potential disputes.

With ThoughtRiver's AI at the helm, legal professionals are equipped to handle a multitude of tasks that once demanded exhaustive manual oversight. From the meticulous drafting of contracts to monitoring compliance and analyzing terms, the platform's robust capabilities mean that the large volumes of contracts that law firms and legal departments handle are managed with heightened effectiveness. Machine learning algorithms enable ThoughtRiver to not just perform tasks but to learn from them, evolving to become an even more powerful tool in the legal tech arsenal.

AI's impact on the legal profession extends into the realm of predictive risk analysis. The proactive insights provided by ThoughtRiver's platform allow for the early identification of contractual risks, ensuring that potential issues can be mitigated before they evolve into significant problems. This preemptive approach is not just about avoiding risks; it also streamlines the negotiation process, reducing the time and resources traditionally spent on due diligence.

By integrating ThoughtRiver's capabilities into their practices, legal professionals not only safeguard their organizations but also enhance the level of service offered to clients. The data-driven insights gleaned from AI analysis afford lawyers the opportunity to provide more informed, strategic advice, which is invaluable in an industry that is becoming increasingly competitive and fast-paced. The adoption of such AI-powered tools is fast becoming a standard, a necessary step for those who wish to remain at the forefront of the legal field in an era marked by rapid digital transformation.

AI in Contract Analysis: Efficiency and Accuracy

The realm of contract management is experiencing a monumental shift, driven by the advent of artificial intelligence (AI). In legal practices, where managing and analyzing contracts is both crucial and labor-intensive, AI offers a promising avenue to streamline these processes, enhancing both efficiency and accuracy. The integration of AI into contract management represents not just a trend, but a fundamental transformation in how legal professionals handle contracts.

Traditionally, managing contracts has been a manual, tedious process, fraught with exhaustive reviews and a high potential for human error. However, with the advent of AI, it's clear that these conventional methods are increasingly impractical in the fast-paced legal world. AI's ability to process and analyze large volumes of data at unprecedented speeds signifies a major leap forward in contract management. This leap is not about replacing human effort but augmenting it, allowing legal professionals to focus on more strategic and analytical tasks.

A prime example of this transformation is Evisort, a leading force in revolutionizing contract management through AI. Evisort's advanced, award-winning machine learning software excels at searching and analyzing contract text. Originally developed for M&A due diligence, Evisort has expanded its capabilities across various legal domains. By turning complex legal language into structured data, Evisort enhances consistency and insight in contract reviews, setting a new standard for legal tech innovation.

In the automation of contract creation, AI-powered systems like Evisort can generate contracts from predefined templates and rules, ensuring legal compliance while drastically reducing drafting time. This automation goes beyond simple drafts to complex, tailored contracts with minimal human intervention.

AI's impact on contract review is profound. Systems equipped with natural language processing, such as Evisort, can meticulously analyze contracts, highlighting issues, deviations, and risks that might be overlooked by human reviewers. This meticulous analysis significantly reduces risk and secures more robust contractual agreements.

Version control and contract lifecycle management benefit from AI's ability to track document changes, maintain version integrity, and ensure access to the latest document versions, thus mitigating confusion and potential conflicts.

In compliance monitoring, AI systems continually assess contracts against changing regulations, flagging potential compliance issues early and preventing legal mishaps. This proactive stance is crucial in navigating today's intricate regulatory environment.

AI also revolutionizes data extraction and analysis in contract management. Systems can quickly identify and organize critical contract information, streamlining data retrieval and enhancing decision-making with comprehensive data insights.

The predictive nature of AI in contract management, analyzing historical data to forecast outcomes like contract breaches or execution bottlenecks, enables preemptive measures that smooth contract execution and minimize risks.

Integration with other legal tech solutions is a significant benefit of AI-driven contract management systems. They can integrate seamlessly with other legal software, forming a cohesive ecosystem that boosts overall operational efficiency and ensures that contract management is an integrated, streamlined process within the larger legal workflow.

However, the transition to AI-driven contract management is not without challenges, requiring technological upgrades and a shift in legal practice culture. Successful AI implementation in contract management demands an openness to new technologies and operational changes.

Training and education are key to overcoming adoption barriers. Legal professionals must understand how to effectively utilize AI, going beyond mere functionality to strategic application in enhancing contract management.

The benefits of AI in contract management are undeniable: reduced turnaround times, increased accuracy, improved compliance, and predictive insights lead to a more efficient, secure, and strategic contract management process, yielding cost savings and enhanced client satisfaction.

Yet, the adoption of AI in contract management also raises ethical and data privacy concerns. Ensuring responsible AI use and data protection is essential to maintaining trust and integrity in legal practices.

As we look ahead, the evolution of AI in contract management promises further advancements, with technologies like Evisort at the forefront, enhancing contract analysis and facilitating real-time collaborative contract management. The future of contract management with AI is not merely about automation but about elevating the role of legal professionals to strategists and advisors, empowered by AI to deliver unparalleled value.

In sum, the revolution in contract management facilitated by AI, exemplified by pioneers like Evisort, represents a significant leap forward for legal practices. Embracing AI enables legal professionals to transform contract management from a cumbersome manual process into a strategic advantage, paving the way for operational excellence and redefining the role of legal professionals in the digital era.

AI Automation in Contract Drafting

The domain of contract management, a core facet of legal practice, has traditionally been fraught with complexity and inefficiency. However, the advent of Artificial Intelligence (AI) is driving a transformative shift. This chapter explores the use of AI technologies in greatly enhancing the efficiency of the contract lifecycle, providing legal professionals with powerful tools to navigate this changing landscape.

Contract lifecycle management involves the creation, execution, and analysis of contracts to optimize operational and financial performance while mitigating risk. Traditionally, these tasks have required considerable time and effort from legal teams. AI's ability to automate and optimize processes offers an opportunity to redefine these norms.

AI’s role in enhancing contract lifecycle efficiency is profoundly evident in its automation capabilities. AI-driven tools, such as those developed by Icertis, can automate routine tasks like contract drafting and employ Natural Language Processing (NLP) to comprehend and generate complex legal terms. This not only speeds up the drafting phase but also reduces errors.

Icertis's AI-powered platform, the Icertis Contract Management (ICM), takes this further by streamlining the entire contract lifecycle from initiation to execution and analysis. It provides visibility and compliance across contracts, leading to more strategic and agile business decisions. Through its Contract Intelligence (ICI) platform, Icertis turns contracts from static documents into strategic assets, structuring and connecting the critical contract information that defines organizational operations. Today, the world's most iconic brands and disruptive innovators trust Icertis to fully realize the intent of their contracts, totaling more than 10 million across 40+ languages in 90+ countries.

Beyond drafting, AI, as utilized by Icertis, significantly streamlines the review process. Machine learning algorithms, trained on vast datasets of legal documents, can quickly identify key clauses and potential issues. This enables legal professionals to concentrate their review efforts where they matter most, thus reducing cycle times.

Furthermore, AI contributes to negotiation efficacy. Predictive analytics, a feature of Icertis's platform, can inform negotiation strategies by analyzing historical data to predict the opposing party's responses or to suggest advantageous terms. This intelligence allows lawyers to approach negotiations with greater confidence and strategic acumen.

Once a contract is executed, AI offers unparalleled advantages in management and analysis. Icertis's system can continuously monitor contractual obligations, compliance requirements, and performance metrics. Doing so provides real-time insights and alerts to ensure any deviations are quickly managed.

The capabilities of AI also extend into the renewal and termination phases of contracts. By evaluating contract performance data, tools like Icertis can advise whether to renew, renegotiate, or terminate agreements. This data-driven approach ensures that decisions are both strategic and aligned with the organization's objectives.

In terms of data security and privacy, AI plays a crucial role in maintaining the integrity and confidentiality of sensitive information. Icertis contributes to this by employing advanced encryption technologies and access control mechanisms that are consistent with legal standards.

AI also excels in integration. AI-powered contract management systems, including Icertis, can integrate seamlessly with other enterprise software, enhancing data sharing and process connectivity across departments. This holistic view supports better organizational coordination and decision-making.

The collaborative aspect of AI in contract lifecycle management is also significant. By enabling shared access to contracts and related data, AI platforms like Icertis encourage a cooperative environment where attorneys, paralegals, and stakeholders can work together effectively, irrespective of their physical locations.

Continual learning and improvement are intrinsic to AI systems. They evolve with each contract processed, learning from feedback and outcomes to increase their accuracy and efficiency over time. Despite initial resistance and skepticism towards technology within the legal profession, the tangible benefits AI brings in boosting contract lifecycle efficiency make a compelling case for its adoption. Legal professionals who embrace AI technologies like Icertis can gain a competitive edge in operational efficiency and service quality.

Implementing AI in contract lifecycle management requires careful planning and a strategic approach. Legal teams must assess their specific needs, choose the right AI tools, and ensure proper training and support for users. Ethical considerations regarding AI use must also be meticulously addressed to ensure compliance with legal and professional standards.

In conclusion, AI's role in enhancing contract lifecycle efficiency is not just an incremental improvement but a paradigm shift in legal practice. By adopting AI, and specifically Icertis's platforms, legal professionals can not only streamline operational processes but also provide greater strategic value to their clients and organizations. The journey toward AI integration may be complex, but the transformative benefits are clear and indisputably valuable.

Scalable Risk Mitigation through AI

Within the rapidly transforming legal sector, the influence of Artificial Intelligence (AI) is increasingly becoming integral, particularly in the realms of contract analysis and management. This comprehensive exploration underscores AI's significant contributions to mitigating contract risks, propelling traditional practices into a new era marked by heightened efficiency, precision, and cost-effectiveness.

At the heart of commercial exchanges, contracts compel adherence to mutually agreed terms. Nonetheless, the intricacy and sheer volume inherent in contracts can spawn considerable challenges, such as the inadvertent disregard of pivotal clauses or duties that might precipitate legal feuds or pecuniary losses. AI transcends a mere auxiliary role, offering a holistic approach to these perennial challenges.

Harnessing machine learning and natural language processing, AI scrutinizes contracts with a speed and exactitude that was previously unattainable. Legal practitioners, therefore, can swiftly pinpoint terms and stipulations that represent potential hazards, such as indeterminate language, disadvantageous conditions, and infractions concerning prevailing laws and regulations. AI thus reduces the duration devoted to manual examination while augmenting the caliber of scrutiny.

AI's role is extensive, encompassing the identification of issues and categorizing and ascertaining the gravity of risks based on their potential repercussions. Consequently, legal squads can allocate their resources judiciously, assuring the considerable threats are not dismissed due to human lapse or inattention.

A notable merit of AI is its scalability. As enterprises burgeon and contract quantities surge, AI proffers a scalable remedy capable of managing the growing load without a commensurate increment in time or resources dedicated to manual examination.

Invaluable, too, are AI's predictive faculties in mitigating risks. By assessing historical data and contractual trends, AI can prognosticate impending risks and advocate strategies to circumvent them in upcoming contracts. This forward-looking approach shifts risk mitigation from a reactive undertaking to a strategic organizational function.

The integration of AI across the contract lifecycle is indispensable. From composition and negotiation to performance monitoring and renewal, AI utilities identify and allay risks at every juncture, ensuring contracts remain advantageous and in compliance throughout their term.

AI also bolsters collaboration among legal divisions and stakeholders. By offering a unified platform for contract appraisal and risk assessment, AI ensures comprehensive access to requisite data, thereby optimizing and informing the decision-making process.

With the ascendance of data privacy and security in contract administration, AI frameworks imbued with robust security protocols are essential in protecting sensitive contract information against illicit access or breaches.

Furthermore, AI's incorporation into contract risk mitigation has profound compliance implications. As regulations undergo continuous evolution, AI systems can adapt to reflect legal changes, ensuring enduring contractual compliance — a critical aspect in regulatory-intensive sectors.

The initial commitment in terms of time and data to train AI systems for contract risk mitigation is substantial. Nevertheless, once primed, these systems become increasingly adept and precise, perpetually learning from novel contracts and judicial precedents. This continuous educational process assures the enduring relevancy and efficacy of AI tools in pinpointing and mitigating risks.

AI's significance in contract risk mitigation further spans dispute avoidance.

By predicting potential friction points within contracts, AI can guide parties in rectifying these issues preemptively, thus averting litigation and conserving resources.

Despite these advantages, AI's application in contract risk mitigation is not devoid of hurdles. Legal professionals are tasked with the ethical and responsible utilization of AI, acknowledging its potential and confines. Moreover, the adoption of AI necessitates a cultural metamorphosis within organizations, embracing innovation whilst upholding the rigorous standards intrinsic to the legal profession.

In sum, the transformative role of AI in contract risk mitigation equips legal practitioners with a potent instrument to manage risks with unprecedented effectiveness and efficiency. As AI continues to advance, its amalgamation into contract analysis and management is poised to further revolutionize the field, endowing legal practitioners with unparalleled abilities to safeguard their clients' interests amid the convoluted realm of contracts.

At this juncture, the case of LawGeex is illustrative, underlining the instrumental role of AI in contract risk mitigation. LawGeex exemplifies the strategic deployment of AI, employing its technology to review and harmonize contracts with corporate policies and benchmarks. Its platform stands as a bulwark against potential contractual pitfalls and deviations, epitomizing the vital role technology plays in proactively diminishing contract-related risks.

LawGeex, a pioneering legal technology entity, introduces an AI-powered platform that automates the scrutiny and negotiation of contracts. The platform is a boon for in-house legal teams inundated with voluminous low-stakes contracts, such as procurement agreements, service contracts, non-disclosure agreements, and SaaS accords. LawGeex's AI engine is capable of evaluating contracts and signaling issues within an hour, thereby enabling legal teams to concentrate on moments of active review time, a quantum leap from traditional contract management.

The forward-thinking integration of AI into legal practices, especially for contract examination and administration, is an unambiguous stride towards enhancing operational proficiency, client contentment, and the attainment of optimal case results. As the legal field adopts AI's capabilities, it opens up innovative pathways for excellence and innovation in legal services.

LawGeex's AI engine, specifically, is a testament to this advancement. By accelerating the contract review process, it exemplifies how AI can condense what once took hours into minutes without sacrificing the meticulous attention to detail that is the hallmark of legal work. This shift not only conserves time but also reallocates the invaluable expertise of legal professionals to more strategic, high-impact tasks. Moreover, LawGeex's system is calibrated to align with company-specific policies and standards, ensuring that all contracts adhere to internal requirements and best practices. This alignment is crucial as it directly influences the risk profile of each contract and, by extension, the organization.

The platform's capabilities are not limited to risk identification; it extends to risk prevention. By incorporating a database of legal knowledge and precedents, LawGeex's AI can provide suggestions for alternative phrasing and clause structuring that mitigate risks before they materialize. This preventive measure is critical in maintaining the integrity of the contract and the protection it offers both parties.

However, the integration of AI like LawGeex's into the risk mitigation strategy is not simply a matter of installing software; it involves a rethinking of the entire contract management process. It requires training the AI with a wealth of data, which in turn demands a strategic approach to data collection and curation. The data must be representative, comprehensive, and, most importantly, continuously updated to reflect the latest legal developments and organizational changes.

The dynamic nature of AI-powered tools like LawGeex's platform means that they improve with use. As more contracts are analyzed, the system refines its understanding of risk factors and enhances its predictive analytics, offering even more accurate assessments over time. This self-improving aspect of AI ensures that the system evolves in step with the legal and business landscape, a feature that is particularly vital in an environment as fluid and fast-paced as the modern marketplace.

Moreover, the adoption of such AI systems influences not only the efficiency of risk mitigation but also the culture of legal teams. Embracing AI requires an openness to change, a willingness to rely on technology for critical tasks, and a commitment to redefining the role of the legal professional. In this new paradigm, lawyers become supervisors of technology, directing its focus and interpreting its findings, rather than sifting through documents themselves.

Despite these profound benefits, it is crucial to navigate the potential ethical and practical implications of AI deployment with caution. Questions about bias, transparency, and accountability in AI decision-making necessitate rigorous oversight. Legal professionals must maintain control over AI systems to ensure that they do not perpetuate existing biases or introduce new ones into the contract review process. This means that while AI can and should be used to streamline and enhance risk mitigation, it cannot replace the nuanced judgement of experienced legal practitioners.

The cultural shift towards AI in legal practices also involves reshaping client expectations. As clients become more aware of AI's capabilities, they seek faster, more cost-effective services without a drop in quality. Legal teams equipped with tools like LawGeex are well-positioned to meet these demands, delivering high-quality contract analysis at a pace that matches the speed of business today.

Ultimately, the integration of AI into contract risk mitigation is not an endpoint but a beginning. It represents the start of a continuous journey towards greater efficiency, higher quality legal services, and more strategic use of human expertise. As AI tools become more sophisticated, their role in legal practice will only deepen, offering unprecedented opportunities to enhance the practice of law and the administration of justice.

In embracing AI and platforms like LawGeex, the legal industry takes a significant leap towards a future where technology and human expertise combine to produce the best possible outcomes for all parties involved. This synergy is the cornerstone of modern legal practice, where AI's role in contract risk mitigation is just one aspect of a broader movement towards innovation, efficiency, and excellence in law.

Thus, as the legal sector forges ahead, the incorporation of AI in contract risk mitigation exemplifies a paradigm shift, enabling legal professionals to transcend traditional boundaries and redefine the essence of effective contract management and client service. With AI's continued evolution, the potential for its impact on the legal profession is boundless, heralding a new era of precision, foresight, and strategic acumen in the world of legal contracts.

Chapter 5:

AI-Powered Paralegals: Transforming Legal Workflows for Maximum Efficiency

The legal industry, traditionally perceived as conservative and resistant to change, is currently undergoing a significant transformation. This shift is primarily driven by the rapid advancement and integration of Artificial Intelligence (AI) and Legal Technology (Legal Tech). The implications of these technologies are profound, not just in terms of enhancing operational efficiencies but also in reshaping the roles and capabilities of legal professionals, particularly paralegals. As we stand on the brink of this technological revolution within the legal sector, it is crucial to understand how these tools are not only augmenting existing workflows but also empowering paralegals to redefine their roles within legal practices.

Paralegals have traditionally been the backbone of legal operations, handling a myriad of tasks that range from document management and research to supporting lawyers in trial preparation and client communications. However, the integration of AI and Legal Tech into their daily routines is paving the way for a shift from manual, repetitive tasks to more strategic and impactful roles. This transition is marked by both opportunities and challenges as paralegals navigate through learning new technologies and adapting to enhanced roles that these technologies facilitate.

The evolution of AI in the legal field is being driven by a combination of factors, including the exponential growth of data, the increasing complexity of legal cases, and the pressing need to reduce legal costs while improving service delivery. AI and Legal Tech offer tools that can automate routine tasks such as sorting through documents, identifying relevant case laws, and drafting standard legal correspondence. These tools leverage machine learning algorithms and natural language processing to perform tasks at a speed and accuracy that is unattainable by human paralegals.

Moreover, the advent of Legal Tech has catalyzed a more collaborative and integrated approach within law firms. Advanced communication tools and document management systems facilitate better coordination among legal teams, allowing for more streamlined processes and efficient handling of cases. Paralegals, equipped with these technological tools are finding themselves at the center of this transformation, playing pivotal roles in bridging the gap between traditional legal practices and modern, tech-driven methodologies.

This technological empowerment of paralegals is enhancing not only the productivity and efficiency of legal teams but also contributing to a more dynamic legal environment where strategic thinking and data-driven decision-making are at the forefront. However, the transition is not devoid of challenges. The adoption of new technologies requires not only a significant investment in terms of time and resources but also a cultural shift within organizations that values innovation and continuous learning.

The role of AI and Legal Tech in transforming paralegal roles also extends to client interactions. In an era where clients are more informed and expect quick and efficient service, the ability of paralegals to utilize AI tools to provide prompt and accurate information can enhance client satisfaction and loyalty. Moreover, AI-driven analytics can provide insights into client preferences and behaviors, enabling law firms to tailor their services to meet the specific needs of their clients more effectively.

In this context, law firms that are proactive in integrating AI and Legal Tech into their operations are likely to gain a competitive edge. These firms benefit not only from increased operational efficiency but also from the ability to attract and retain top talent who are eager to work in an innovative and technologically advanced environment. Conversely, firms that are slow to adopt these technologies risk falling behind, not just in terms of operational efficiency but also in their ability to attract new clients and top talent.

As we delve deeper into the integration of AI and Legal Tech in the legal sector, it becomes evident that the role of paralegals is becoming more crucial than ever. They are not just users of technology but are also becoming key players in shaping how technology is adopted and utilized within law firms. Their deep understanding of legal processes and their ability to adapt to new tools make them indispensable in the modern legal ecosystem.

The continued advancement of AI and Legal Tech is set to create more opportunities for paralegals to expand their roles and contribute more significantly to their organizations. As we explore the specific impacts of these technologies on paralegal work, it is important to consider both the opportunities they present and the challenges they pose. This exploration not only sheds light on the current state of the legal profession but also helps us envision the future of legal practices where efficiency, innovation, and human expertise converge to create a more effective legal system.

With this comprehensive overview of the transformative impact of AI and Legal Tech on paralegals, the following sections will delve deeper into specific aspects of this transformation, beginning with an exploration of how AI is reshaping paralegal work in "The Rise of AI in Paralegal Work."

Advancements in AI: Redefining Paralegal Tasks

In the domain of contemporary legal practice, the ascendance of Artificial Intelligence (AI) is manifesting a paradigm shift in the operational fabric of law firms, particularly in the sphere of paralegal work. This phenomenon is not an exercise in technological substitution but a significant enhancement of the paralegal profession, augmenting their capacity to execute their responsibilities with greater efficiency and precision. AI's foray into legal workflows heralds an era where paralegals transcend conventional task execution, venturing into realms that necessitate a nuanced understanding of legal intricacies and sophisticated data management. This transformative phase redefines the traditional roles of paralegals, pivoting from administrative tasks to becoming integral components in the strategical and analytical undertakings of legal services.

The digitization of document management signifies one of the most substantial influences AI has exerted in legal operations. Law firms, as repositories of voluminous legal documents, have conventionally grappled with the manual labor of organizing and retrieving vast amounts of paperwork, a process fraught with inefficiencies and predisposed to human inaccuracies. AI-driven platforms like Luminance disrupt this status quo through the deployment of advanced algorithms that autonomously orchestrate the classification, sorting, and indexing of legal documents. This technological prowess not only curtails the expenditure of time but also escalates the accuracy of document processing, ensuring immediate and reliable access to pivotal information. Moreover, these intelligent systems discern patterns and irregularities within swathes of data, facilitating an analytical lens for paralegals that might otherwise be mired in the granularity of manual document scrutiny.

In conjunction with document management, the sphere of legal research has witnessed a substantial reformation at the hands of AI. The exigencies of legal practice command a swift and precise accumulation of legal texts, case precedents, and statutory provisions, a demand that AI-equipped legal research instruments address with remarkable aptitude. These tools transcend the rudimentary keyword search, embracing natural language processing to interpret legal inquiries and provide bespoke results that cater to the unique demands of legal cases. Platforms such as Everlaw do not merely furnish advanced search functions but integrate predictive analytics that intuit potential documents and judicial precedents relevant to ongoing cases. This not only optimizes the research trajectory but amplifies the scope and depth of legal research, fortifying the foundation upon which legal strategies are constructed.

The automation capabilities of AI extend to the realm of routine paralegal tasks that historically consume substantial professional bandwidth. Scheduling, invoicing, and client communication, once ensnared in the web of manual labor, are now elegantly managed by AI applications. Tools such as virtual assistants and chatbots assume these responsibilities with a high degree of accuracy and efficiency, liberating paralegals to devote their intellect and time to the essence of legal work. This liberation of human resources enhances the client experience, fostering a responsive and smooth interaction with legal services, which in turn bolsters client trust and satisfaction.

AI's reconfiguration of paralegal duties sets the stage for deeper involvement in strategic facets of legal practice. Liberated from the shackles of routine tasks, paralegals are afforded the latitude to engage intimately with case strategies, trial preparations, and client advocacy. This transition from administrative support to strategic partnership not only enhances the intrinsic value paralegals contribute to their legal firms but also enriches their professional journey with intellectually stimulating and rewarding experiences.

In contemplation of the metamorphosis ushered in by AI, it is imperative to recognize that the transformation extends beyond the mere enhancement of individual tasks. It signals a recalibration of the paralegal role itself, aligning it more closely with the strategic core of legal practice—an alignment that will be further explored in the ensuing discourse on the "Transition from Routine to Strategic Roles."

Evolving Roles: Paralegals as Strategic Partners

The integration of Artificial Intelligence (AI) into the paralegal domain signifies a profound evolution, transcending mere alteration of tasks to a complete redefinition of the paralegal role within the legal ecosystem. This evolution is pivotal, heralding a new epoch where paralegals emerge as strategic partners, their contributions echoing profoundly through the corridors of their firms and resonating in the heightened satisfaction of the clients they serve.

AI's encroachment into the paralegal sphere has liberated these professionals from the confines of routine task execution, propelling them into a landscape brimming with complex, high-value activities. Paralegals, historically relegated to the background tasks of document filing, data entry, and preliminary client interactions, are now engaging with the very heart of legal operations. AI-driven automation endows them with the opportunity to delve into profound legal research, participate actively in strategic discussions, and contribute directly to the architecture of case strategy. This participatory shift in their roles not only inflates their job satisfaction but also augments their prominence within the legal firm's architecture, fostering newfound visibility and significance.

Illustrative of this transition is the litigation sphere, where paralegals, now unshackled from the drudgery of administrative tasks, allocate their expertise to scrutinizing case law, crafting intricate reports, and collaborating closely with attorneys on the nuances of case management. This paradigmatic shift in operational roles engenders a symbiotic environment, enhancing the collective efficacy of legal teams in securing favorable outcomes for clients.

Concomitantly, the transformation of paralegal roles is manifest in the sphere of client relations. With AI assuming the helm of administrative functions, paralegals are afforded the bandwidth to forge and fortify client relationships. Their role evolves to encompass a spectrum of client-centric interactions, from the provision of consistent updates to the gathering and analysis of comprehensive information, which allows for a more profound resolution of client concerns. Such intensified engagement with clients does not merely uplift client satisfaction; it nurtures a foundation of trust and fidelity, imperative for the enduring success of any legal establishment.

The ascension to strategic roles mandates that paralegals partake in deliberate training and professional development. Recognizing this, law firms are increasingly channeling resources into educational programs, designed to equip paralegals with advanced competencies in legal analytics, sophisticated case management, and refined client communication strategies. Such investment in professional development is quintessential, ensuring that paralegals navigate their expanded roles with proficiency and assurance.

Nevertheless, the journey from routine to strategic roles is fraught with challenges. Paralegals confront the daunting task of acclimatizing to novel operational modalities, a stark departure from the familiarity of conventional methodologies. This metamorphosis necessitates a cultural recalibration within firms, recognizing and valuing the strategic contributions of paralegals. Moreover, this technological reliance introduces challenges of data security and the imperative for robust IT support, demanding paralegals to be proactive in safeguarding the sanctity of client information.

In summary, as the narrative on the transformation from routine to strategic roles concludes, it is evident that the function of AI and Legal Tech transcends the realm of efficiency tools; they act as catalysts for a profound redefinition of roles within the legal sector. This evolution necessitates a holistic embrace of change, commitment to continuous learning, and a nimble adaptation to the shifting landscapes of legal work.

We now segue into the subsequent analysis, "Challenges of Non-Adoption," wherein we scrutinize the consequences faced by law firms that are reticent or fail to integrate AI and Legal Tech, emphasizing the risks and the forfeiture of competitive edge inherent in such resistance. This introspection affords a comprehensive insight into the imperative of adopting AI and

Legal Tech in the contemporary, highly competitive legal milieu.

Consequences of Stagnation: The Impact of Non-Adoption of AI and Legal Tech

In a profession as dynamic and competitive as the law, the inertia of non-adoption of AI and Legal Tech bears the onus of inefficiencies, detriments to competitive edge, and discord with the evolving paradigms of legal service delivery. Law firms that abstain from embracing these seminal technologies risk falling into obsolescence, as their operational models clash with the exigencies of modern legal practice and client expectations.

Reduced Competitive Edge

In the legal battlefield, agility and responsiveness are the hallmarks of success. Firms reluctant to employ AI and Legal Tech find themselves encumbered by sluggish operational processes and protracted service delivery times. Paralegals tethered to manual methodologies are ensnared in time-intensive activities, precipitating case delays and elongated client wait periods. This lethargy in service is antithetical to client retention, particularly as clients' predilections shift towards firms offering prompt and effective legal resolutions.

Increased Risk of Human Error

Traditional, manual legal processes are fraught with the potential for error. Bereft of the precision and consistency offered by AI tools, the risk of inaccuracies in data entry, document analysis, and elements of legal research is amplified. These human-induced errors can cascade into a series of deleterious consequences: misfiled documents, overlooked pivotal information, and erroneous case filings, each undermining the firm's repute and potentially engendering legal setbacks for clients. This underscores the indispensability of AI and Legal Tech as bastions of reliability in the contemporary legal arena.

Inability to Leverage Big Data

AI's proficiency in dissecting and assimilating large data sets stands as one of its cardinal advantages. Firms bypassing AI forfeit the profound insights yielded by big data analytics—insights that could illuminate patterns in case outcomes, nuances in client behaviors, and strategies for resource optimization. The deprivation of this data-centric acumen stymies firms, constraining their strategic agility and impeding potential growth trajectories.

Cost Inefficiencies

Contrary to prevailing concerns that technological integration is cost-prohibitive, it is the expense of non-adoption that looms larger. Traditionalist firms grapple with inflated operational costs attributable to the extended hours necessitated by manual task execution and the need for an expansive workforce to shoulder these demands. Moreover, inefficiencies and manual errors induce a spectrum of collateral expenses—legal liabilities, corrective measures, and client compensations—bolstering the economic case for the adoption of technological aids.

Talent Retention and Attraction Challenges

The contemporary legal labor market, especially the emergent cohorts of the profession, exhibit a pronounced inclination towards technologically enriched work environments. They are drawn to roles that promise substantive engagement rather than the tedium of administrative monotony. Firms that forgo technological advancements suffer a drain of talent, ceding potential innovators to more progressive practices. Simultaneously, their allure wanes in the eyes of prospective talent, who prefer employers that invest in technology and foster professional development.

Conclusion of Non-Adoption

The exploration of the repercussions of non-adoption elucidates a stark reality: the embracement of AI and Legal Tech is not an elective strategic maneuver but an imperative for survival and prosperity in the legal domain.

As we proceed to the ensuing discourse on "Tools Enhancing Paralegal Productivity," we shall delve into the specific technologies that are redefining the operational landscape for paralegals, revealing how these tools not only circumvent the aforementioned challenges but elevate legal practice to unprecedented pinnacles of efficacy and sophistication.

Leveraging Technology for Paralegal Efficiency

In the swiftly evolving landscape of legal practice, the advancements in AI and Legal Tech have unveiled a suite of tools explicitly architected to escalate the productivity of paralegals. These tools, by mechanizing a spectrum of legal processes, free paralegals to allocate their expertise to more cerebral and rewarding tasks, heralding a new era of efficiency and effectiveness within modern legal practices.

Document Automation and Management Systems

Amongst the spectrum of legal technologies, document automation and management systems stand out for their revolutionary impact. Platforms such as Luminance and Logikcull epitomize innovation by harnessing AI to automate and refine the creation, organization, and analysis of legal documents. Utilizing the dual strengths of machine learning and natural language processing, these systems deftly navigate vast document landscapes, pinpointing essential information and orchestrating data in an intuitively accessible manner. Such automation is not merely a conduit to the time economy; it is a bulwark against the foibles of human error, ensuring that legal professionals can rely on a foundation of accurate and readily available data.

Consider Luminance's AI-driven document analysis, which transcends traditional review methodologies to furnish paralegals with a profound comprehension of legal documents' content, structure, and embedded risks. This insight empowers paralegals to expedite case material preparation, thereby augmenting their support for attorneys in strategic decision-making processes.

Legal Research Tools

AI's incursion into legal research has recast this once arduous task as an exercise in efficiency, critical to the fruition of successful legal outcomes. Platforms like Everlaw emerge as quintessential in this transformation, proffering advanced search functionalities that surmount the limitations of conventional keyword searches. By understanding the context enveloping user queries, these tools permit paralegals to uncover relevant case laws, precedents, and legal arguments with enhanced precision and speed.

Furthermore, the integration of predictive analytics in some legal research tools proposes probable outcomes predicated on historical data, thereby equipping paralegals and attorneys with strategic foresight. This analytic prowess is instrumental in formulating case strategies, as it offers a predictive panorama of how analogous legal scenarios have previously unfolded.

Client Relationship Management (CRM) Software

In the realm of client engagement, CRM software tailored for legal operations, such as Clio, is instrumental. These systems centralize client communications and case particulars, streamlining the maintenance of comprehensive client records. Enhanced scheduling capabilities, prompt follow-ups, and personalized service—facilitated by such CRMs—escalate client contentment. Furthermore, the seamless integration of these CRMs with various office management applications crystallizes a workflow conducive to efficient billing, document management, and overarching communication.

Communication and Collaboration Platforms

Amidst a tide shifting towards remote or hybrid work modalities, communication and collaboration platforms such as Microsoft Teams or Slack are indispensable. Adapted by myriad law firms, these platforms ensure seamless document collaboration, secure information sharing, and consistent team engagement, regardless of geographical dispersion. They are pivotal in preserving the integrity of team dynamics and guaranteeing the alignment of all case stakeholders with the latest developments.

AI-Powered Contract Review and Analysis Tools

AI-powered contract review and analysis tools, like LawGeex and ContractPodAi, are redefining contract management within corporate law sectors. These tools scrutinize contracts with unprecedented speed and accuracy, identifying non-standard clauses and potential discrepancies, thereby prompting amendments that align with legal best practices. This automation expedites contract review procedures and fortifies the consistency and accuracy of legal documents.

Conclusion

The suite of tools outlined herein represents just a fraction of the extensive arsenal by which AI and Legal Tech are fortifying paralegals, enabling them to divert their focus from rote tasks to endeavors of strategic significance. By embracing these technologies, paralegals enhance their intrinsic value to their firms, championing a new paradigm of legal practice.

As we pivot from the examination of discrete productivity-enhancing tools to the subsequent section, "Beyond Automation: AI's Broader Implications," we will explore the expansive influence of AI as it permeates beyond task automation to inform strategic decision-making and predictive analytics within legal practice. This foray will underscore AI's transformative impact throughout the legal sector, highlighting its instrumental role in not merely optimizing operations but in reconceptualizing the very fabric of legal strategies and outcomes.

Beyond Task Automation: AI's Role in Legal Strategy

As the legal industry harnesses AI and Legal Tech with increasing fervor, it is not merely routine tasks that are being reimagined; the technology is substantially recalibrating the legal landscape itself. These advancements bolster strategic decision-making, enable predictive analytics, and fundamentally revise the methodologies by which legal professionals approach their vocations. The implications of AI within the legal sphere underscore how the technology buttresses and enriches the strategic and analytical acumen of law firms.

Enhancing Strategic Decision-Making

AI's incursion into legal decision-making is a tour de force. It amalgamates and deciphers vast data troves, equipping paralegals and attorneys with insights that historically eluded capture due to logistical constraints. AI's prowess in discerning trends from legal case histories empowers legal teams to architect strategies grounded in data-driven projections rather than reliance on precedent or intuition alone. Moreover, AI's vigilant monitoring of legislative and jurisprudential evolutions ensures that legal strategies remain in lockstep with contemporary legal standards—an indispensable advantage in rapidly mutating domains like intellectual property or environmental law.

Predictive Analytics in Legal Practice

The domain of predictive analytics, too, has been touched by AI's transformative hand. Applying sophisticated algorithms to historical case data, AI furnishes forecasts of case outcomes, client dispositions, and even judicial predilections. This prescient ability emboldens law firms to assess case risks with greater precision, selectively pursue promising cases, and optimize resource allocation. For instance, platforms like Lex Machina distill legal analytics, granting firms a strategic edge through insights into the behavioral patterns of opposition, judicial track records, and jurisdictional ruling histories.

Automating Due Diligence and Compliance Checks

In the domain of due diligence, essential in corporate transactions, AI expediently sifts through documentary mountains, spotlighting salient information and potential anomalies. This alacrity not only expedites transactions but also bolsters the integrity and thoroughness of due diligence. Similarly, AI's surveillance of regulatory changes across jurisdictions is relentless, safeguarding law firms and their clientele against inadvertent non-compliance and the specter of legal penalties.

Reshaping Client Interactions and Services

AI is also reinventing client-firm interactions. AI-facilitated platforms proffer regular case updates, prognostic insights, and tailored advice—enhancements that engender deeper client engagement and satisfaction. What's more, AI's efficiency allows firms to adopt competitive pricing structures, broadening the accessibility of legal services and potentially expanding market reach.

Conclusion

The ramifications of AI in legal practice transcend the enhancement of efficiencies. By bolstering strategic deliberations, predictive analytics, and client services, AI is redefining the essence of legal practice in the digital epoch. Yet, amidst these seismic shifts, it remains imperative to recognize the undiminished significance of the human element within legal practice.

The next discourse, "Maintaining the Human Touch in a Digital Age," shall contemplate the symbiosis of AI's technological muscle with the essential, nuanced empathy that characterizes the legal profession. This equilibrium is vital, ensuring that as firms grow more efficient and data-centric, the soul of legal service—its inherent humanity—persists unobscured.

Preserving Humanity in the Age of Digital Law

In the intersection of the digital revolution and legal practice, AI and Legal Tech are potent forces of change, but it is the nuanced human touch that remains the beacon of the legal profession. As AI refines the efficiency and analytical prowess of legal operations, the craft of law continues to rely on the discernment and empathy inherently human. This section considers how legal professionals can harness AI to amplify their services while steadfastly preserving the soul of legal practice.

The Complementarity of AI and Human Insight

AI boasts the ability to process and analyze information at a superhuman pace and precision, yet it lacks the capacity for deep, empathetic comprehension that defines the legal professional's craft. The nuanced understanding of human emotion, the subtleties of ethical considerations, and the fine-grained judgment of complex legal matters remain the exclusive domain of human intelligence. Consequently, the most effective deployment of AI in law is one that supplements rather than supplants human expertise, ensuring that while AI provides a robust analytical foundation, the final judgment is rendered with a comprehensive understanding of the unique human elements at play.

Legal professionals, in leveraging AI-generated insights, must consider these insights as components within a broader, more complex decision-making framework. AI's data-driven predictions are valuable, but they must be weighed alongside the full spectrum of a client's personal circumstances and the ethical dimensions of each case.

Enhancing Client Relationships

In the digitized age, the importance of fostering robust client relationships cannot be overstated. Clients seek counsel that transcends transactional outcomes—they value understanding and support throughout their legal journey. Herein lies the irreplaceable value of the human touch. Paralegals, utilizing AI to efficiently manage routine tasks, are thereby empowered to invest more deeply in client interactions. CRM systems and other AI tools can streamline the management of client relationships, but they cannot replicate the empathetic connection, the reassurance, and the personalized care that a human paralegal offers. By prioritizing these human interactions, paralegals can significantly enhance client satisfaction and foster long-term loyalty.

Training and Development for Empathetic Practice

To marry technological adeptness with empathetic client service, law firms must be committed to ongoing training and development. It is essential to educate legal staff not only in the operational use of AI tools but also in the soft skills that foster compassionate client interactions—emotional intelligence, ethical practice, and nuanced communication. Such education ensures that legal professionals are equipped to navigate the complexities of legal issues with sensitivity and to provide support that is understanding and thoughtful.

A legal culture that equally values technical proficiency and interpersonal skill engenders a more holistic legal practice, one that recognizes AI as a tool for enhancing service delivery without diminishing the importance of human interaction.

Ethical Considerations in the Use of AI

As AI becomes increasingly integrated into legal workflows, ethical considerations must take center stage. It is imperative to ensure that AI systems are employed with a sense of responsibility, vigilance against biases, and an unwavering commitment to maintaining client confidentiality. Legal professionals must stay informed about the ethical dimensions of AI and are tasked with deploying these technologies in ways that adhere to the highest standards of legal practice.

Conclusion

To conclude, the essence of maintaining the human touch in a digitized legal age lies not in drawing lines between technology and personal interaction but in finding the harmony that enhances both. As legal professionals skillfully navigate this confluence of technology and humanity, they affirm the sanctity of their practice—ensuring it remains as humane as it is proficient.

The subsequent and final discussion, "The Indispensable Role of Paralegals in Modern Legal Practice," will encapsulate the significance of paralegals in today's tech-forward legal milieu. Amidst swift technological progress, the personal skills paralegals possess are not only invaluable but also imperative, safeguarding the quintessence of legal service—its unwavering commitment to humanity—even as it evolves in efficiency and scope.

The Vital Role of Paralegals in a Digital Legal Landscape

In the innovative expanse of legal technology, paralegals stand at the vanguard, navigating the transformative tide that AI and Legal Tech have ushered in. These technological advances confer remarkable efficiency and analytical prowess, yet the indispensable human core that paralegals embody remains the fulcrum of effective legal practice. This conclusive section amplifies the critical role that paralegals fulfill as they meld advanced technology with the nuanced, empathetic essence of law, spotlighting their ever-essential skills and contributions in the modern legal tapestry.

Paralegals as Strategic Assets

Amidst the mechanization of routine tasks and the unveiling of new data analytic horizons, paralegals are emerging as strategic assets within their firms. Possessing a profound grasp of legal procedures coupled with a keen sensitivity to client needs, paralegals adeptly navigate AI-generated data streams. This dual adeptness fosters a capacity to distill strategic insights that synthesize the raw computational output of technology with the rich tapestry of human experience.

Moreover, paralegals are pivotal in fostering client relationships and upholding the ethical deployment of technology, tasks quintessential in preserving the trust and integrity that underpin the legal profession. Through their stewardship, paralegals not only bolster the stature of their firms but also anchor legal practice to the bedrock principles of equity and justice.

Continuous Learning and Adaptation

The brisk pace of technological innovation within the legal field necessitates an ethos of incessant learning and adaptation from paralegals. To maintain relevance and efficacy, paralegals must actively refine their skills, embracing advancements in legal technology and fortifying their competencies in critical thinking, effective communication, and client service.

Progressive law firms that dedicate resources to the continuous education and development of their paralegals position themselves to harness the full potential of AI and Legal Tech. Such strategic investment cultivates a dynamic and versatile workforce, poised to spearhead the legal industry in a market increasingly defined by technological advancements.

The Human Touch in Legal Services

In the milieu of legal services, AI's capabilities, although extensive, cannot supplant the human touch that is the hallmark of the profession. Paralegals stand as guardians of this personal dimension, ensuring that technological tools enhance, rather than eclipse, the intrinsic aspects of legal practice. Their innate ability to resonate with clients, comprehend the intricacies of their concerns, and dispense bespoke counsel renders them irreplaceable in the delivery of holistic legal services.

This human aspect assumes heightened significance in complex legal scenarios that demand emotional and ethical discernment. The presence of an astute and compassionate paralegal can significantly sway the client experience and materially influence case outcomes.

Conclusion

Reflecting on the evolving paradigm of the paralegal role in an age of AI and Legal Tech, it becomes manifestly clear that paralegals are not merely enduring but flourishing. By assimilating novel technologies, magnifying their strategic involvement, and preserving the indispensable human touch, paralegals affirm their critical place within the legal ensemble.

Their readiness to adapt and their zest for embracing emergent challenges render paralegals invaluable in contemporary legal practice. As the technological wave progresses, the paralegal's role will invariably expand and adapt. Yet, anchored by a commitment to ethical standards, client service, and the human dimension of the law, legal practices continue to be defined as much by the people at their core as by the technology they employ.

In summation, the integration of AI and Legal Tech into legal practices is redefining not only the mechanics of work but also the very skills and roles of legal professionals. Paralegals, central to this evolution, are instrumental in guiding the legal profession towards a future where the symbiosis of technology and humanity coalesces to dispense justice and legal services with unprecedented efficacy and compassion.

Chapter 6:

AI Chatbots in Law: Bridging the Gap in Legal Consultations

Transformative and tireless, AI chatbots are reengineering the client experience in law with remarkable efficiency and reach. Among the myriad of applications, AI chatbots stand out for their capability to automate legal advice, a development that promises to enhance efficiency and accessibility in legal consultations. These AI assistants are designed to interpret and respond to legal queries in real-time, offering preliminary advice and guidance to those who might otherwise face barriers in accessing legal support. By offloading routine inquiries to AI legal chatbots, firms are able to allocate their human resources to more complex, higher-value tasks, thus optimizing their service delivery model.

However, the embrace of AI chatbots in legal practices isn't without its challenges. Concerns regarding the accuracy of automated advice, confidentiality of client data, and the potential for misunderstanding nuanced legal problems are valid. It's imperative, then, that legal professionals thoroughly evaluate these tools, ensuring they're implemented in ways that amplify the strengths of human lawyers rather than dilute the quality of legal advice. Crafting detailed protocols for chatbot interactions and continuously monitoring these AI assistants for learning and improvement are crucial steps in mitigating risks associated with automation. Importantly, these measures also ensure that the ethical standards of the profession are maintained, even in the face of rapid technological advancement.

The potential of AI chatbots in legal settings is undeniably vast. When used judiciously, they can make legal advice more accessible, improve client satisfaction through quick responses, and even educate the public on basic legal matters. Nonetheless, they are not a replacement for the nuanced judgment and empathetic understanding that experienced legal professionals provide. Thus, the goal is not to replace lawyers with machines but to leverage AI to enhance the legal profession's ability to serve and advocate. As we navigate this integration, it's incumbent upon legal practitioners to stay informed about advancements in AI, ensuring that technology serves the noble aim of justice rather than undermining it.

AI Legal Assistants: Harnessing AI Chatbots for Legal Task Automation

The legal profession, a bastion of tradition, now teeters on the edge of a revolution, with artificial intelligence (AI) poised to catalyze an unprecedented transformation. As the intricate tapestry of legal innovation unfurls, AI legal assistants stand out as pivotal actors in the redefinition of legal service delivery. Not simply ornamental, these digital cohorts—infused with cutting-edge AI and machine learning—are indispensable for legal practitioners determined to excel in a rapidly evolving landscape.

These AI legal assistants are engineered to automate a plethora of tasks that typically dominate a lawyer's agenda. The drafting of legal documents, and the sifting through of case law—activities once burdensome—are now within the effortless purview of these intelligent systems. This automation heralds a paradigm shift, promising not just efficiency but a radical enhancement in the caliber of legal service.

A deep dive into the underpinnings of AI legal assistants reveals a foundation of complex algorithms adept at navigating massive datasets with speed and precision. Their use of natural language processing (NLP) allows them to comprehend and generate text that approximates human conversation, making them indispensable for document drafting and review.

However, efficiency is not the sole benefit of AI legal assistants. They stand as champions of legal advice quality. With their access to fresh case law, statutes, and scholarly commentary, they ensure legal counsel is built upon the most recent and relevant legal authority. This diminishes human error and guarantees that legal guidance is as thorough and current as it can be.

Client service is another realm ripe for revolution by AI legal assistants. In an era where immediacy is king, AI-driven chatbots deliver swift answers to a spectrum of queries, establishing a new standard for client satisfaction while liberating lawyers to address more nuanced legal challenges.

The true disruption, however, lies in the adaptability and learning prowess of AI legal assistants. Tailorable to a firm's or lawyer's specificities, they refine their utility over time, aligning closely with the particular legal nuances of their users, thus offering an ever-more personalized form of assistance.

While the ascendancy of AI in legal practices signals a new dawn, it is not without its complexities. The effectiveness of AI legal assistants is tethered to the expanse and integrity of the data they are fed. Moreover, legal practitioners must acclimate and learn to trust and interpret the results produced by their digital aides.

The ethical domain is just as critical. The deployment of AI legal assistants must respect legal advice's sanctity and client confidentiality. This ethical navigation demands deliberate policy crafting and constant vigilance.

For those in the legal field embracing this new technological epoch, the rewards are vast. AI legal assistants not only pare down time spent on routine tasks but also augment the accuracy and pertinence of legal research, thereby lifting the quality of legal counsel. Yet, harvesting these rewards calls for an investment in technology, training, and the management of change.

The integration of AI legal assistants into day-to-day practice is challenged by resistance to change, fears of job displacement, and the practicalities of melding new systems into established IT frameworks. Nevertheless, these challenges are surmountable. With strategic planning, thorough training, and a dedication to continual evaluation and improvement, the legal sector can deftly manage this transition.

In this vein, DoNotPay serves as a paragon in the vanguard of AI legal assistants. Born from the simple premise of contesting parking tickets, it has burgeoned into an all-encompassing legal tool that streamlines a wide gamut of legal issues for the layperson. Its interface—intuitive in design—and its automated processes stand as a testament to AI's power to democratize access to legal resources, establishing a benchmark for others in the space.

Coined as "The World's First Robot Lawyer," DoNotPay epitomizes the blend of innovation and user empowerment. Despite legal threats to its creator, including jail time for potential courtroom appearances by the AI, DoNotPay persists, exemplifying the mission "to level the playing field and make legal information and self-help accessible to everyone." In light of shifting focuses, DoNotPay has recently pared down its offerings, concentrating solely on "non-consumer legal rights products," underscoring a commitment to the everyday individual's legal challenges.

AI legal assistants are set to redefine the essence of legal practice. They offer a transformative potential in the realm of legal services, shaping the future of law with each task automated and each process streamlined. The legal sector must now pivot to embrace and capitalize on these technologies to remain relevant and competitive in a digitized world.

As the legal community perches on the precipice of this new era, it is not a question of whether AI will revolutionize the industry but rather of how swiftly and adeptly legal professionals will adapt to the impending changes. The dawn of AI legal assistants is merely the commencement, and their potential to overhaul the legal industry is bound only by our collective willingness to embrace and exploit this transformative technology.

The course ahead is explicit. By embedding AI legal assistants within their practices, legal professionals are not merely mechanizing tasks—they are redefining the essence of law practice in the 21st century. The future of law has arrived, with AI as its bedrock.

This journey towards the adoption of AI legal assistants is undoubtedly intricate, peppered with obstacles yet rich with opportunity. By leveraging AI, legal professionals can unlock unprecedented levels of proficiency, precision, and client contentment, propelling the evolution of the legal field.

Standing at the threshold of this novel epoch in legal services, the pivotal query is not if AI will transmute the legal profession, but rather how swiftly and efficaciously legal professionals will assimilate these advancements. The inception of AI legal assistants represents just the initial step, with their potential to revolutionize the legal sector being constricted solely by our readiness to espouse and utilize this burgeoning technology.

The path forward is lucid. By integrating AI legal assistants into their modus operandi, legal practitioners do more than streamline tedious tasks—they pioneer a renaissance of what it means to practice law in the contemporary world. AI is not merely an addition to the legal arsenal; it is the cornerstone of a new legal reality.

The metamorphosis of law by AI is an inexorable tide. With entities like DoNotPay leading the charge, legal professionals have tangible exemplars of how AI can be wielded to extend legal aid's reach and calibrate its delivery. DoNotPay's trajectory from a simplistic parking ticket dispute app to a robust legal aid tool illustrates the vast scope of AI's application. It has redefined what constitutes a legal assistant, melding technological sophistication with a staunch dedication to public service.

Joshua Browder, the architect behind DoNotPay, encapsulates the spirit of innovation that drives the legal tech industry. Despite facing legal intimidations, Browder's steadfast vision of a robot lawyer accessible to all underscores a deep-seated belief in equity and accessibility. DoNotPay's recent streamlining initiative, focusing on the enhancement of consumer legal rights, manifests a strategic reorientation towards empowering the individual against a backdrop of legal complexity.

As legal professionals consider the incorporation of AI legal assistants like DoNotPay into their practices, it is critical to acknowledge the broader implications. These tools not only reshape how tasks are performed but also amplify the potential for consumer empowerment. They make legal assistance more approachable and decipherable to the general populace, thereby democratizing legal knowledge.

However, the march towards AI integration is not a solo venture. It requires a collaborative ethos, where legal minds and tech innovators converge to sculpt tools that are not only technologically advanced but also ethically sound and user-centric. The result is a synergistic blend of legal expertise and AI efficiency that expands the frontiers of legal service delivery.

Moreover, the adoption of AI legal assistants is a dynamic process, necessitating a cultural shift within the legal community. It demands a willingness to evolve, to reassess traditional workflows, and to envision a legal future that is inclusive, intelligent, and interconnected.

DoNotPay's story, emblematic of this shift, serves to remind us of the transformative power of AI. It is a call to action for the legal profession to not only adopt AI but to adapt to it—to harness its capabilities to enhance access to justice, streamline the mundane, and ignite a reformation of legal processes.

As we stand at this juncture, the role of legal professionals is more pivotal than ever. It is not simply to adopt AI legal assistants but to become the stewards of this technological renaissance. To guide it with foresight and a principled commitment to the greater good, ensuring that as the wheels of justice turn, they do so with greater speed, accuracy, and fairness.

In sum, AI legal assistants mark the commencement of a new chapter in legal practice. As we embrace the profound capabilities of DoNotPay and its ilk, we step into a future where legal services are more accessible, efficient, and equitable. The legal profession must not only welcome AI but must also actively shape its trajectory, ensuring that this technological boon realizes its full potential in serving the cause of justice. The future of law is now, and AI legal assistants are at the forefront, beckoning us towards an era of unrivaled innovation and opportunity.

Client Service Innovations: The Impact of AI Chatbots

The legal profession, traditionally seen as a bastion of human expertise and personalized service, is undergoing a seismic shift with the advent of artificial intelligence (AI), particularly AI chatbots. These chatbots, exemplified by the innovative Law Bot Pro, are revolutionizing the way client services are rendered in law firms. The integration of AI chatbots like Law Bot Pro into legal practices is not just an enhancement of existing processes but a monumental step towards creating more efficient, accessible, and client-focused services.

AI chatbots, including Law Bot Pro, are engineered to simulate human-like interactions, providing clients with instant responses to their queries. This capability marks a significant improvement in client service, offering 24/7 availability that surpasses the limitations of human staff. For example, when clients have urgent questions outside of regular business hours, AI chatbots can supply them with timely and relevant information, a feat previously unattainable.

Furthermore, these chatbots are equipped with advanced natural language processing skills, enabling them to comprehend and respond to a broad spectrum of legal inquiries. This is particularly advantageous for addressing routine questions related to legal procedures, documentation requirements, or general legal advice. By automating these responses, law firms can liberate valuable time for their attorneys to concentrate on more complex and nuanced legal matters.

A significant benefit of AI chatbots in legal practices is their capacity to provide personalized recommendations based on a client's specific situation. With machine learning algorithms, chatbots like Law Bot Pro can analyze a client's query and past interactions to deliver tailored advice. This level of personalization enhances the client experience, making it more akin to interacting with a human legal advisor.

Addressing potential concerns about the reliability and accuracy of legal advice provided by AI chatbots is essential. To alleviate these worries, law firms must ensure their chatbots, including Law Bot Pro, are regularly updated with the latest legal information and trained using a comprehensive legal knowledge dataset. Additionally, there must be a clear process for escalating complex or sensitive queries to human attorneys, thereby maintaining the standard of legal advice.

Privacy and confidentiality are of utmost importance in the legal field. Thus, implementing AI chatbots necessitates stringent data protection measures. Law firms must adhere to strict data privacy regulations and ensure their chatbots are designed with security as a foundational element. This includes encrypting client communications and protecting against unauthorized data access.

The introduction of AI chatbots like Law Bot Pro offers law firms an opportunity to gather valuable insights into their clients' needs and preferences. By analyzing interactions between clients and chatbots, firms can identify common concerns, trending legal issues, and areas where additional client education might be beneficial. This data can guide the development of new services, legal content, and targeted marketing strategies.

Careful planning and consideration are required for implementing AI chatbots. Law firms must select the right technology partners, invest in training their teams, and establish clear protocols for chatbot deployment, including educating clients on interacting with chatbots and setting realistic expectations about the issues these automated systems can address.

The economic rationale for adopting AI chatbots is strong. By automating routine inquiries and client interactions, law firms can achieve significant cost savings. These savings can then be redirected towards more strategic investments, like expanding legal services or enhancing client support infrastructure. Moreover, the scalability of AI chatbots allows firms to manage growing client bases without proportionately increasing their staff count.

However, the adoption of AI chatbots in legal practices, such as Law Bot Pro, faces challenges. Concerns about the impersonal nature of chatbots, potential resistance from clients accustomed to traditional service models, and the ongoing need for technical support and upgrades must be addressed. Successful integration hinges on finding the right balance between automation and human interaction, ensuring that clients feel valued and supported throughout their legal journey.

In conclusion, AI chatbots represent a substantial opportunity for law firms to upgrade their client services. By offering timely, accessible, and personalized legal assistance, these automated systems can improve client satisfaction and loyalty. The success of these initiatives depends on careful implementation, continuous monitoring, and a commitment to maintaining the high standards of legal advice and confidentiality that clients expect. As the legal landscape continues to evolve, embracing AI chatbots may become a necessity for law firms aiming to stay competitive and meet the changing needs of their clients.

The decision to integrate AI chatbots like Law Bot Pro into legal services is strategic, aligning with the firm's broader goals and a commitment to future investment. Doing so allows law firms to not only enhance operational efficiency but also deepen client relationships, laying the groundwork for long-term success in an increasingly digital environment.

Moving forward, AI's role in enhancing client services within the legal profession will undoubtedly expand. Law firms that recognize this potential and act decisively, like those adopting Law Bot Pro, will lead in setting new standards for excellence and innovation in client service. Thus, the journey toward integrating AI chatbots is not just an upgrade; it's a leap into the future of legal practice.

Law Bot Pro, developed by Rare Labs, serves as a prime example of how chatbots are transforming client-lawyer interactions. With its conversational AI capabilities, Law Bot Pro discerns client needs and offers clear legal pathways, merging efficiency with personalized care. Its adoption serves as a model for how legal practices can maintain ethical standards while embracing cutting-edge technology.

Rare Labs, an Indian legal tech company, launched Law Bot Pro as a pro-bono initiative, making AI-powered tools accessible to the legal community. The creation of Law Bot Pro was driven by the acute need for informative legal aid, which was previously accessible to only a few, and limited to specific areas where legal aid drives were conducted. Since its inception, Law Bot Pro has become a household name that provides free legal informative aid to Indian citizens. With over 130,000 users as of July 2023, the platform has proven the necessity and impact of such innovative solutions in meeting the legal needs of the populace.

Mandaar Mukesh Giri, a recent law graduate and an Advocate based in New Delhi, India, founded Rare Labs. Under his leadership, Law Bot Pro has established itself as India's own legal AI companion, offering assistance specifically tailored to Indian laws. This localized approach ensures that the legal advice provided is relevant and accurate, reflecting the specific legal framework and nuances of Indian law.

Law Bot Pro's integration into legal services highlights the broader trend of digital transformation in the legal sector. By offering a user-friendly, accessible, and efficient legal aid tool, Law Bot Pro addresses the gap in legal services accessibility, democratizing legal information for a wider audience. Its success story underscores the potential of AI in legal tech to bridge the gap between legal professionals and the public, enhancing the overall efficacy and reach of legal services.

The chatbot's impact extends beyond individual client interactions. It serves as a tool for legal education and empowerment, helping users to understand their legal rights and options. By providing immediate, reliable legal information, Law Bot Pro helps to reduce the intimidation factor often associated with legal processes and encourages more people to seek legal assistance when needed.

Furthermore, the data collected through interactions with Law Bot Pro offers invaluable insights into the legal needs and concerns of the population. This data can be leveraged to identify trends, inform policy decisions, and drive legal research, contributing to the overall improvement of the legal ecosystem.

The story of Law Bot Pro exemplifies the transformative power of AI in the legal industry. It shows how technological innovation when aligned with the mission to serve and educate, can have a profound impact on society. For law firms and legal professionals, adopting AI chatbots like Law Bot Pro is not just about upgrading client services; it's about participating in a larger movement towards a more inclusive, efficient, and responsive legal system.

In embracing AI chatbots, law firms must navigate the complexities of integrating new technologies with traditional legal practices. They need to consider not only the technical and operational aspects but also the ethical implications of using AI in legal contexts. Ensuring transparency, maintaining client trust, and upholding legal standards are paramount in this process.

As AI continues to evolve and become more sophisticated, its role in the legal field will expand, offering new opportunities and challenges. Law firms that are early adopters of this technology, like those using Law Bot Pro, will be at the forefront of a new era in legal services characterized by increased accessibility, efficiency, and client satisfaction.

In conclusion, the integration of AI chatbots like Law Bot Pro into legal practices signifies a pivotal shift in how legal services are delivered. It reflects a broader trend towards digitalization and innovation in the legal sector, offering promising prospects for enhancing client services and advancing the cause of justice. As law firms and legal tech companies continue to explore and harness the potential of AI, the future of legal practice looks increasingly dynamic and client-centered, heralding a new chapter in the ongoing evolution of the legal profession.

AI-Driven Excellence in Client Satisfaction

In the landscape of legal services, the pursuit of client satisfaction is paramount. As law firms strive to not only meet but exceed client expectations, the advent of Artificial Intelligence (AI) has emerged as a cornerstone of client-centric innovation. LegalMation, at the forefront of this technological renaissance, is redefining the benchmark for client service in the legal industry.

LegalMation's AI technology presents a transformative approach to the drafting of legal documents, a cornerstone of legal practice. It is not merely an incremental improvement but a complete reimagining of the process. By harnessing the power of AI, LegalMation automates the production of responsive pleadings, discovery requests, and various legal documents. This capability is transformative, allowing attorneys to produce documents at a speed and accuracy that were previously unattainable. The system takes into account the intricate details of each case, ensuring that every document is tailored specifically to the nuances and demands of the situation at hand while also aligning with jurisdictional variances.

This level of detail and customization enhances the overall quality of legal work presented to clients. It instills a deep sense of confidence in clients, who can clearly see that their legal matters are being handled with the utmost care and precision. This directly correlates with elevated levels of client satisfaction, as they recognize the value being added through the use of advanced technology in their legal representation.

Beyond streamlining document creation, LegalMation redefines the concept of time within the legal profession. Traditionally, the creation of legal documents is a time-intensive task, often consuming hours of meticulous work. By automating this task, LegalMation frees up invaluable time for attorneys, allowing them to invest more in their client relationships. With more time available, lawyers can delve deeper into the intricacies of each case, providing more comprehensive and strategic advice. This transition from task-oriented work to a focus on strategic legal counsel adds a new layer of value to client services, as clients receive not just a service, but a dedicated legal partnership.

Moreover, the swift response facilitated by LegalMation's AI technology aligns perfectly with the expectations of today's clientele, who are accustomed to immediacy in service delivery. When legal professionals leverage LegalMation, they are able to provide prompt document turnaround, thereby laying a solid foundation of trust and reassurance from the very beginning of the client relationship.

LegalMation's AI solutions signify a law firm's commitment to embracing innovative solutions. Clients are increasingly tech-savvy and seek out firms that demonstrate a forward-thinking approach. By incorporating LegalMation, a law firm showcases its dedication to modernization and efficiency. This commitment resonates with clients, who feel assured that their legal representatives are at the forefront of industry advancements.

One of the most impactful features of LegalMation is the empowerment it offers to clients. By enabling more efficient processes, the platform offers clients greater transparency and involvement in their legal matters. They gain an enhanced understanding of the legal process, which not only reinforces their trust in their legal advisors but also encourages a deeper engagement with their own cases.

Looking to the future, LegalMation's role in the industry transcends operational improvements, marking a shift towards comprehensive client service enhancements. This is not a matter of AI taking over the role of legal advisors; it is about AI facilitating lawyers to become better advisors. LegalMation enables legal professionals to practice at the zenith of their capabilities—providing sophisticated, individualized, and attentive service.

The contributions of LegalMation to the sector demonstrate a bold move towards redefining client service in the legal industry. This AI system does not merely support legal practice; it enriches it by blending the precision of technology with the personalized touch of human counsel. This symbiosis is the future of law practice—a future where technology elevates the client experience and where satisfaction is the linchpin.

By adopting LegalMation, law firms are not just keeping pace with technology—they are leading a transformation towards a more accessible, responsive, and client-centric legal service delivery. They are reimagining what it means to serve clients in the legal field, opening a new chapter in the narrative of client satisfaction, where superior service is not just expected but delivered consistently and with distinction. LegalMation stands as a beacon of this new era, demonstrating that the future of client satisfaction in legal services is here, and it is being shaped by the intelligent application of AI.

Chapter 7:

Data-Driven Judgement: AI in Case Strategy and Outcome Forecasting

Dramatic shifts in courtroom strategy unfold as artificial intelligence steps into the spotlight, altering the landscape of support and prediction. Traditionally, litigation outcomes were forecast based on the lawyer's experience, historical precedent, and instinctual judgment. However, with the advent of AI technologies, legal practitioners now have access to tools that can analyze vast datasets, recognize patterns, and predict outcomes with a level of accuracy previously deemed impossible. This paradigm shift offers an unparalleled advantage in strategic planning, resource allocation, and overall case management.

One of the key applications of AI in litigation is its capacity for legal outcome predictions. By leveraging historical data, AI algorithms can identify trends and probabilities associated with different legal strategies, judge decisions, and jury reactions. This predictive capability enables attorneys to make more informed decisions, tailor their approaches, and set realistic expectations with their clients. It's a sophisticated tool that, when used correctly, has the potential to significantly increase the chances of a favorable outcome. However, this technology is not without its challenges, including issues related to data bias and the ethical use of predictive information.

In contemplating the ethical dilemmas that accompany AI in the courtroom, it's essential to engage in a nuanced discussion on the balance between technological advancement and ethical responsibility. The accuracy and impartiality of AI predictions depend heavily on the dataset it learns from, which can inadvertently reflect historical biases. Furthermore, the question of whether AI should influence legal decisions raises concerns about transparency and the essence of justice. These ethical considerations necessitate a comprehensive framework that governs the use of AI, ensuring that its employment enhances the legal process without compromising the core values of justice and equity. As we move forward, the legal profession must tread carefully, embracing the benefits of AI while vigilantly guarding against its potential pitfalls.

Data-Driven Defense: AI as the Modern Litigator's Tool

The frontier of legal technology has been irrevocably altered by the advent of artificial intelligence (AI), with its influence particularly pronounced in litigation support and the anticipation of legal outcomes. The essence of this chapter is to examine the monumental shift brought forth by AI, which is facilitating a new era of legal strategy and decision-making through predictive analytics.

AI's pivotal role in this realm is underscored by machine learning technologies, which imbue computer systems with the ability to learn from data, refining their capabilities with each new case. Delving through extensive legal data—ranging from filings to court decisions—these algorithms discern intricate patterns that are too subtle for the human eye, laying the groundwork for litigation strategies deeply rooted in empirical analysis. This nuanced understanding of past legal outcomes paves the way for more strategic planning and improved advisories.

Beyond mere historical analysis, AI in the legal domain also critically evaluates the merits of legal arguments, offering a prognostic evaluation of their likely success. This is not simply a passive analysis but a dynamic, predictive model that contrasts against vast databases of legal history to forecast potential outcomes. This surge towards empirically supported lawyering is redefining the preparation and presentation of legal cases.

In practice, AI's benefits are manifold, offering lawyers the tools for intricate scenario analysis. With AI, attorneys can model diverse case trajectories, modifying key variables to assess their impact. This enables a more robust and strategic representation, empowering clients with forward-thinking legal counsel.

However, the utility of AI does not signify an encroachment on the lawyer's traditional role; rather, it is a tool that augments the legal expert's analytical capacity. It serves to enhance—rather than replace—the discerning judgement that only experienced legal professionals can provide.

The ethical integration of AI in law also demands attention. With the power of predictive analytics comes the responsibility to uphold transparency and mitigate biases that could otherwise compromise the justice system's integrity. Legal professionals must engage with the methodologies of AI, ensuring they understand and can counteract any embedded biases.

The implementation of Lex Machina's Legal Analytics platform illustrates the transformative impact of AI. By leveraging a colossal repository of legal data with cutting-edge machine learning, Lex Machina grants legal professionals unprecedented strategic insights. This includes detailed analyses of judicial behavior, case histories against specific defendants, and much more. This confluence of quantitative data and analytics fosters a predictive capability that empowers lawyers to craft well-informed, tailored litigation strategies with heightened predictability.

Lex Machina: The Pinnacle of Legal Analytics

With an unrivaled depth of accurate, comprehensive, and transparent data, Lex Machina stands out as the epitome of legal analytic tools. Our approach—melding advanced machine learning with the expertise of in-house legal analysts and ensuring data currency with daily updates—sets a new industry standard. We highlight several unique features:

- **Unique Outcomes:** Lex Machina's Outcome Analytics™ offers exclusive insights into damages, resolutions, and remedies—vital to comprehending the trajectory and conclusion of a case.

- **Valuable Insights**: Our Case List Analyzer™ employs exclusive filters to quickly identify relevant precedents, based on actual case documents, facilitating a comprehensive and expedient case assessment.

- **Instant Results**: Lex Machina's Quick Tools enable immediate comparative analyses of judges, law firms, and other entities, and facilitate the development of compelling motions and the evaluation of patent portfolios or expert witnesses.

- **ccurate Data**: Our patented Attorney Data Engine™ amends and completes attorney records, yielding precise representation of legal counsel involvement—a granularity unmatched by other platforms.

- **Transparent Sources**: Lex Machina encourages demonstrable analysis, with the ability to delve from overarching trends to the granular level of dockets and documents, coupled with clear definitions and sources.

This technological confluence underscores the necessity for cross-disciplinary collaboration, merging legal acumen with data science expertise. Law firms are increasingly forming alliances with AI specialists, capitalizing on the comprehensive potential of these technologies.

Clients, too reap the rewards of AI's integration, gaining more precise outcome forecasts that inform their litigation and resolution strategies. Such clarity can revolutionize litigation approaches, economizing time, resources, and reducing the emotional burden on all parties.

Nevertheless, a judicious approach to AI predictions is warranted. Algorithms offer probabilities, not certainties, and external variables can always alter a case's direction. Hence, AI should complement, not dictate, a lawyer's strategic arsenal.

As AI's predictive proficiency continues to mature, we anticipate increasingly sophisticated understandings of legal outcomes. The legal sector is set for a continued evolution, with AI as a catalyst for innovation and scholarly expansion. Education and ongoing training in AI will be pivotal for the legal community, with academic and professional programs incorporating this knowledge into their offerings.

The democratization of AI tools is levelling the playing field, offering smaller practices access to resources that were traditionally exclusive to larger firms. This not only boosts competition but also fuels innovation, benefiting the legal field at large.

The forward momentum of AI in law is presenting intriguing possibilities. With AI's deeper entrenchment in legal procedures, we can expect it to spur further advances, open new scholarly avenues, and transform legal service delivery.

In summation, the utilization of AI for legal outcome predictions denotes a significant shift in litigation and client advisory. By grounding their counsel in data-driven insights, legal professionals can offer advice with enhanced accuracy and assurance. As the legal landscape undergoes this transformative shift, it is crucial for the profession to harness the potential of AI while maintaining a vigilant stance on its ethical implications and innate limitations.

Lex Machina's Distinction in the AI Legal Landscape

Lex Machina's pioneering Legal Analytics platform exemplifies the apex of leveraging AI in the legal field. Its precise, comprehensive, and transparent analytic data, augmented by proprietary outcome analytics, positions it uniquely in the marketplace. The data, meticulously curated and constantly refined by a blend of sophisticated algorithms and legal expertise, is regularly updated, ensuring lawyers have access to the most current information. Lex Machina's distinct features include:

- **Exclusive Outcome Analytics**: The platform's Outcome Analytics™ delve into the specific details of damages, case findings, resolutions, and remedies, offering a clear view of what transpired and determining the victor with precision.

- **Unparalleled Insights**: Lex Machina's Case List Analyzer™ facilitates the rapid discovery of similar cases through practice-specific filters and tags, grounded in the realities of the court's documents, allowing for the swift and comprehensive assembly of case lists.

- **Swift, Actionable Results**: Quick Tools offer on-the-spot analysis, whether comparing judicial tendencies, evaluating parties, or assessing law firms. These tools streamline the creation of compelling motions and enhance the investigation of expert witnesses or patent portfolios.

- **Enhanced Data Accuracy**: The patented Attorney Data Engine™ of Lex Machina amends and enhances counsel data, ensuring an accurate portrayal of lawyer appearances, outperforming other systems that rely on less detailed document-based analysis.

- **Unmatched Transparency**: Lex Machina's platform stands alone in its ability to allow users to trace their analytical journey from broad patterns down to individual dockets and documents, ensuring complete understanding and transparency of the data's origin.

As we navigate this new era, the promise of AI to reshape the judicial process is palpable. AI's predictive capabilities will only sharpen with time, promising even more profound insights into the complex web of legal outcomes. This inevitable advancement beckons a new breed of legal professionals, skilled not only in law but in the language of AI and analytics.

To leverage AI to its fullest potential in predicting legal outcomes, education and continual learning are essential. The legal profession must ensure practitioners are well-versed in the nuances of AI interpretations and can apply these insights with a critical eye.

The democratization of AI also creates a more egalitarian legal sector. Now, even solo practitioners have at their disposal the analytical prowess that once required the resources of a large firm. This shift fosters greater innovation and a more dynamic legal market.

Looking to the future, the assimilation of AI in legal prediction heralds an exciting chapter in the practice of law. As AI technologies become more entrenched in day-to-day legal operations, they will likely be a beacon for further ingenuity, initiate new legal research areas, and recast the landscape of legal services.

In closing, embracing AI for legal outcome prediction is a move towards a new strategic paradigm in the legal arena. It equips legal professionals to guide their clients with an enhanced level of foresight and specificity. As we embrace this new phase, the legal community must recognize and engage with AI's capabilities, always aware of its ethical considerations and the limits within which it operates.

Empirical Insights: Testing AI’s Litigation Efficacy

In the intricate dance of litigation, where every step is calculated, and every move can tip the scales of justice, the emergence of Artificial Intelligence (AI) has introduced a new rhythm. The real-world accuracy of AI within the confines of legal disputes is a profound concern for those who wear the robe and wield the gavel, as well as for the advocates arguing before them. In this exploration, we shall navigate the depths of AI’s efficacy in litigation, a journey through a landscape where legal precedents intertwine with algorithms.

The precision and dependability demanded by the legal profession are well-known. The entrance of AI has been heralded with a mix of excitement and wariness. AI's capacity to forecast litigation outcomes, dissect intricate legal documents, and pinpoint relevant precedents isn't just a triumph of technology; it's a bedrock for legal professionals to construct innovative, robust strategies. Yet, the empirical validation of AI in these arenas demands scrutiny.

Studies have shown AI, particularly when wielding machine learning algorithms, to have achieved commendable accuracy in various litigation-associated tasks. In predictive analytics, AI tools like Lex Machina have furnished lawyers with the ability to anticipate case outcomes, drawing on historical data from past case law and judges' rulings. This remarkable ability to sift through immense data sets to discern patterns offers lawyers a formidable tool to refine case strategies and set realistic expectations for their clients.

AI’s prowess extends to the jury selection process, where tools such as Voltaire have revolutionized the selection strategy. By analyzing potential jurors’ digital footprints across social media and public records, Voltaire equips legal teams with insights into behavioral tendencies, helping to construct a jury composition that could prove favorable to the case at hand. This edge in strategy not only underpins the objectives of a fair trial but also enhances the likelihood of successful litigation outcomes.

The labyrinth of legal language, with its nuances and jargon, presents a formidable challenge for AI, especially in the realm of natural language processing (NLP). Yet, companies like Lionbridge have risen to the occasion, offering AI-powered translation services that cater to the specialized needs of legal document translation. These services ensure that the intricacies of legal terminology are preserved, underpinning the fidelity of multilingual legal proceedings and documentation.

Despite these advancements, AI’s accuracy in litigation support is not uniformly distributed across all areas. While predictive analytics and jury selection analysis show promise, other tasks, like the comprehension of legal nuances in document analysis, may encounter hurdles due to the contextual complexity. These challenges are a clarion call for continual development to refine AI technologies for better accuracy.

Another critical factor in AI's performance is the integrity of the data it learns from. Data riddled with biases or gaps can lead to flawed predictions, calling into question the fairness and objectivity of AI-supported legal processes. It falls upon legal professionals to exercise caution, employing AI as a complement to human discernment, not as a blind substitute.

The interpretability of AI's suggestions also poses a quandary. The 'black box' phenomenon – where the reasoning behind AI's conclusions is obfuscated – clashes with the legal profession's core tenets of transparency and explicability. This necessitates more interpretable AI models, ensuring that legal experts can trust and comprehend the advice tendered by their digital assistants.

The practical application of AI in litigation has nonetheless shown encouraging signs. Insights gleaned from AI can streamline case management, inform strategic direction, and offer predictions on judicial tendencies. These benefits are capable of improving case resolutions and enhancing client satisfaction. Yet, the effective integration of AI within the legal arena demands a delicate equilibrium – capitalizing on AI's strengths while critically appraising its boundaries and connotations.

The pedagogical landscape for legal practitioners is evolving, increasingly spotlighting AI and technological fluency to equip the new cadre of lawyers. This shift is crucial for fostering a profound comprehension of AI's possibilities and potential pitfalls, enabling practitioners to wield AI judiciously and ethically.

As we move forward, it is imperative that legal experts, technologists, and legislators work in concert. This synergy is vital for sculpting AI tools that are not only precise but also consonant with the ethos and aims of the judicial system. Comprehensive regulatory frameworks can guide the ethical deployment of AI in litigation, mitigating misuse and assuring that AI advances the quest for justice.

Case studies showcasing AI in legal settings serve as a rich repository of knowledge, highlighting conditions under which AI fortifies legal operations and where human supervision is indispensable. Examining these instances grants legal professionals valuable insights into the effective and ethical incorporation of AI in their practice.

As AI's evolution marches on, its role in litigation will surely grow, presenting novel tools and approaches for legal analysis. Yet, this progression also brings challenges, necessitating ongoing education and flexibility among legal professionals. With an attitude of innovation and vigilance, the legal sector can surmount these challenges, harnessing AI's potential while preserving the integrity of the legal process.

The real-world accuracy of AI in litigation is a dynamic metric, an ever-shifting balance that mirrors technological progress, the caliber of the data at hand, and the capacity of the legal field to assimilate AI responsibly. In this ever-evolving arena, it's incumbent upon legal professionals to remain attuned to AI's advancements, critically assessing its role and impact on the mechanisms of litigation.

In conclusion, the precision of AI in legal proceedings encapsulates a complex interplay between technological innovation, legal acumen, and moral judgment. AI presents a panoply of tools that promise to enhance legal practices, yet its utmost value does not lie in supplanting human discretion but in amplifying it. By nurturing a sophisticated understanding of AI's virtues and limitations, the legal fraternity can employ AI to attain heightened efficiency, exactitude, and equity in litigation.

The path of integrating AI into legal practice is charted by continuous education, adaptation, and cooperation. It heralds not just the transformation of legal procedures but also the advancement of our shared quest for justice amidst the intricacies of the modern age.

By fostering these partnerships and pursuing these developments, the legal profession can secure AI's position as a beacon of support in the pursuit of justice. It is a journey that requires balancing the scales of technological possibility with the weight of legal tradition, ensuring that each new tool, each new algorithm, enhances rather than detracts from the human element at the core of the law.

Lawyers of the future may find themselves not only as practitioners of the law but as stewards of a new legal order, one where AI serves the cause of justice without overshadowing the human wisdom that has long been its guardian. In this pursuit, the legal community must tread with both caution and curiosity, embracing the new while respecting the enduring principles that constitute the bedrock of legal practice.

As AI becomes a more ubiquitous part of the legal landscape, its real-world accuracy will be tested and proven in the crucible of the courtroom. It will be measured against the yardstick of human experience and judged not only by its ability to predict outcomes or sift through data but also by its capacity to enhance the human pursuit of fairness and truth.

Thus, as we stand at the intersection of a traditional field and a transformative technology, the challenge will be to ensure that AI's role in litigation is constructive, clear, and, above all, just. This is not the end of the story but rather a new chapter in the annals of law and technology—a chapter that we must write with care, with insight, and with an unwavering commitment to the ideals that underpin the rule of law.

The Morality of Machines: Ethical Considerations for AI in Law

The integration of Artificial Intelligence (AI) into the courtroom is a paradigm shift, redefining the contours of legal proceedings. This technological renaissance, while laden with potential, introduces complex ethical dilemmas that necessitate diligent scrutiny and thoughtful navigation.

AI systems, like Premonition, employ sophisticated algorithms to analyze colossal datasets, predicting case outcomes with astonishing accuracy. Such systems can pinpoint which attorneys frequently succeed before specific judges, offering an unparalleled strategic edge in litigation. Yet, this predictive prowess brings forth critical ethical considerations. The risk of reinforcing systemic biases looms large; historical legal data, tainted by prejudicial undercurrents, may serve as a skewed foundation for AI decisions, amplifying disparities rather than attenuating them.

The 'black box' nature of AI algorithms presents a formidable ethical quagmire. The obscure logic that underpins AI conclusions obscures the judicial decision-making process, clashing with the tenets of open justice that demand transparency and accountability. This opaqueness hinders the ability to contest AI-influenced decisions, casting a shadow over the right to a fair trial and the due process rights of individuals.

AI's increasing role in the courtroom could also potentially encroach upon the sacrosanct domain of judicial discretion. The essence of legal adjudication lies in the moral and empathetic discernment of human judges, an element AI cannot replicate. Delegating critical decisions to algorithms strips the legal process of its humanistic core, raising questions about the role and relevance of empathetic jurisprudence in an AI-driven era.

Data privacy concerns are amplified by the integration of AI into judicial processes. AI's reliance on extensive data hoards heightens the vulnerability of sensitive personal information, increasing the risk of breaches and misuse. This intrusion on privacy rights is a stark reminder of the ethical responsibility to protect individual liberties in the digital age.

Furthermore, the ethical quagmire deepens when considering the potential impact of AI on judicial independence. Overreliance on AI predictions could inadvertently influence or bias judicial decisions, undermining the bedrock principle of an autonomous judiciary, unswayed by external or algorithmic influences.

The democratizing potential of AI to broaden access to justice is commendable. Yet, it also ignites ethical debates over the equitable distribution of such technology. The dichotomy between those who can leverage AI's capabilities and those who cannot threaten to widen the chasm of justice availability, thereby challenging the very notion of equality before the law.

Assigning responsibility and accountability in AI-assisted legal decisions is a labyrinthine task. The blurred lines of culpability, whether it falls on the AI developers, the legal practitioners, or the judiciary, complicate the foundational legal principles of fault and redress.

To navigate these ethical mires, a comprehensive and multifaceted strategy is imperative. Crafting ethical guidelines tailored to the unique context of AI in legal settings is paramount. Such standards should enshrine the principles of fairness, transparency, accountability, and respect for privacy, safeguarding against the subversion of justice by technological means.

Collaboration between technologists, legal professionals, and ethicists is essential in this ethical odyssey. Joint efforts can cultivate nuanced AI systems that are attuned to the ethical complexities of legal practice, ensuring that such technologies are allies rather than adversaries of justice.

The importance of ongoing vigilance cannot be overstated. Regular assessments of AI in legal contexts can preemptively address any detrimental repercussions, aligning AI deployment with societal ethical norms.

Education for legal professionals on the ethical ramifications of AI is critical. A well-informed legal community can steer AI applications responsibly, fostering an environment where technology is a boon, not a bane, to justice.

Civic engagement on AI's role in the judiciary is equally vital. Public discourse can elucidate the societal impacts of AI, engendering a collective wisdom that will guide the ethical integration of these technologies in the courtroom.

In essence, while AI heralds a new dawn of judicial efficiency and insight, it is accompanied by a host of ethical challenges. These challenges must be met with a steadfast commitment to ethical principles, ensuring AI's role in the courtroom strengthens rather than undermines the pillars of justice. By addressing these ethical conundrums head-on, the legal profession can wield AI's power with integrity, reinforcing the equitable administration of justice for all.

AI’s Role in Evolving Litigation Strategies

In the realm of legal technology, the importance of enhancing litigation strategy cannot be overstated. As the legal landscape becomes increasingly complex, the need for advanced tools that can analyze and interpret vast amounts of data grows. This is where AI Legal Tech, specifically court analytics platforms, comes into play, offering groundbreaking solutions that transform judicial decision-making.

The integration of AI Legal Tech into litigation strategy is paramount for several reasons. First, it allows legal professionals to harness the power of data, providing insights that were previously unattainable. By leveraging machine learning and analytics, legal teams can predict outcomes, understand trends, and make informed decisions that significantly impact the success of their cases. This data-driven approach not only enhances the efficiency of the litigation process but also contributes to a more transparent and fair judicial system.

Moreover, AI Legal Tech platforms facilitate a more strategic allocation of resources. With real-time data analysis, courts and legal practitioners can identify bottlenecks and inefficiencies, enabling them to focus their efforts where they are needed most. This not only improves the workflow within legal institutions but also enhances the overall experience for litigants, who benefit from faster and more predictable case resolutions.

A leading example of innovation in this domain is the Tyler Technologies Court Analytics Platform. Tyler Technologies has been at the forefront of the judicial transformation, offering a solution that revolutionizes how data is utilized to enhance litigation strategy. Their Court Analytics Platform is designed to modernize the judicial process by integrating data from various information systems and presenting it in user-friendly dashboards.

The platform's strength lies in its ability to provide comprehensive insights into the judicial process. By drawing data from the Enterprise Case Manager to the Court Analytics dashboard, it presents a clear picture of an organization's performance. These insights are grounded in standardized metrics developed by the National Center for State Courts (NCSC), ensuring that the analysis is both reliable and relevant.

For court administrators and judges, the platform offers a tool that transcends traditional data analysis methods. Instead of sifting through dense and complex data, users are greeted with streamlined, accessible dashboards that highlight key performance indicators. This accessibility of information empowers decision-makers to make strategic choices that can lead to significant improvements in the judicial system.

One of the platform's most impactful features is its ability to facilitate comparative analysis. Courts can benchmark their performance against other institutions using NCSC metrics, providing a clear framework for identifying strengths and areas for improvement. This benchmarking is not merely about introspection but also fosters a learning environment where courts can share best practices and learn from each other's experiences.

A hypothetical scenario illustrates the transformative potential of Tyler's Court Analytics. Imagine a court struggling with case backlogs and perceived delays in justice delivery. Upon adopting Tyler's platform, the court identifies patterns in case durations and pinpoints bottlenecks in the motion-to-trial phase. Through strategic adjustments in staffing and procedures, it manages to reduce the average case duration by 15%, significantly enhancing the perception and reality of judicial efficiency.

Tyler's Court Analytics also prioritizes user experience with dashboards designed to democratize data. These intuitive visualizations ensure that court staff, regardless of their data expertise, can interact with the platform and extract actionable insights. This user-centric approach not only makes data more accessible but also fosters a culture of informed decision-making within the judiciary.

To maximize the platform's impact, Tyler Technologies provides comprehensive training and support, ensuring that users can fully leverage the tool's capabilities. This commitment to user empowerment underlines the platform's role as a catalyst for judicial enhancement, rather than a mere analytical tool.

The benefits of adopting Tyler's Court Analytics extend beyond administrative efficiencies to improve the litigant experience. With more predictable case processing times and increased transparency, litigants gain a clearer understanding of their journey through the legal system, contributing to a perception of fairness and trust in the judiciary.

In conclusion, the Tyler Technologies Court Analytics Platform is not just an innovative tool but a strategic imperative for courts aiming to navigate the complexities of the 21st-century legal environment. By adopting this platform, courts can enhance their litigation strategies, ensuring that they not only meet the current demands of justice delivery but are also well-prepared for future challenges. Through such advanced AI Legal Tech solutions, the judicial system can achieve a balance between efficiency and fairness, heralding a new era in the pursuit of justice.

Chapter 8:

AI-Enhanced Compliance: Transforming Risk Management in Law

In the rapidly evolving legal landscape, artificial intelligence (AI) has emerged as a cornerstone for enhancing compliance and risk management practices. Firms that embrace AI tools find themselves at a significant advantage, navigating the complexities of regulatory requirements with greater precision and efficiency. The integration of AI into compliance strategies isn't just about keeping pace with current trends; it's about setting a new standard for meticulousness and foresight in legal operations. As regulatory frameworks become increasingly intricate and the stakes of non-compliance soar, the utilization of AI to predict, analyze, and mitigate risks stands as a beacon of innovation in legal practice.

Among the most transformative applications of AI in this domain is its role in advancing Anti-Money Laundering (AML) and Know Your Customer (KYC) processes. By harnessing sophisticated algorithms, legal professionals can now sift through vast datasets to identify patterns and anomalies that would be impossible for the human eye to catch in a timely manner. This not only accelerates the verification processes but also enhances the accuracy of background checks and due diligence efforts. Moreover, predictive analytics in risk evaluation enables a proactive approach to compliance. Legal practitioners are equipped with the tools to foresee potential compliance issues before they escalate, allowing for preemptive measures that can save firms from hefty penalties and reputational damage.

Yet, the journey towards the full integration of AI in compliance and risk management is not without its challenges. Concerns surrounding data privacy, algorithmic bias, and the ethical use of AI technology are at the forefront of discussions among legal professionals. Addressing these concerns requires a commitment to transparency, ongoing dialogue with stakeholders, and a keen awareness of the ethical implications of AI deployment in legal settings. Nevertheless, the potential of AI to revolutionize compliance and risk management practices remains undeniable. By embracing AI, the legal profession can not only enhance its compliance efforts but also pave the way for a future where legal practices are more resilient, efficient, and aligned with the demands of an increasingly digital world.

Enhancing Legal Compliance through AI Technologies

In the labyrinthine world of compliance and risk management, the relentless quest for more streamlined and efficient methodologies is a constant challenge for legal professionals. With the dawn of a more technologically empowered era, artificial intelligence (AI) has arisen as a transformative force, poised to redefine and enhance compliance endeavors within legal practices.

Compliance is a field that necessitates not only a profound comprehension of the law but also an acute sensitivity to the ever-changing tides of regulations and societal norms. Within this realm, AI stands out as a formidable ally. Especially with the integration of platforms like Compliance.AI, legal teams are now equipped with the advanced capability to analyze extensive volumes of data with both speed and precision. This is pivotal for monitoring adherence to multifaceted regulatory frameworks and significantly alleviates the workload that often burdens legal teams.

AI's most salient feature in the context of compliance is its ability to automate the traditionally labor-intensive and time-consuming process of routine compliance checks. Where once such tasks would have been carried out manually, they can now be streamlined through sophisticated AI algorithms. Compliance.AI, in particular, is adept at perusing through documents and data with a level of precision that is unparalleled, expediting the compliance process while simultaneously minimizing the potential for human error. This results in a more fortified compliance posture for legal entities.

The analytical strength of AI extends far beyond mere automation. Through the use of historical data and the power of continuous learning, AI systems, such as those developed by Compliance.AI, have the capacity to forecast potential compliance violations even before they manifest. Identifying patterns and anomalies indicative of non-compliance risks enables legal professionals to proactively address and mitigate issues, thus circumventing the hefty penalties and reputational damage that could ensue.

In addition to proactive risk management, AI's role in improving the accuracy and efficiency of compliance reporting is nothing short of revolutionary. Regulatory bodies often demand precise, comprehensive reports delivered within stringent deadlines. With tools like Compliance.AI, much of the reporting process can be automated, resulting in the creation of thorough reports that align with regulatory expectations. This automation allows legal professionals to redirect their focus to more strategic tasks, leveraging their expertise where it is most needed.

A cornerstone of compliance is transparency and accountability, and AI significantly strengthens these elements through its meticulous record-keeping capabilities. Compliance.AI ensures that each decision and action executed by its algorithms is documented, establishing a transparent audit trail that proves invaluable during regulatory examinations or legal reviews. Such comprehensive documentation not only evidences compliance but also cultivates a culture of integrity and ethical behavior within organizations.

The transformative power of AI in compliance is perhaps most evident in the realms of anti-money laundering (AML) and know your customer (KYC) processes. Here, AI is revolutionizing the way legal teams verify and monitor client information. Compliance.AI, with its advanced algorithms and machine learning, can rapidly analyze client data, flagging any anomalies that could suggest financial misconduct. This capability is essential not only for adhering to regulatory mandates but also for significantly enhancing an organization's capacity to thwart financial crimes.

Yet, the integration of AI, including Compliance.AI, into compliance strategies does not come without its challenges. Paramount among these is the issue of data privacy and security, as legal professionals are tasked with ensuring that the AI systems they employ are compliant with strict regulations governing the protection and handling of sensitive information. Additionally, there is a necessity for ongoing oversight and refinement of AI systems to maintain their effectiveness as legal and regulatory environments evolve.

The introduction of AI into compliance has also initiated a shift in the role of legal professionals. Far from diminishing their significance, AI, and platforms like Compliance.AI, empower legal practitioners, enabling them to concentrate on strategic, analytical tasks. This shift heralds a new era where the required skill set is heavily biased towards technological proficiency, in conjunction with a robust legal foundation.

For those law firms and legal departments considering the adoption of AI in their compliance processes, a gradual, phased approach is recommended.

Initiating the use of AI in low-risk areas allows legal teams to grow accustomed to the new technologies and methodologies, incrementally expanding AI's role as confidence in the system is established. This strategy ensures a smooth transition for staff and minimizes disruption to ongoing operations.

Choosing the appropriate AI tool is a critical step for any legal entity. The market is replete with options, yet not all are suited to the unique compliance requirements of every legal practice. A comprehensive evaluation of potential AI solutions, including Compliance.AI, is imperative, examining their accuracy, efficiency, and compatibility with existing systems to ensure a successful integration.

Education and training are the pillars of successful AI integration in compliance efforts. Legal professionals must not only understand the basic mechanics of AI technologies but also possess insights into how these tools, such as Compliance.AI, can be leveraged to bolster compliance operations. Regular training sessions, workshops, and seminars are vital to keeping legal teams up-to-date with the latest developments in AI and familiar with best practices in the field.

The synergy between AI expertise and legal acumen is fundamental for advancing compliance endeavors. Collaboration between technologists and legal professionals ensures that AI systems, including those provided by Compliance.AI, are developed and implemented in ways that are congruent with legal standards and the specific needs of the practice. This partnership not only improves the effectiveness of AI in compliance but also promotes innovation and the ongoing refinement of these tools.

As we look toward the future, the integration of AI into compliance and risk management signifies a substantial advancement for the legal profession. With AI technologies continually progressing, new avenues and benefits for legal practices are sure to emerge. Yet, the successful navigation of this terrain necessitates a dedicated effort to comprehend, adopt, and adapt to these technological strides, all the while maintaining a vigilant focus on ethical and responsible AI utilization.

AI's potential to transform the compliance landscape within the legal domain is immense. By automating routine tasks, forecasting potential compliance risks, enhancing the precision of reporting, and fostering a proactive compliance culture, AI is set to elevate the efficiency and efficacy of legal practices to new heights. Compliance.AI serves as a prime example of this potential, offering a suite of tools that can help legal professionals not only keep pace with but also stay ahead of the curve in regulatory compliance.

However, the journey toward fully realized AI-enabled compliance is multifaceted. It requires careful strategizing, ongoing education, and a steadfast commitment to ethical principles in AI usage. Legal entities must navigate this journey with a clear vision and an open mind, ready to leverage the power of AI while ensuring that the integrity of the profession and the rights of individuals are preserved.

In conclusion, while AI, embodied by platforms like Compliance.AI, presents a promising horizon for transforming compliance efforts, the legal profession stands at a crossroads. It can either continue down traditional paths or choose to embrace the new world of possibilities that AI offers. By doing so, legal professionals can not only achieve heightened levels of efficiency and effectiveness in their practices but also contribute to the pioneering spirit of innovation that defines our times. The future of compliance in the legal sector is not just about adapting to change—it's about leading it. With AI, and specifically with Compliance.AI, legal teams are well-equipped to do just that, navigating the complexities of compliance with a newfound agility and foresight that will serve the industry well into the future.

As the landscape of legal compliance becomes increasingly complex, the role of AI will undoubtedly become more central. The ability to quickly adapt to new regulations, maintain meticulous records, and provide strategic insights will be the hallmarks of successful practices. With the right tools, training, and ethical framework in place, the potential for AI to revolutionize the legal industry is limitless, and Compliance.AI is at the forefront of this transformative journey. By harnessing the power of AI, the legal profession is poised to meet the challenges of compliance with confidence, ensuring that it not only responds to the demands of the present but also shapes the standards of the future.

Transforming AML and KYC: AI's Strategic Impact

AI's Impact on AML and KYC Processes

In the fast-evolving world of legal compliance and risk management, the emergence of artificial intelligence (AI) has marked a significant transformation, especially within Anti-Money Laundering (AML) and Know Your Customer (KYC) protocols. The innovative integration of AI technologies is redefining traditional approaches to crucial compliance tasks such as client identity verification, risk profile assessment, and monitoring of financial transactions for illicit activities.

AML and KYC, cornerstones in the global fight against financial crime, compel institutions to establish robust customer due diligence systems. Historically, these processes have been beleaguered by their manual, labor-intensive nature, leading to inefficiencies, delays, and susceptibility to human error. AI has disrupted this status quo by introducing levels of efficiency and accuracy that were once thought unattainable. Sophisticated AI algorithms have the capacity to analyze extensive datasets at unprecedented speeds, detecting potential risks and anomalies that might elude human scrutiny.

The significance of AI extends well beyond the realm of simple automation. Machine learning, a key subset of AI, leverages historical data to identify complex patterns indicative of money laundering or other fraudulent activities. These dynamic algorithms enhance their detection capabilities continuously as they ingest new data, making them increasingly effective in flagging suspicious transactions.

In the context of KYC, AI has been instrumental in refining process accuracy. Utilizing advanced data analytics and natural language processing (NLP), AI systems can efficiently parse through sprawling databases and public records. This facilitates a more rapid and accurate client identification and verification process, significantly reducing the risk of errors that could lead to regulatory non-compliance, thus optimizing the client onboarding journey.

Moreover, AI is invaluable for ongoing due diligence and monitoring. These intelligent systems track and adapt to changes in client profiles and behaviors, swiftly identifying and mitigating risks. This level of proactive monitoring is crucial in sustaining compliance and safeguarding institutions against the complexities of financial crime.

One of the most laudable advantages of AI integration into AML and KYC workflows is the potential for substantial cost reduction. By automating routine tasks and refining the accuracy of risk assessments, organizations can redistribute their human capital towards more strategic and analytical endeavors. This not only bolsters productivity but also translates into considerable savings on operational expenditures.

Despite its myriad benefits, the implementation of AI in compliance processes is not without its challenges. Data privacy stands as a predominant concern, as the technology inherently requires access to sensitive personal and financial information. Legal professionals are tasked with navigating a complex landscape, where they must weigh the benefits of AI-driven efficiency against the imperative to protect client privacy in accordance with stringent data protection laws.

Furthermore, questions of transparency and accountability loom large, particularly with AI algorithms founded on deep learning. This 'black box' phenomenon, where the decision-making process of algorithms is not easily interpretable, can complicate the understanding of how certain assessments are made, raising valid concerns about potential biases and fairness. Ensuring the explicability of AI models and the justifiability of their decisions is paramount for maintaining trust and adhering to legal and ethical standards.

Overcoming these challenges necessitates a concerted collaborative effort that spans legal experts, data scientists, and technology professionals. It is imperative that this collaboration is informed by a deep understanding of both the legal regulatory landscape and the technical intricacies of AI. It also demands continuous vigilance to stay abreast of evolving legal standards and technological advancements.

The role of legal professionals is particularly pivotal in this integration. They must serve as custodians of these technologies, providing informed guidance on compliance issues, aiding in the design of AI systems that respect legal and ethical boundaries, and ensuring that the application of AI in AML and KYC processes is congruent with broader goals of justice and fairness.

The proactive development of AI governance frameworks tailored for financial institutions is a commendable stride toward addressing the nuanced challenges of AI integration. These frameworks are designed to facilitate the ethical design, deployment, and monitoring of AI systems within the sensitive realm of compliance. Such initiatives are instrumental in promoting transparency, ensuring accountability, and building trust in AI applications.

As we cast our gaze forward, the trajectory of AML and KYC in the legal profession seems indelibly linked to AI. With an increasingly intricate regulatory environment and the advancing sophistication of financial crimes, there is a burgeoning demand for more effective, accurate, and scalable compliance solutions. AI, with its unparalleled analytical capacity and potential for continuous innovation, is at the vanguard of meeting this demand.

Yet, the embracement of AI in compliance processes signifies more than mere adoption of new technologies. It represents the adoption of a forward-thinking mindset—one that comprehends the transformative potential of AI while remaining acutely conscious of its ethical, legal, and societal ramifications. It is about forging a symbiosis between innovation and responsibility, ensuring that our pursuit of technological progression remains aligned with our commitment to upholding legal integrity and the preservation of human values.

In summation, AI's burgeoning impact on AML and KYC processes signals a landmark evolution in the sphere of legal compliance and risk management. When implemented with care and precision, AI technologies empower legal professionals to enhance operational efficiency, elevate the accuracy of verifications and risk assessments, and significantly cut costs. However, realizing these benefits depends on a nuanced understanding of AI's vast potential coupled with an awareness of its inherent pitfalls, all guided by a steadfast commitment to upholding stringent legal and ethical standards.

Amidst this backdrop, the incorporation of technologies like BehavioSec's by LexisNexis® Risk Solutions heralds a new chapter in innovation. BehavioSec's foray into behavioral biometrics adds a fresh layer of security to the domain, facilitating real-time identity verification that is both seamless and robust, thereby greatly reinforcing the AML and KYC defenses without undermining the user experience. This continuous and unobtrusive authentication represents a crucial step in adjusting to the evolving tactics of fraudsters, ensuring that only legitimate users gain access and maintaining the trust of all stakeholders.

The strategic acquisition of BehavioSec by LexisNexis demonstrates a commitment to enhancing their AML and KYC offerings, positioning them at the forefront of market leaders, and providing customers with a layered security approach that incorporates state-of-the-art AI technology. Consequently, institutions are positioned to reap the rewards in the form of reduced fraud-related losses, a decrease in the rate of false positives commonly produced by traditional verification methods, and an optimized user experience that judiciously balances convenience with security.

As the era of digital transformation deepens, AI's role in legal compliance is expected to not just grow but flourish. The increasing complexity of regulatory frameworks and the continuous evolution of financial crime strategies demand advanced, scalable, and intelligent compliance solutions. AI, with its sophisticated analytical capabilities and adaptability, is well-suited to rise to these challenges.

Nevertheless, the full potential of AI can only be harnessed when coupled with a cultural shift within organizations. This shift requires not just an embrace of AI's transformative capabilities but also a strategic approach to its implementation. Organizations must cultivate a culture of innovation that readily adopts AI while also embedding a strong sense of ethical responsibility in its usage.

The path forward is intricate and multifaceted—balancing the strengths of AI against the backdrop of ethical, legal, and societal questions it elicits. It is a journey punctuated by exciting developments in the field of legal technology, demanding a consistent and resolute commitment to preserving the integrity of our legal frameworks and ensuring the protection of individual rights and privacy.

In light of these considerations, AI's influence on AML and KYC processes marks not just a significant shift but a revolutionary stride in legal compliance and risk management. Legal professionals have the opportunity, through the astute application of AI, to not only streamline compliance procedures but also to advance the fight against financial crime. The success of this advancement, however, will require an equilibrium between leveraging AI's strengths and conscientiously navigating the ethical, legal, and societal considerations it presents.

In conclusion, the profound impact of AI on AML and KYC processes is shaping a new era in the field of legal compliance. The judicious application of AI technologies stands to significantly enhance operational efficiencies, improve the accuracy of compliance tasks, and reduce overhead costs. Yet, the true measure of success in these endeavors will be determined by a deep understanding of AI's capabilities, informed by an unwavering commitment to legal and ethical principles. As the legal landscape evolves alongside these technological advancements, the indispensable role of legal expertise in guiding the future of AI in compliance becomes ever more apparent.

Forecasting Compliance and Risks: The Power of Predictive Analytics

The legal industry's transformation is underway, significantly propelled by the integration of advanced technologies, among which predictive analytics is a game-changer for risk evaluation. This innovative approach transcends the traditional boundaries of risk assessment, offering a proactive lens through which legal practitioners can preemptively engage with potential issues. Neota Logic emerges at the forefront of this technological evolution with its no-code platform, empowering legal professionals to construct robust predictive models without the need to delve into the complexities of coding.

The essence of predictive analytics lies in its ability to parse through historical data, employing machine learning to discern patterns and predict potential compliance violations or other risks before they escalate. This technique is pivotal for legal practitioners, including lawyers, paralegals, and attorneys, as it revolutionizes risk management, transforming it from a reactive task to a strategic imperative.

The value of predictive analytics extends beyond risk identification. It reshapes the entire framework of legal defense and compliance strategies. Legal challenges, once only reactively addressed, can now be anticipated with a degree of precision previously unattainable. This forward-thinking approach can forestall legal disputes and the financial and reputational damages they might inflict.

For example, in the complex and ever-changing realm of regulatory compliance, predictive analytics acts as a sentinel. It can alert firms to upcoming legislative changes, providing the luxury of time to adapt and strategize. This aspect alone can confer a significant competitive advantage, as it translates into a more agile and responsive practice that can maintain compliance and avoid the pitfalls of non-adherence.

Predictive analytics also facilitates a deeper understanding of operational risks. It enables firms to spot inefficiencies or procedural flaws that could precipitate legal vulnerabilities. This insight is instrumental in optimizing operations and shoring up defenses, ensuring that firms are not just reacting to risks but actively preventing them.

In the financial domain, predictive analytics is equally transformative. It allows legal professionals to anticipate and mitigate potential financial risks by analyzing historical data trends. This could lead to adjustments in client agreements or the fortification of fiscal policies, precluding risks that could otherwise severely impact a firm's financial health.

Neota Logic's platform plays a crucial role in these scenarios, offering a way for legal entities to automate the detection and analysis of risks. This automation is achieved through no-code solutions that allow firms to scale their knowledge, optimize processes, and automatically generate necessary documentation. This innovative approach enables legal professionals to harness the power of predictive analytics quickly and efficiently, leading to strategic, data-driven decisions that underpin risk management.

Neota Logic's collaboration with its sister company, ComplianceHR, combines the legal expertise of Littler Mendelson with a powerful no-code automation platform to deliver revolutionary employment law solutions. This synergy exemplifies how legal tech can drive innovation in legal practices, especially in areas as critical as employment law where compliance is paramount.

Yet, the deployment of predictive analytics is not without challenges. The accuracy of such analytics depends heavily on the quantity and quality of historical data. Firms must commit to establishing comprehensive data collection and management systems to underpin their predictive efforts. Moreover, predictive analytics necessitates a cultural transformation within the legal sector—a shift from traditional methods to a more data-driven, innovative mindset. This shift is critical for the legal sector to realize the full potential of predictive analytics.

Ethical concerns also surface with the use of predictive analytics, particularly around client data. The balance between the advantages of data analytics and respecting individual privacy must be meticulously managed. Legal practitioners are tasked with navigating these ethical waters, ensuring compliance with legal standards and societal expectations of privacy and consent.

Furthermore, the nuanced and intricate nature of legal data requires a close collaboration between legal professionals and data scientists. The interplay between these two disciplines is vital for the creation of predictive tools that are not only accurate but also contextually aware and aligned with the nuances of legal practice.

As we peer into the future of the legal sector, the role of predictive analytics is undoubtedly expanding. Its capacity to foster innovation in compliance and risk management is vast. Legal practitioners who adeptly incorporate predictive analytics into their practice will set a new standard in legal services, enhancing their risk management capabilities and solidifying their competitive position in the marketplace.

In sum, the implementation of predictive analytics heralds a substantial leap forward in how the legal industry evaluates and manages risk. Its impact is profound, offering a method that proactively anticipates and prepares for future challenges. With the continuous evolution of legal practice in this technological age, the adoption of predictive analytics will serve as a pivotal differentiator, augmenting operational efficiency, compliance, and client contentment.

Legal professionals stand at the threshold of a new era in legal risk management. By embracing predictive analytics and incorporating it into their workflows, they can tackle the complexities of modern legal environments with greater confidence, ensuring compliance and securing a competitive edge in an ever-evolving legal landscape. This strategic integration of technology will not only elevate their risk management strategies but will also redefine excellence and innovation in the legal field.

Chapter 9:

Legal Tech and AI: Efficiency in Legal Operations

In an era where the legal profession is continuously evolving, the introduction of Artificial Intelligence (AI) into legal operations emerges as a transformative tool, enhancing efficiency and productivity in unprecedented ways. Legal firms are now positioned to innovate billing, timekeeping, and the overall management of their practices through AI's capabilities. By automating mundane and repetitive tasks, AI enables legal professionals to reallocate their focus towards more complex, value-added activities. This shift not only optimizes operations but also significantly reduces the margin of error inherent in human intervention, ensuring a higher degree of accuracy in tasks such as billing and documentation.

Further, the enhancement of team productivity through AI-integrated systems cannot be overstated. By leveraging tools that streamline communication and project management, legal teams can operate with greater cohesion and flexibility. This entails a more agile response to client needs and legal challenges, bolstered by data-driven insights and predictive analytics that inform decision-making processes. AI's integration into legal operations thus plays a pivotal role in a law firm's competitiveness, enabling a leaner, more strategic approach to practice management that aligns with the modern legal landscape's demands.

Moreover, the implementation of AI in legal firms is not merely about enhancing current processes but fundamentally rethinking the approach to legal operations. It demands a culture shift towards continuous improvement and innovation, encouraging legal professionals to embrace new technologies and methodologies. As firms navigate this transition, the focus must remain on leveraging AI to foster a more dynamic, efficient, and client-centered practice. Ultimately, optimizing legal operations through AI is not just a pathway to operational excellence but a strategic imperative for firms aiming to thrive in the digital age of law.

The AI Advantage in Legal Billing Precision

In the current landscape of legal operations, the tide of digital transformation is not merely an emerging trend but an essential force of change. This shift is deeply influenced by the potent capabilities of Artificial Intelligence (AI). Within this transformation, legal billing and timekeeping stand out as prime candidates for innovation. The intricacy and specificity required for effective legal billing necessitate not only precise attention to detail but also a robust understanding of legal procedures and the expectations of clients. AI is at the forefront of redefining this space by integrating efficiency, precision, and transparency into these critical processes.

The old-school methods of billing and timekeeping, deeply rooted in the legal profession, have long been a source of inefficiency and frustration. Manual time entry is not just arduous; it is fraught with the potential for error. Additionally, the subjective determination of what constitutes billable hours often results in inconsistency and disputes with clients. AI disrupts this outdated model with its capability to process immense amounts of data swiftly, ensuring that every minute of billable work is meticulously captured and accurately documented.

A significant innovation that AI introduces to legal billing is the application of predictive analytics. Through the examination of historical billing data, AI provides law firms with insights into their billing patterns, pinpointing areas of inefficiency and offering suggestions for process enhancement. This predictive prowess is a valuable asset for law firms aiming to refine their billing accuracy and streamline their operations.

AI-driven legal billing systems also incorporate Natural Language Processing (NLP) technology. This feature enables the system to understand and categorize the textual descriptions of tasks completed by attorneys. It facilitates the generation of automated invoices that are not only precise but also rich in detail and easily understandable by clients. This ensures that clients are no longer puzzled by obscure billable entries, as AI-generated invoices provide comprehensive descriptions of the services provided.

One of the key players in this space is SimpleLegal, a modern legal operations platform that has incorporated AI to revolutionize billing and timekeeping. SimpleLegal's machine learning algorithms meticulously analyze time entries, enhancing accuracy and adherence to predefined billing guidelines. This not only streamlines the billing process but also reduces manual errors and increases transparency. The result is a fortified trust between law firms and their clients, and an organization that operates with peak efficiency.

SimpleLegal stands as a testament to the modernization of in-house legal operations. The platform enables legal teams to effortlessly track, manage, and report on legal expenses, accruals, budgets, and vendor relationships. Its intelligent legal operations solution, trusted by hundreds of companies, represents a monumental shift in managing legal matters, spend, vendors, documents, and reporting.

Through SimpleLegal, firms can bridge communication gaps between departments, becoming a strategic asset to the organization with insights gleaned from the platform's smart reporting and AI-powered features. Following its acquisition by OnIt, an AI-enabled workflow hub, SimpleLegal has further strengthened its position in the market. OnIt connects teams with how they wish to operate, both today and in the future, creating premier workflow automation experiences for sophisticated, legally-related business processes from a singular, integrated platform.

Furthermore, the compliance aspect of AI in legal billing cannot be overstated. Law firms frequently wrestle with the various and often intricate billing guidelines set by their clients. AI systems can be programmed to internalize these rules, systematically flagging entries that fall outside of compliance. This pre-emptive compliance ensures that invoices meet client expectations right from the beginning, thus minimizing the time invested in resolving invoice disputes and making revisions.

The embrace of AI in legal billing and timekeeping goes beyond the immediate improvements in operations. It signifies a shift towards a more transparent and trust-based dynamic between law firms and their clients. With AI-enabled oversight, clients have confidence in the fairness and accuracy of billing, nurturing a relationship grounded in reliability and trust. This transparency is vital in a sector where the very foundation of client relationships is built on trust.

Moreover, the introduction of AI in timekeeping liberates attorneys from the monotonous task of logging hours, freeing them to dedicate more of their time to the essence of their work: the practice of law. This liberation not only propels productivity forward but also enhances job satisfaction among legal professionals, fostering a healthier work-life balance.

Nevertheless, the integration of AI into legal billing and timekeeping is accompanied by its own set of challenges. Issues concerning data privacy, security, and the potential displacement of jobs are at the forefront of concerns. Law firms must tackle these concerns directly, initiating transparent conversations with their teams and maintaining a rigorous adherence to data protection regulations. Importantly, the implementation of AI should be perceived not as a replacement for human roles but as an enhancement of human capabilities.

As we look towards the full integration of AI into billing and timekeeping, the need for legal professionals to develop a thorough understanding of AI technologies becomes increasingly apparent. Education and training on the applications of AI in legal billing are crucial to enable a smooth transition and to fully harness the benefits of AI in this field.

In light of these developments, SimpleLegal has emerged as a pivotal technology in the legal tech landscape. Its AI-powered features provide a comprehensive solution for law firms and legal departments seeking to enhance their billing processes. Through SimpleLegal, firms can experience the advantages of machine learning for time entry analysis, ensuring accuracy and compliance with client-specific billing guidelines. This leads to a more streamlined billing process, reducing the time spent on manual entry and the potential for human error.

SimpleLegal's platform extends its capabilities to the management of legal spend, accruals, budgets, and vendor relations, which is indispensable for in-house legal teams. It empowers these teams with the ability to make data-driven decisions, backed by intelligent reporting and analytics. The ability to track and manage legal matters, spend, and vendor performance through a single platform streamlines operations and enhances overall efficiency.

Furthermore, SimpleLegal's integration into OnIt's ecosystem signifies a strategic move that broadens its scope. OnIt provides a hub for AI-enabled workflow automation, which facilitates the design of sophisticated, legal-related business processes. This acquisition implies not only a synergy between SimpleLegal's robust legal ops solution and OnIt's workflow automation capabilities but also the potential for expanded innovation in legal tech.

The introduction of such AI-driven tools in the legal industry underscores the importance of upskilling legal professionals. The continuous evolution of technology necessitates ongoing education to keep pace with the advancements and to effectively leverage the tools at their disposal. Law firms must prioritize the training of their teams on AI applications, ensuring that their workforce is equipped to utilize these innovations to their fullest potential.

Looking toward the future, the integration of AI in legal billing and timekeeping stands as a testament to the transformative power of technology within the legal profession. It's not merely an evolutionary step but a reimagining of operational workflows that can lead to significant gains in efficiency, accuracy, and client service.

To summarize, AI's role in legal billing and timekeeping is a transformative force that enables law firms to enhance their service delivery while focusing on core legal work. SimpleLegal's sophisticated platform exemplifies the practical application of AI in improving the efficiency and transparency of legal operations. As AI continues to make inroads into the legal industry, the onus is on legal professionals to adapt and adopt these technological advancements. The integration of AI in legal billing and timekeeping is not just an innovation; it's an essential evolution to stay ahead in a competitive and ever-changing legal landscape.

In a profession where time is of the essence, AI provides a critical advantage, allowing lawyers to invest their time where it matters most: in serving their clients effectively and advancing the practice of law. The journey toward integrating AI into legal billing and timekeeping is one of continuous improvement and adaptation, promising a more dynamic, responsive, and client-centric approach to legal services.

In conclusion, the promise of AI in legal billing and timekeeping is immense, offering a future where the routine is automated, the complex is simplified, and the relationship between law firms and clients is strengthened through trust and transparency. Law firms that embark on this journey of innovation with partners like SimpleLegal and OnIt are positioning themselves at the forefront of a legal services revolution, ensuring they remain competitive and relevant in an increasingly digitized world. The path ahead is clear: for law firms to thrive, embracing AI in billing and timekeeping is not just an option but a strategic necessity.

Streamlining Operational Excellence and Team Productivity with AI

In the ever-evolving realm of legal practice, the imperative for heightened efficiency and enhanced productivity is a resounding call to action, one that the advent of artificial intelligence (AI) and machine learning technologies answers with robust solutions. These technological advancements promise an era of unprecedented streamlining and optimization, ushering in new levels of operational excellence and team productivity for legal firms and departments.

Clio, a trailblazer in legal practice management software, stands at the forefront of this transformation. By leveraging AI, Clio integrates case management, document organization, and team collaboration into a singular, cloud-based platform, epitomizing the efficiency potential AI brings to legal operations. With Clio, routine tasks such as document sorting, billing, and scheduling are automated, allowing legal professionals to channel their expertise into substantial legal work—where the irreplaceable value of human intelligence and the personal touch are most impactful.

This chapter explores the transformative role of AI in revolutionizing the operational infrastructure of legal practices. AI's integration into legal operations paves the way for a metamorphosis in procedural tasks, driving accuracy and reliability to new heights and freeing up lawyers to focus on strategy, analysis, and client relations. The manual and often monotonous tasks that once consumed substantial time are now swiftly handled by intelligent systems, affording legal teams the capacity to invest in areas that significantly benefit from human acumen.

Moreover, the deployment of AI-driven analytical tools within case management has introduced a paradigm shift in legal strategy formulation. Through the analysis of vast datasets, AI algorithms can discern patterns and precedents, equip legal teams with predictive insights, and contribute to more robust case strategies. Such tools not only amplify success rates but also curtail the time traditionally dedicated to research and case preparation.

The synergy between AI and human expertise is perhaps nowhere more evident than in the realm of team collaboration. Legal projects, often characterized by their complexity and the necessity for precise coordination among numerous stakeholders, are streamlined through AI-powered platforms. These platforms facilitate seamless communication and project management, ensuring that all team members are in lockstep, thereby eliminating duplicative efforts and circumventing miscommunication.

AI also lends a personal touch to client interactions. By tailoring communications and advice based on specific client data and preferences, AI elevates the client service experience. This capability highlights the imperative for a deep understanding of client needs and the customization of legal services to meet those needs, thereby reinforcing client satisfaction and loyalty.

Document management, a conventionally arduous task, has been redefined by AI. Lawyers now utilize AI for swift document retrieval, categorization, and analysis, markedly reducing the hours previously poured into these tasks. This acceleration of document-related activities not only expedites legal processes but also drastically lowers the potential for human error.

In billing and timekeeping, AI ushers in an era of exactitude and transparency once thought unachievable. Automated tracking of billable hours coupled with the creation of detailed invoices ensures that billing practices are fair and transparent, cultivating a foundation of trust with clients.

AI's implications extend beyond process efficiency, playing a vital role in talent acquisition and retention. In today's competitive landscape, legal professionals increasingly gravitate towards work environments that minimize rote work and foster engagement with meaningful and innovative legal practices. AI-enabled environments are attractive to top-tier talent, offering a stimulating work atmosphere conducive to professional growth and satisfaction.

Despite the clear benefits, integrating AI into legal practices presents its own set of challenges, necessitating not only technological adoption but also a cultural shift within organizations. Legal professionals must be willing to engage in lifelong learning, adjusting to a landscape where AI is not a mere tool but a collaborator that amplifies their expertise.

The successful incorporation of AI into legal operations hinges on continuous training and education. Training programs must extend beyond the mere mechanics of technology, exploring its implications for legal practice and ethics. Furthermore, a rigorous data governance framework is crucial to ensure the security and privacy of sensitive information—a non-negotiable in the legal domain.

In selecting AI tools and partners, legal firms must undertake a comprehensive evaluation process to ensure the chosen solutions are fitting, reliable, and scalable to the unique needs of their practice and clientele.

In closing, the integration of AI into legal operations represents a watershed moment for the sector. Legal practices that adopt AI stand to redefine the boundaries of efficiency, accuracy, and personalized service in legal services. Overcoming the challenges of adopting these technologies requires a strategic approach, a dedication to continuous learning, and a steadfast commitment to ethical practice.

This chapter not only charts a course towards an AI-enhanced future for legal practices but also reinforces the critical role that technology plays in the optimization of legal operations. The partnership between human expertise and artificial intelligence is the crucible from which unprecedented levels of productivity and service excellence will be forged.

Legal professionals are at the threshold of a new era, one where the integration of AI into their practice is not just a possibility but an imperative. Embracing this change is more than streamlining operations; it is about reimagining the possibilities of the legal profession and setting new benchmarks for innovation and excellence within the industry.

AI and legal technology not only streamline tasks but also herald a new paradigm in legal service delivery. By automating processes that traditionally consumed significant manpower, technology like Clio enables legal teams to concentrate on more strategic and engaging work, thereby enhancing overall productivity. This transition allows for a redefined professional landscape, one in which legal practitioners can excel in the aspects of the job that most benefit from human insight and ingenuity.

The adoption of AI technology like Clio for case management simplifies every aspect of a law firm's operation. By providing everything a law firm needs in one place, from billing and document management to communication and case management, Clio exemplifies the seamless integration of multiple functions into a cohesive whole. This all-in-one approach removes the friction and fragmentation that can slow down legal work, enabling a smoother, more streamlined workflow.

Clio's impact is profound, with the software being lauded as the number one legal practice management solution. Its widespread acceptance and the high volume of positive reviews highlight the system's effectiveness and the tangible benefits it delivers to legal professionals. Clio caters to law firms of all sizes, from large to small and solo practices, with flexible solutions that support a variety of legal specializations. Its centralization of work into one system ensures that everything a legal professional requires is at their fingertips, regardless of where they are or what device they use.

In terms of talent acquisition and retention, AI's influence is significant. Legal professionals are increasingly seeking out innovative work environments that leverage technology to maximize efficiency and reduce the burden of repetitive tasks. Firms that adopt AI-driven platforms like Clio are more likely to attract and retain the best talent in the industry. These professionals are drawn to environments that promise a forward-thinking, technologically-advanced approach to legal practice.

However, the path to integrating AI into legal practice is not devoid of obstacles. It necessitates a willingness within the organization to embrace change and evolve. Legal practitioners must be proactive in their learning, continuously updating their skills to stay abreast of technological advancements. A comprehensive understanding of how AI tools work, their potential impact on legal processes, and the ethical considerations they raise is essential.

Training and development play a critical role in this transition. Effective training programs are essential for ensuring that legal teams are proficient in using AI tools and understanding their broader implications. It's not enough to know how to use the technology; legal professionals must also comprehend how it affects legal practices, client relationships, and the ethical framework of the profession.

Furthermore, a strong data governance strategy is indispensable in the age of AI. Legal firms must ensure the security and confidentiality of the sensitive information they handle. This includes implementing robust measures to protect data against breaches and ensuring compliance with data protection regulations.

The choice of AI tools and solutions requires careful consideration. Legal firms must rigorously evaluate AI products to ensure they are suitable for their specific needs. Factors such as reliability, scalability, and compatibility with existing systems must be thoroughly assessed before making a decision.

In conclusion, the integration of AI and legal technology like Clio into legal operations has the potential to revolutionize the industry. It promises not only to enhance efficiency and productivity but also to redefine the standards of legal service delivery. The successful adoption of AI requires overcoming various challenges, but the rewards—increased efficiency, improved client satisfaction, and the attraction of top legal talent—are well worth the effort.

As we forge ahead, the legal profession must be willing to embrace the changes that AI and legal technology bring. It's not just about adopting new tools; it's about embracing a new mindset that sees technology as an ally in the quest to deliver superior legal services. The law firms that successfully navigate this transition will set themselves apart as leaders in a new, technology-driven age of legal practice.

Legal Tech Transformations: AI Adoption in Law Firms

In today's rapidly evolving legal landscape, the implementation of AI and

Legal Tech has transitioned from a competitive edge to an absolute necessity. Law firms that have yet to integrate these technologies face a stark predicament: modernize or risk obsolescence. The importance of AI in the legal sector is underscored by the transformative effect it has on every facet of practice management—from predictive analytics and legal research to client service and operational efficiency. Legal Tech's role in automating routine tasks is not merely about streamlining processes; it's a fundamental shift towards augmenting the intellectual capacity of legal practitioners. Firms that ignore this shift are poised to find themselves outmatched, as their AI-empowered counterparts can offer faster, more accurate, and cost-effective services. The risk for traditional firms lies not just in falling behind in efficiency but in the possibility of entirely missing the next wave of legal services innovation, which could effectively push them out of the competitive landscape.

As we delve into the importance of embracing AI and Legal Tech, we encounter Rocket Matter—a beacon of modern legal software and a paragon of what AI integration can achieve for today's busy law firms. With over a decade of trust from thousands of attorneys, Rocket Matter stands as a testament to the power of legal practice management when harnessed through intuitive, cloud-based software. This platform embodies the transformation that AI brings to the legal domain, ensuring that firms remain not just relevant but dominant in their field.

Rocket Matter revolutionizes legal practice management with its comprehensive suite of easy-to-use legal software, built explicitly for the demanding needs of contemporary firms. The platform provides a holistic approach to managing a law firm's affairs, enabling quick invoice generation, prompt payment processing, and deep business intelligence. By incorporating AI, Rocket Matter enables legal professionals to manage their law firm's matters with unparalleled efficiency. From case initiation to resolution, every step is optimized for peak performance, granting firms the ability to grow and scale without the traditional increase in overheads.

The essence of Rocket Matter's impact lies in its ability to integrate intelligent time tracking, analytics, and advanced reporting. These features provide law firms with vital insights into productivity and financial health, allowing them to undertake digital transformation with confidence and strategic acuity. The platform's AI-driven analytics offers a window into the future of law practice, with predictive capabilities that help firms anticipate trends and adapt accordingly. This is not just about survival in the digital age—it's about thriving, expanding, and setting new standards for legal service delivery.

Moreover, Rocket Matter's AI capabilities extend to the realm of client engagement and satisfaction. In an era where client expectations are ever-increasing, the platform's AI tools help firms tailor their services to the unique needs of each client, fostering stronger relationships and improving client retention. With features that enable automated client communication and personalized legal advice, law firms can ensure that their services are not just efficient but also client-centered.

Despite the clear advantages, the path to AI integration is laden with challenges that law firms must navigate with prudence. One of the principal dangers is the potential loss of the human touch that is intrinsic to the practice of law. Law is not just about processing information; it's about understanding the human elements of each case. Therefore, while AI can enhance the capability to process legal information, firms must ensure that they do not become overly reliant on technology at the expense of their core values and the personal touch that clients cherish.

Another peril in the realm of AI is the ethical and privacy concerns associated with handling sensitive data. Law firms are custodians of confidential information, and any breach can have severe consequences. As such, the integration of AI must be accompanied by stringent cybersecurity measures to protect against data breaches and ensure the integrity of client information.

The introduction of AI in legal firms also presents an opportunity for a more inclusive legal system. By reducing the costs associated with legal services, AI has the potential to make legal advice more accessible, thereby promoting justice and equality before the law. Rocket Matter's automation of administrative tasks means that lawyers can offer their services at a more affordable rate, expanding their client base to include those who might otherwise be unable to afford legal counsel.

To conclude, the importance of implementing AI and Legal Tech in law firms cannot be overstated, and the risks associated with neglecting this advancement are significant. Firms that do not embrace these technologies may very well find themselves sidelined in an industry that is becoming increasingly tech-driven. Rocket Matter leads the charge in this respect, offering law firms the tools they need to become more automated, profitable, and attuned to the business of law. As the legal industry continues to transform, those who adopt AI and Legal Tech will be the ones who set the pace for the future, defining what it means to be a modern and successful law firm.

Chapter 10:

Navigating the AI Landscape in IP Law

As the legal profession navigates the transformative wave of artificial intelligence, Intellectual Property (IP) Law, standing at the confluence of creativity, technology, and legal protection, faces a unique set of challenges and opportunities. The deployment of AI in patent law and research, for instance, represents a paradigm shift in how legal professionals approach IP management and defense. AI-powered tools have significantly reduced the time and resources required for patent searches and analysis, enabling lawyers and paralegals to focus on more strategic aspects of patent prosecution and litigation. However, this shift also requires a deep understanding of the underlying technology to effectively leverage these tools in practice. This chapter aims to demystify the application of AI in IP law, guiding legal professionals through the intricacies and pointing out both the immediate benefits and the long-term considerations of integrating AI into their IP practices.

The landscape of IP law is notoriously complex, with nuances that can dramatically affect the outcome of patent applications, copyright registrations, and trademark disputes. AI's entrance into this realm introduces another layer of complexity, particularly concerning authorship and creativity in the age of machine learning algorithms capable of generating original works. Legal frameworks around the globe are grappling with these novel concepts, striving to adapt traditional notions of intellectual property to accommodate the capabilities of AI systems. This chapter not only navigates the current state of AI's intersection with IP law but also critically examines the evolving legal and ethical perspectives related to AI-generated content. Through a balanced discussion, it aims to equip practitioners with the knowledge to skillfully navigate AI's IP law challenges and prospects, forging pathways that respect both innovation and established IP legal principles.

Moreover, the exploration of AI in IP law isn't merely an academic exercise; it's a practical guide to navigating a future that has already begun. The trajectory of technological advancement suggests that AI will not only become an integral part of the practice of IP law but might also redefine the very essence of what constitutes intellectual property. The chapter concludes by presenting a forward-looking perspective, contemplating the role of AI in the future of IP law. By providing legal experts, from the seasoned attorney to the curious paralegal, with a comprehensive overview of AI's impact on IP law, this chapter serves as an essential component of a roadmap for incorporating AI and machine learning into legal practices. It underscores the importance of understanding and adapting to these changes to improve operational efficiency, enhance client satisfaction, and positively impact case results in the realm of intellectual property law.

Transforming Patent Law with AI Innovation

In the intricate world of patent law and research, the evolution of artificial intelligence (AI) heralds a new dawn of possibilities and challenges. LexisNexis IP stands at the vanguard of this revolution, offering a suite of AI-enhanced tools that are reshaping the strategies and efficiencies of legal practitioners. This chapter delves into the multifaceted ways in which AI is infused into the domain of patent law, spotlighting LexisNexis IP's role in this transformation while providing an in-depth analysis of the opportunities and challenges inherent in such technological advancements.

Patent law is underpinned by an extensive, intricate repository of global patent data—a vast and ever-expanding universe of innovation. With millions of patents filed each year, the herculean tasks of parsing through prior art, assessing patentability, and monitoring for infringement are daunting. AI, as developed and refined by LexisNexis IP, presents an indispensable solution. Its advanced algorithms can rapidly analyze large datasets, revolutionizing the way patent professionals conduct comprehensive and efficient research.

At the core of LexisNexis IP's tools is an advanced application of natural language processing (NLP). NLP allows these tools to interpret the complex lexicon of patent documents meticulously. This level of understanding is critical when identifying relevant prior art, as it directly impacts the evaluation of a patent's novelty and non-obviousness—two pillars essential for the granting of a patent. The time savings and improved accuracy of searches using AI are undeniable, representing a paradigm shift in patent research methodologies that traditionally demanded much more time and resources.

Predictive analytics is another arena where AI, as wielded by LexisNexis IP, is proving indispensable. Through the analysis of historical patent filings and current technological trends, AI is able to anticipate future innovations and shifts in the technological landscape. This prescient capability enables patent professionals to craft more informed strategies for their clients' patent filings and R&D investments. It also equips them with strategic insights that are invaluable in carving a niche in competitive marketplaces.

Drafting patents is an art that requires precision and a deep understanding of both legal and technological nuances. LexisNexis IP's AI-driven tools aid this process by utilizing machine learning algorithms to suggest specific technical language and legal phrasings that are in line with the most current and relevant patents. This assistance not only streamlines the drafting process but also ensures that new applications are solidified against future legal challenges and positioned for success in the patent office's review process.

Enforcement of patent rights is a critical aspect of intellectual property law, and here too, AI tools have begun to play an essential role. LexisNexis IP provides sophisticated monitoring capabilities that enable patent holders to identify potential infringements with unprecedented speed. This kind of proactive approach to enforcement is indispensable in the modern marketplace, where a rapid response can be the difference between safeguarding one's intellectual property and losing it to infringers.

Despite the clear advantages, the integration of AI into patent law is not without significant challenges. The use of AI raises complex questions about the privacy of data, the reliability of AI-generated outcomes, and the future role of human professionals in the field. Additionally, the legal industry faces a profound ethical debate regarding the status of AI as an inventor and whether AI-generated inventions can, or should be patented.

To effectively address these challenges, legal entities must implement stringent policies and guidelines for the ethical use of AI in patent law. Firms are tasked with establishing procedures for the integration of AI-generated content into patent applications while also maintaining vigilant oversight to mitigate biases that may arise from AI algorithms.

The legal profession must embrace continuous education and training to stay abreast of the rapid developments in AI tools and applications. A thorough understanding of AI's capabilities and limitations is critical for legal professionals to maximize the benefits of AI while avoiding potential pitfalls.

Furthermore, collaboration between AI technology developers, patent professionals, and regulatory authorities is essential in setting the standards for the application of AI in patent law. By working together, these groups can help ensure that AI is used in ways that uphold fairness, accuracy, and transparency.

The future of AI in patent law promises to bring even more sophisticated tools for claim analysis, predictive modeling for litigation outcomes, and automated systems for patent management. The ability of the legal sector to adapt to and embrace these changes will shape the evolution of patent law practice.

In conclusion, the intersection of AI with patent law and research represents a remarkable shift in the legal profession, characterized by a balance between embracing technological innovation and addressing the profound challenges it brings. As AI continues to mature, the strategies and practices of patent professionals will evolve correspondingly. While the journey towards fully leveraging AI's potential in patent law is fraught with complexity, the potential rewards are immense. Through the use of AI tools like those provided by LexisNexis IP, practitioners can navigate the intricacies of patent law with greater clarity and precision than ever before, enhancing the protection and management of intellectual property in the information age.

In embracing AI, legal professionals are not merely adopting a new set of tools; they are partaking in a cultural and methodological shift that redefines the essence of patent research and law practice. As AI becomes more deeply integrated into the fabric of legal operations, it demands a recalibration of roles, responsibilities, and strategies within firms. Legal practitioners must be at the forefront of this wave, ensuring that their use of AI aligns with the core values of the legal system, including justice, fairness, and human dignity.

AI's influence extends beyond streamlining existing processes; it opens doors to novel approaches to legal reasoning and problem-solving. With AI, patterns imperceptible to the human eye can be uncovered, and predictions about legal outcomes can be made with a higher degree of confidence.

LexisNexis IP's predictive analytics serve as a crystal ball, providing foresight into examiner behaviors and the likelihood of patent grant success, thereby equipping patent attorneys with a strategic advantage in prosecution proceedings.

The role of LexisNexis IP in this AI-driven revolution is multifaceted. It serves not only as a tool provider but also as a thought leader and innovator, continuously pushing the boundaries of what's possible in patent law. Its contributions are shaping how AI is perceived and used within the industry, setting a benchmark for innovation and excellence. As AI technology continues to evolve, LexisNexis IP remains committed to enhancing its offerings, ensuring that its clients are always at the cutting edge of patent law and strategy.

However, as legal professionals and firms increasingly rely on AI, there is a pressing need for vigilance against overreliance on technology. The human element of legal work—ethical considerations, professional judgment, and creative legal thinking—remains indispensable. AI is a powerful assistant, but it does not replace the nuanced understanding and ethical compass that experienced legal professionals bring to their work.

In terms of IP workflow optimization, LexisNexis IP's solutions play a pivotal role. By streamlining the search and application process, enabling more efficient monitoring of patent landscapes, and facilitating better strategic decision-making, LexisNexis IP is not just a tool but a strategic partner for businesses and law firms. This partnership is crucial as clients demand more than just transactional legal services; they seek a strategic ally to navigate the complex world of intellectual property.

The expectations placed upon law firms in this new era are unprecedented. Clients are more knowledgeable and discerning, and their choice of legal representation hinges not only on cost but also on the strategic value the firm can provide. LexisNexis IP's data-driven solutions enable law firms to meet these heightened expectations, offering reliable metrics that aid in stakeholder management and process optimization.

In addition, as firms adapt to this new AI-enhanced landscape, they must also consider their role in shaping the broader dialogue about the future of patent law. How AI-related legal challenges are addressed in courts and through legislation will have far-reaching implications. Firms using LexisNexis IP's tools can contribute valuable insights and expertise to these discussions, ensuring that the legal framework surrounding AI and patents remains robust and fair.

Ultimately, AI in patent law is more than just a series of technological advancements; it is a pivotal shift in the way intellectual property is created, protected, and leveraged. As we look to the future, it is clear that AI will continue to transform the patent landscape in profound ways. Legal professionals, empowered by LexisNexis IP's innovative tools, stand ready to navigate this evolving terrain. They will write the next chapter of patent law, one in which AI and human expertise converge to create a more dynamic, just, and efficient system for managing the lifeblood of innovation—intellectual property.

Futurizing IP Law: The Impact of AI Technologies

Navigating the complex landscape where artificial intelligence (AI) converges with intellectual property (IP) law requires not only legal acumen but also a prescient understanding of the technological transformations at play; it is a dance on a mosaic floor that is constantly shifting beneath our feet, a dynamic interplay of evolving patterns that reflect the relentless march of innovation and the ensuing legal quandaries that accompany such progress. Within this domain, AI is not merely a tool; it is a quasi-creator, a force that prompts us to question the very foundations of IP law, to differentiate between the outputs driven by human creativity and those generated by algorithms, and to consider the adequacy of our existing legal frameworks when faced with the nontraditional nature of AI-generated works. The pertinence of AI in IP law is particularly palpable in the realm of patentability, where the traditional patent system, constructed with the human inventor in mind, now encounters the novel phenomenon of AI systems conceiving innovations autonomously, thereby necessitating a recalibrated understanding of what constitutes an 'invention' and potentially prompting a reformulation of the patent law to embrace AI-produced innovations. This conundrum is mirrored in copyright law, where the works created by AI software—be they musical compositions or literary pieces—challenge our notions of originality and authorship, blur the distinction between tool and creator, and push us to reconsider the bounds of copyright law.

Trade secrets, too, are in a state of flux due to AI's formidable data analytical capabilities, which can unravel patterns and insights that might otherwise remain obscured; the traditional methods to safeguard trade secrets might fall short in the face of AI's analytical prowess, urging us to forge stronger defenses and innovative strategies to maintain competitive edges in a rapidly changing marketplace. Trademark law is similarly undergoing a transformation as AI-generated logos and machine-learned brand names compel us to reevaluate established principles of distinctiveness and the likelihood of confusion among consumers, requiring a reimagined legal approach that keeps pace with technological advancements. Yet, amidst these challenges, AI harbors considerable potential to revolutionize IP law by streamlining complex and laborious processes such as patent searches, copyright registrations, and trademark applications, thereby enhancing operational efficiency and precision across the board. In the context of combating counterfeiting and piracy, AI's potential is particularly noteworthy; with its ability to analyze patterns and track digital footprints, AI technologies can detect infringement activities at an unprecedented scale, providing rights holders with a powerful tool in their arsenal against IP violations.

When considering licensing agreements, the increasing integration of AI into products and services demands a nuanced approach to licensing frameworks that can accommodate the complexities of AI while ensuring fair use and fostering an environment conducive to innovation. The international dimension of AI in IP law is another critical aspect that cannot be ignored; the borderless nature of the digital domain and AI technologies requires harmonized international standards and collaborative efforts to effectively manage IP rights in the AI sphere. Moreover, the ethical implications of AI's expanding role in IP law must be thoughtfully addressed; the implementation of AI must be underpinned by robust ethical principles that prevent misuse, bias, and infringement upon privacy, thereby ensuring that technological advances align with societal values and norms.

To successfully navigate these waters, an interdisciplinary dialogue among legal practitioners, policymakers, and technologists is essential, facilitating the development of legal frameworks that are resilient and agile enough to accommodate rapid technological changes. Legislative bodies and international organizations are instrumental in this process, as their forward-thinking legislation and policy initiatives can lay the groundwork for effectively tackling the IP challenges presented by AI while simultaneously nurturing an ecosystem that encourages creativity and innovation. IP professionals, for their part, must be committed to continuous learning and skill enhancement to stay ahead of the curve in understanding AI's implications for IP law; by becoming proficient in the subtleties of AI technology, they can offer their clients enlightened guidance and innovative solutions that address the unique IP issues of the digital age.

As we traverse the path of AI's IP law challenges and prospects, it is abundantly clear that AI is set to redefine not only the practice of IP law but also our conceptualization of intellectual property itself; this journey is characterized by a constant flux of learning and adaptation. The path ahead is strewn with both hurdles and opportunities, necessitating a balanced approach that honors the legacy of IP law while wholeheartedly embracing the potentialities wrought by an AI-driven era. In this evolving narrative, the role of entities like Clarivate becomes ever more consequential; as a beacon of innovation, Clarivate leverages AI to reshape our approaches to copyrights and trademarks. With its sophisticated predictive analytics and conflict detection tools, Clarivate provides a strategic advantage in managing intellectual assets, epitomizing the progressive synthesis of AI with IP law.

Clarivate, as a global leader in providing trusted insights and analytics, stands as a testament to the transformative power of AI in the legal domain. By amalgamating enriched data, astute analytics, and specialized workflow solutions, Clarivate exemplifies the vanguard of AI's application in intellectual pursuits, offering services that are deeply rooted in extensive domain expertise and an unwavering commitment to innovation. Clarivate's history, which commenced in 1864 with the establishment of the Zoological Record, underscores a tradition of pioneering information services that have continuously evolved to address the ever-changing needs of their diverse clientele, which includes not just law firms but also a multitude of stakeholders in the broader knowledge economy.

With over 11,000 professionals worldwide, including legal experts, consultants, and data scientists, Clarivate is uniquely positioned to guide its clients through the labyrinth of AI's intersection with IP law. Their collective expertise and experience serve as the cornerstone for the successful navigation and utilization of AI in the legal sphere, providing insights that are not merely reactive but predictive, enabling clients to anticipate and adapt to the legal challenges ahead. As we cast our gaze forward, the impact of AI on the practice of IP law is undeniable. It is a force that compels the legal profession to reassess and reformulate traditional doctrines to re-envision the mechanisms of protection, enforcement, and exploitation of intellectual property in a way that accommodates the realities of the digital age.

AI's capabilities extend beyond mere automation; they offer a paradigm shift in the detection, analysis, and resolution of IP-related issues, thus allowing the legal profession to transcend traditional barriers and achieve greater effectiveness and precision. The continuous evolution of AI demands that IP law not only responds to current challenges but also anticipates future developments, thus calling for a proactive and prescient legal framework that is capable of adapting to the rapid pace of innovation. This framework must be both robust and flexible, providing a foundation that is solid yet capable of evolving in tandem with the technological advancements that characterize the age of AI. It must also be internationally cohesive, as the digital economy does not recognize national boundaries and IP rights often extend across borders, necessitating a global perspective and cooperation to effectively safeguard these rights in an interconnected world.

The ethical considerations surrounding AI's deployment in the field of IP law also require vigilant attention. As AI systems become more sophisticated and autonomous, the potential for misuse, bias, and infringement upon individual rights escalates, making it imperative that ethical guidelines are not only established but rigorously enforced. These guidelines must ensure that AI is utilized in a manner that is consistent with the public interest, respecting individual privacy, promoting fairness, and preventing discrimination. Legal professionals, along with technologists and policymakers, must work collaboratively to craft these ethical standards, embedding them into the fabric of legal practice to ensure that AI serves the greater good while advancing the field of IP law.

In the face of these multifaceted challenges and prospects, Clarivate's AI-driven technologies and services offer a beacon for the future, illuminating the path forward for IP law. Their tools for predictive analysis, copyright and trademark assessment, and conflict identification are not merely functional; they are transformative, enabling a proactive approach to managing intellectual property that is aligned with the evolving landscape of AI. This approach is not only strategic but also ethical, reflecting Clarivate's commitment to fostering an environment where innovation flourishes while rights are respected and protected.

The convergence of AI and IP law represents a formidable frontier for the legal profession. It is a horizon brimming with complex challenges but also abundant with exhilarating prospects. By confronting these issues with strategic foresight and nurturing a culture of adaptability and innovation, the legal profession can lay the foundation for a future that harmonizes AI with IP law to foster creativity, protect rights, and drive progress. This is a future where the synthesis of technology and legal expertise catalyzes the advancement of both fields, ensuring that the march of progress is both responsible and aligned with the foundational principles of intellectual property law. In this respect, the role of Clarivate, and similar pioneers at the nexus of AI and IP is crucial. They are not just witnesses to the unfolding narrative of AI in IP law; they are active participants and shapers of that narrative, providing the tools, insights, and expertise that will enable the legal profession to harness the full potential of AI in safeguarding intellectual property for generations to come.

In conclusion, the interface between AI and IP law is replete with nuances that demand a nuanced understanding and approach. Through a concerted effort that combines the foresight of entities like Clarivate, the wisdom of the legal fraternity, and the ingenuity of technologists, this complexity can be navigated successfully. It is through such collaboration and continuous innovation that the legal profession can ensure a future where AI not only transforms IP law practice but does so in a way that upholds the integrity of intellectual property rights and fosters an enduring culture of creativity and progress.

Chapter 11:

Artificial Intelligence: Reshaping International Legal Practices

The advent of Artificial Intelligence (AI) has ushered in a new epoch in the realm of international law, signaling a shift toward a more streamlined and sophisticated legal order. At the heart of this transformation is the power of AI to enhance the efficiency and efficacy of legal processes in an increasingly interconnected world. With the complexity of international regulations, the myriad of legal instruments, and the intricate weave of jurisdictional issues, AI has emerged as a vital instrument in the arsenal of legal practitioners. Its capability to process, analyze, and interpret colossal datasets is not merely an addition to the legal toolkit; it represents a paradigm shift in how legal professionals approach the very concept of international law. LexisNexis, as a leader in legal research and technology, exemplifies the integration of AI into the services it provides, offering a vanguard platform for the application of AI in this expansive field.

AI stands at the forefront of innovation in international law, bringing precision and depth to a field characterized by its diversity and complexity. Platforms like LexisNexis leverage AI to parse through the labyrinth of international treaties, conventions, case law, and myriad legislative texts. It acts as a magnifying glass over the fine print of global legal frameworks, providing clarity and insights that are critical in shaping international legal practice. With the aid of AI, practitioners can swiftly navigate through historical documents, extracting essential data that informs current legal strategies and compliance. The synergy between AI and comprehensive legal databases transforms the nature of legal research, enabling a more strategic and informed approach to international legal issues.

The significance of AI extends to the sphere of international dispute resolution. Here, it functions as an invaluable tool for legal analysis and strategy. Predictive analytics empower legal professionals with a data-driven forecast of case outcomes, potentially revolutionizing arbitration and mediation processes. AI provides a dispassionate, evidence-based perspective in disputes, which is invaluable for reaching fair and well-informed resolutions. Whether addressing trade conflicts, intellectual property disputes, or human rights violations, AI stands as a nonpartisan arbiter, equipped to deliver rational and data-backed solutions.

In the crucial area of compliance, AI is a game-changer for multinational corporations and international organizations alike. The constant flux of international regulatory environments poses a formidable challenge for entities striving to maintain legal and ethical operations across various jurisdictions. LexisNexis harnesses AI to offer a real-time, predictive, and comprehensive compliance toolkit. AI's role in compliance is twofold: it ensures adherence to current regulations and provides strategic foresight into upcoming legislative changes. By preemptively adjusting to the evolving legal landscape, organizations can avoid the risks associated with non-compliance, such as legal sanctions or reputational damage.

However, the adoption of AI within international law is not without its challenges. Ethical considerations regarding the decision-making processes of AI, the accountability for actions taken based on AI-driven advice, and the equitable distribution of AI technology are pressing issues. The potential for a digital divide—where resource-rich nations have disproportionate access to AI—is a significant concern that threatens the fair and unbiased application of international law. It is imperative that there be concerted efforts to democratize access to AI technologies, ensuring that all nations, regardless of economic standing, can partake in the benefits brought about by AI. As a global entity, LexisNexis stands as a beacon for the potential democratization of access to AI in legal research and practice.

Beyond functional roles, AI acts as a catalyst for systemic transformation within international law. It is not only a tool for optimizing processes but also a harbinger of methodological evolution, shaping how international law is conceptualized, interpreted, and applied. The true potential of AI lies in its capacity to facilitate more equitable, transparent, and accessible legal practices on a global scale. For legal practitioners, mastering the capabilities and limitations of AI is crucial. This involves integrating legal proficiency with an understanding of AI to navigate this emerging landscape effectively.

Embracing AI's transformative potential mandates a balanced approach—leveraging its power to innovate while ensuring it aligns with ethical standards and contributes positively to the global legal system. The promise of AI in international law is not confined to enhancing the performance of existing systems; it encompasses a broader vision of advancing legal practices to be more inclusive, predictive, and attuned to the principles of international governance. The integration of AI into international law must be guided by a commitment to uphold the rule of law, protect human rights, and promote global cooperation.

The responsibility of adopting AI in international law extends beyond legal practitioners to the broader community of nations and international organizations. It requires a collaborative approach to policy-making, standard-setting, and the development of frameworks that govern the use of AI. This collaborative spirit is encapsulated in platforms like LexisNexis, which provide a nexus for legal professionals worldwide to engage with AI technology in their practices. The collective intelligence that AI embodies must be matched by collective responsibility in its application.

In practice, AI can assist in the drafting of international treaties, the automation of compliance checks, the facilitation of legal education, and the dissemination of legal aid across borders. Such applications reflect the multifaceted nature of AI's influence. In treaty drafting, AI can analyze existing treaties to identify common clauses and language patterns, making the process more efficient and helping to ensure consistency and coherence across international documents. For compliance, AI not only streamlines the verification of regulations but also predicts future regulatory trends, allowing organizations to prepare for and adapt to changes proactively. In legal education, AI-powered platforms can tailor learning materials to the needs of students from diverse legal systems, enhancing their understanding of international law. Furthermore, AI can facilitate access to legal assistance, breaking down barriers to justice by providing low-cost or free services to those in need, particularly in underserved regions.

As we proceed into the AI-integrated future of international law, it's important to consider the role of entities like LexisNexis. With its comprehensive databases and AI-driven tools, LexisNexis not only provides the raw materials for AI's operation in the legal field but also serves as an exemplar of how technology can be leveraged to enhance the practice of law. Their commitment to continuous innovation reflects the dynamic nature of both AI and international law.

Moreover, the standardization of AI applications in international law is crucial for maintaining consistency and reliability. Standard-setting bodies, legal institutions, and international organizations must work together to develop guidelines that ensure AI operates within ethical boundaries and legal frameworks. These standards should promote transparency, accountability, and fairness, ensuring that AI's contributions to international law reflect the values and principles that underpin the legal system.

The transformative impact of AI on international law also extends to the development of new legal norms and the interpretation of existing ones. As AI systems become more adept at processing complex legal information, they could contribute to shaping the evolution of international legal norms, suggesting new ways to address emerging global challenges such as cyber warfare, space exploration, and environmental protection. By offering new perspectives and innovative solutions, AI can play a pivotal role in the progressive development of international law.

In navigating the challenges associated with the integration of AI into international law, the legal community must be proactive in addressing the potential for misuse and unintended consequences. This includes safeguarding against biases embedded in AI algorithms, ensuring that the data used for training AI systems is diverse and representative, and maintaining vigilance against the manipulation of AI tools for unlawful or unethical purposes. The development of robust legal and ethical frameworks to govern AI's application in international law is paramount.

Additionally, as AI tools become more prevalent in international legal practice, continuous education and training for legal professionals become imperative. Lawyers, judges, and policymakers must be equipped with the knowledge and skills to understand and engage with AI technology effectively. This will require a concerted effort to integrate technology education into legal training, fostering a generation of tech-savvy legal professionals capable of harnessing AI's potential responsibly.

In conclusion, the integration of AI into international law is more than a mere technological upgrade; it is a transformative process that holds the promise of more efficient, fair, and accessible legal practices. The role of AI in international law is multifaceted, ranging from enhancing the precision of legal research to redefining the methods of dispute resolution and compliance. Platforms like LexisNexis serve as critical enablers of this transformation, providing the necessary resources and tools for legal professionals to navigate the AI-enhanced legal terrain. As we embrace AI's potential, it is incumbent upon the global legal community to ensure that the deployment of these technologies adheres to the highest standards of ethics and justice, contributing to a more equitable and harmonious international order.

Strategic AI: Mastering Cross-Border Legal Hurdles

The sphere of international law is witnessing a remarkable transformation, thanks in large part to the integration of artificial intelligence (AI). In an era where legal challenges transcend national boundaries and become intricately global, AI emerges as a pivotal force, offering groundbreaking solutions and reshaping the approach to complex legal issues. This exploration focuses on AI's vital contribution to the legal domain, scrutinizing its impact on cross-border disputes, regulatory compliance, and enforcement of international accords, while incorporating insights on the capabilities of Thomson Reuters in the field.

The implementation of AI in legal research across various jurisdictions marks a significant leap forward. With the ability to analyze and interpret large volumes of legislation, case law, and legal precedents in a plethora of languages, AI stands as an indispensable asset for legal professionals. It refines the research process, which historically has been labor-intensive and time-consuming, enhancing it with accuracy and breadth of understanding previously unattainable. Legal strategies are now informed by a more comprehensive and current compendium of legal information, crafted to stand resilient across diverse legal landscapes. The AI solutions offered by Thomson Reuters epitomize this advancement, as they present lawyers with robust, data-driven insights, enabling them to conduct predictive analytics and streamline their work to an unprecedented extent. Their tools afford lawyers strategic foresight in international litigation and arbitration, magnifying their capacity to predict outcomes and tailor their approaches accordingly.

AI's predictive analytics are a cornerstone of its value proposition, particularly in litigation and arbitration on an international scale. By scrutinizing historical data and discerning patterns, tools developed by entities such as Thomson Reuters forecast case outcomes with remarkable precision. This foresight equips legal practitioners with the information needed to make informed decisions about case progression, settlement opportunities, or the pursuit of alternative dispute resolution mechanisms, particularly in jurisdictions where they may lack extensive experience. The predictive edge afforded by these intelligent tools can significantly alter the course of international litigation, providing strategic insights that were previously inaccessible.

Compliance with international regulations presents a formidable challenge, especially for multinational corporations that navigate a complex web of varying legal requirements. AI excels in this arena by perpetually monitoring and analyzing the flux of regulatory frameworks, thereby enabling corporations to anticipate changes and adapt proactively. Automation of compliance processes reduces the risk of infractions and the severe penalties they incur, which is a quintessential advantage in the dynamic and often volatile global market.

The verification and enforcement of international treaties and trade agreements are also areas where AI exhibits formidable capabilities. Through automation and sophisticated algorithms, AI facilitates a more efficient, unbiased verification process, ensuring that all parties adhere to their commitments. Such capabilities strengthen the integrity of international accords and foster a spirit of mutual trust and accountability among nations. This aspect of AI is not just a technological advancement; it's a catalyst for more reliable and transparent international relations.

Human rights initiatives benefit profoundly from AI's surveillance and analytical abilities. AI tools scrutinize various forms of digital data, from satellite imagery to the vast networks of social media, to uncover evidence of human rights violations that might otherwise elude detection. This vigilance amplifies the voices of those suffering oppression and facilitates the pursuit of justice against violators on a global scale, marking a significant stride in human rights advocacy.

Despite these advancements, the integration of AI into the international legal arena is accompanied by notable challenges. The digital divide, a chasm that exists between developed and developing nations, threatens to create an imbalance in the access to and benefits of AI technologies. Addressing this gap is critical to ensuring the equitable application of international law and leveraging AI's full potential across all nations.

Another critical area of concern surrounds data privacy and ethical AI use. Legal data is inherently sensitive, necessitating stringent safeguarding of confidentiality and integrity when processed by AI systems. Legal professionals are tasked with the delicate balance of leveraging AI's power while maintaining the trust of their clients and the public at large by adhering to the highest ethical standards.

The resolution of these challenges necessitates a concerted, interdisciplinary effort. Collaboration between legal scholars, technologists, and policymakers is crucial in creating solutions that optimize AI's strengths and minimize its risks. This cooperative approach can lead to the formulation of standardized guidelines for AI use in international law, fostering global best practices.

Educational initiatives are also essential in realizing the potential of AI in law. As the landscape of legal practice evolves, it is imperative that legal professionals are well-versed in AI technologies. Continuous learning and professional development in AI must become integral to legal education and practice, preparing lawyers to deftly wield the tools AI offers.

While AI augments the efficiency and effectiveness of legal processes, the unique qualities of human lawyers remain indispensable. The nuanced understanding of ethical considerations, the art of persuasive argumentation, and empathetic client relations are attributes where AI complements rather than replaces the human element.

The trajectory of AI in international law is set for continued expansion and innovation. As AI technologies advance, they promise to introduce new solutions to enduring legal challenges. The prospect of AI contributing to a fairer and more equitable international legal system is immense, yet such advancements must be pursued with both caution and intentionality.

In light of these developments, it is evident that the legal profession must adapt, embracing AI as a powerful ally in the quest for justice across nations. The successful integration of AI into international legal practice hinges on the profession's ability to navigate the ethical, educational, and technological challenges that accompany its adoption. As AI becomes increasingly sophisticated, it holds the potential to enhance the legal profession, provided it is implemented with a conscientious and principled approach.

AI's assistance in overcoming global legal challenges signals a watershed moment in the practice of international law. Its capacity to efficiently process vast quantities of information, forecast case outcomes, ensure regulatory compliance, and enforce international agreements equips legal professionals with an advanced toolkit that was once the stuff of science fiction. Yet, it is the legal community's response to the ethical and educational demands of this technology that will determine its ultimate efficacy.

Thus, the intersection of AI and international law presents not merely an option for legal practitioners but an imperative. Those committed to the rule of law and the betterment of global legal practices must not only adopt AI but also actively engage in shaping its influence. It is through this proactive engagement that AI can be steered to serve the greater good, enhancing the capabilities of legal professionals without supplanting the essential human elements of the profession.

Incorporating AI into the fabric of international legal practices offers a wealth of benefits, including the streamlining of processes that have traditionally been cumbersome and prone to human error. AI's ability to provide consistent, reliable analysis across vast arrays of legal documents and precedents is revolutionizing the way legal professionals approach case preparation, due diligence, and risk assessment. This is especially pertinent in an era where international law encompasses a growing array of issues, such as cyber law, environmental law, and intellectual property rights, each with its own complex web of regulations and norms.

Moreover, the continual evolution of AI technology portends the emergence of even more sophisticated tools that can assist in drafting legal documents, negotiating contracts, and facilitating communication across different legal systems. As these technologies mature, they will undoubtedly become integral to the day-to-day operations of law firms and legal departments around the world.

However, the embrace of AI also necessitates a robust framework for its governance. The potential for misuse or misinterpretation of AI-generated data is a valid concern that must be addressed through stringent guidelines and oversight. The legal community must work in tandem with technologists to establish protocols that ensure AI is used responsibly and ethically. Additionally, lawyers and legal scholars have the responsibility to engage in discourse surrounding the implications of AI for privacy, confidentiality, and the preservation of client-lawyer privilege.

As we look to the horizon, the potential for AI to assist in overcoming global legal challenges is limitless. Yet, this potential will only be realized through deliberate and informed action by the legal profession. By embracing AI with an eye towards innovation, ethics, and continuous improvement, legal practitioners can forge a new path for international law. In doing so, they will not only enhance their own practices but also contribute to the development of a more just, efficient, and accessible legal system for all.

The call to action is clear: the integration of AI into international legal practices must be approached with diligence, foresight, and a commitment to upholding the highest standards of the profession. With these guiding principles, AI will not only assist in overcoming global legal challenges but also elevate the practice of law to new heights, benefiting societies and nations across the globe.

The Effect of AI on Worldwide Compliance

In an era where the pace of change in international law accelerates daily, artificial intelligence (AI) stands as a powerful partner in the quest for worldwide compliance. With the proliferation of complex regulations, AI introduces innovative tools for ensuring and managing adherence to these multifaceted mandates. The introduction of AI into compliance regimes signifies a paradigm shift, offering not just heightened operational efficiency in legal practices but also a redefinition of the essence of compliance.

AI's pivotal role in global compliance is incontestable. It automates routine tasks, liberating legal professionals to concentrate on the intricate elements of compliance, such as strategic planning and interpretation—capabilities that are invaluable in a world where regulations are in a constant state of flux. The ability of AI to process extensive data rapidly enables real-time oversight of compliance status, representing a milestone achievement in the legal domain.

The influence of AI on worldwide compliance extends further than simple process optimization. It fundamentally augments the precision of compliance operations. While traditional methods are prone to human error, AI can dissect data with an acuity beyond human capability, diminishing the risk of compliance violations and their attendant consequences.

A noteworthy stride made by AI in this sphere is its predictive acumen. AI systems discern data patterns that could signal impending compliance challenges, permitting organizations to act preemptively. This forward-looking stance not only abates risks but ingrains a culture of compliance that becomes integral to an organization's framework.

In the present international business milieu, AI's value is magnified. Firms operating across borders must decipher and conform to an array of disparate regulations. AI's capacity for comprehending and analyzing regulatory divergences internationally is indispensable, assuring uniform compliance standards globally.

Nonetheless, melding AI into compliance structures is replete with hurdles. Data security, for example, is a predominant issue. With AI systems handling vast quantities of sensitive data, safeguarding against breaches is of the essence. It is incumbent upon legal entities to ally with IT departments to fortify data security protocols and preserve client confidentiality.

Reliance on AI for compliance also provokes questions regarding accountability. In the wake of a compliance lapse, attributing liability—whether to the AI system, its developers, or the legal personnel relying on it—becomes complex. This situation calls for lucid regulations and frameworks overseeing AI's application in compliance.

The ethical ramifications of AI within compliance are profound. As decision-making processes increasingly fall within the purview of AI, guaranteeing equitable and unbiased operation is paramount. Ethical AI development, emphasizing transparency and fairness, is vital.

Despite these impediments, AI's potential to transform global compliance is indubitable. Through AI, legal professionals can traverse international law with enhanced exactitude, swiftness, and efficacy. This boosts compliance initiatives and advances the overarching aim of fostering legal integrity and accountability worldwide.

Moreover, AI's integration into compliance is not simply a trend but a digital age imperative. As regulatory complexity and volume burgeon, the old methods of ensuring compliance will grow obsolete. AI provides a path forward, equipping legal professionals to meet contemporary regulatory challenges.

In conclusion, AI's impact on worldwide compliance is monumental. It heralds an era of change for legal practitioners, offering advanced tools for improved efficiency, precision, and foresight. As AI evolves, its influence on international law will surely expand, underscoring the need to embrace this technology in the pursuit of universal compliance.

Legal professionals must grasp AI's potential and limitations within this domain. By doing so, they can optimize their compliance endeavors and contribute to the evolution of a more equitable legal system. The advent of AI-driven compliance is only beginning, with implications for the legal profession and society poised to be significant.

Looking ahead, synergistic efforts among legal specialists, technologists, and policymakers will be pivotal in harnessing AI's full potential in compliance. Jointly, they can surmount challenges and ethical concerns, paving the way for an epoch of compliance that is efficient, effective, and upholds the highest standards of legal and moral integrity.

Wolters Kluwer: A Vanguard of AI-Enhanced Compliance

Central to this AI-driven transformation is Wolters Kluwer, a beacon in the global legal landscape. Its AI-powered tools revolutionize legal research by providing an exhaustive panorama of international legal frameworks. With succinct representations of pertinent treaties, statutes, and case law, Wolters Kluwer aids the complex decision-making processes intrinsic to the practice of international law.

Wolters Kluwer's digital resources are a stepping stone toward a more productive, efficient legal department. With its international English-language legal information and insights, the Wolters Kluwer Legal & Regulatory U.S., International Group, furnishes legal practitioners, corporate counsel, and business executives with the tools necessary for navigation in the global arena of law. Their cross-jurisdictional products, coupled with online resources and influential blogs hosted by legal luminaries, provide a fertile ground for dialogue and collaborative thought, propelling the legal profession into the future.

Wolters Kluwer stands at the forefront, offering a world of expertise in international law. Their services ensure that clients and colleagues receive swift, well-informed guidance on international regulations that have a direct impact on their operations. From intellectual property law to environmental and energy regulations, Wolters Kluwer is the gateway to essential information and analysis needed to tackle critical legal issues worldwide.

In a world that demands accuracy and certainty, Wolters Kluwer is the go-to source for in-depth analysis and user-friendly digital solutions for matters governed by international law. Their commitment to precision and integrity makes them an indispensable ally in the legal industry.

As we stand on the brink of a new compliance paradigm powered by AI, the contributions of Wolters Kluwer are both timely and impactful. They not only facilitate access to crucial legal information but also enhance the ability to digest, interpret, and apply this knowledge to various international compliance scenarios.

The symbiotic relationship between AI and human expertise is epitomized in the Wolters Kluwer offerings. Their AI tools, combined with the acumen of legal professionals, forge a formidable force in compliance. This alliance enables the anticipation of regulatory shifts, the navigation through the complexities of international law, and the provision of strategic insights that are both comprehensive and compliant.

The landscape of worldwide compliance is rapidly transforming, with AI at the helm. This evolution brings about an elevated standard for compliance practices, which is made tangible through the advanced solutions provided by Wolters Kluwer. They not only meet the existing demands of the legal profession but also shape its future trajectory, ensuring that the legal industry remains at the vanguard of innovation and excellence.

Embracing the potential of AI is not an option but a necessity for contemporary legal entities. Wolters Kluwer exemplifies this adoption by offering digital platforms that are robust, reliable, and ready to meet the challenges of the modern regulatory environment. Their expertise in international law is not merely a resource; it is a catalyst for transformation, enabling professionals to excel in their compliance efforts and beyond.

In essence, the effect of AI on worldwide compliance, amplified by the resources of Wolters Kluwer, presents an unparalleled opportunity. It is an opportunity to reshape the modalities of legal compliance, increase the rigor and responsiveness of legal practices, and instill a proactive, predictive, and precise approach to international regulations.

In this age where legal compliance is a global endeavor, the intersection of AI and the rich repository of knowledge housed within Wolters Kluwer creates a synergy that is greater than the sum of its parts. It equips practitioners with the foresight to foresee and adapt to the dynamism of international law, assuring that compliance is not a mere regulatory hurdle but a strategic advantage.

As AI continues to mature and integrate more deeply into the fabric of legal operations, its potential to transform the landscape of compliance will only be magnified. In partnership with Wolters Kluwer, legal professionals are not just keeping pace with this transformation; they are

driving it, leveraging AI to foster a global compliance environment that is more efficient, transparent, and just.

The partnership with Wolters Kluwer provides an edge to those seeking to navigate the often turbulent waters of international compliance. Their cutting-edge AI applications, when deployed alongside the expertise of legal professionals, allow for a nuanced understanding of legal complexities. This combination of technology and human insight is the cornerstone of effective compliance strategies that resonate with the demands of the modern world.

The integration of Wolters Kluwer's resources into compliance strategies is indicative of the future of legal practice. By harnessing their comprehensive databases, analytical tools, and expert commentary, legal professionals can stay ahead of regulatory changes, adapting their compliance measures to reflect the latest developments in international law.

Moreover, the use of AI-powered solutions from Wolters Kluwer in navigating the compliance landscape underscores the shift towards a data-driven legal practice. Data, once a byproduct of legal processes, has now become central to compliance strategy. Through the advanced analytical capabilities provided by Wolters Kluwer, data is transformed into actionable insights, empowering legal teams to make informed decisions and take proactive steps in maintaining compliance.

The breadth of Wolters Kluwer's influence is seen in its ability to unify diverse legal traditions and practices under a coherent framework for international law. As regulations become increasingly interwoven on a global scale, the need for a consistent and harmonious approach to compliance is imperative. Wolters Kluwer rises to this challenge by providing a global perspective that respects the nuances of local jurisdictions while fostering international harmonization.

In embracing the transformative power of AI, Wolters Kluwer is not only redefining the landscape of worldwide compliance but is also reinforcing its commitment to the principles of legal ethics and responsibility. In the pursuit of compliance, it is essential that the tools and technologies employed are aligned with the highest standards of legal practice. Wolters Kluwer's platforms are designed with this in mind, ensuring that the pursuit of compliance is congruent with the ethical obligations of the legal profession.

As AI reshapes the future of worldwide compliance, the importance of education and awareness among legal professionals cannot be overstated. Wolters Kluwer facilitates this by offering not only resources and tools but also training and support to ensure that legal practitioners are fully equipped to utilize AI in their compliance efforts. This commitment to professional development is critical in ensuring that the legal industry remains adept and agile in the face of AI-driven change.

The journey towards an AI-enhanced compliance framework is ongoing. With each advancement in AI technology, new possibilities emerge for improving the speed, accuracy, and effectiveness of compliance practices. Wolters Kluwer remains at the forefront of this journey, continuously innovating and expanding its offerings to meet the evolving needs of the legal profession.

In sum, the impact of AI on worldwide compliance, as facilitated by the resources of Wolters Kluwer, is both significant and transformative. It heralds a new age of legal practice, characterized by heightened efficiency, strategic foresight, and an unwavering commitment to upholding the rule of law across international borders. As legal professionals and organizations embrace these AI-driven tools, they pave the way for a future where compliance is not just a regulatory requirement but a cornerstone of global legal integrity and ethical practice.

Chapter 12:

Worldwide Legal Transformation Through AI

Under the weight of artificial intelligence, legal traditions worldwide are bending and reshaping from the inside out as AI emerges as the architect of a profound metamorphosis across continents. This transformation isn't limited to automated document analysis or improved client service platforms; it represents a fundamental shift in how legal processes are conceptualized and executed globally. As AI transcends borders, its integration into legal systems worldwide presents an intricate mosaic of adaptation and innovation. In this extensive examination, we delve into the nuanced ways in which AI is being woven into the very fabric of different legal traditions and practices across the globe, spotlighting the unique contributions and pioneering advancements that are setting the benchmarks for the future of the legal profession.

The evolution of AI in the legal field within different jurisdictions paints a rich tapestry of the intersection between technology and law. In Europe, the General Data Protection Regulation (GDPR) presents a dual-edged sword for AI deployment, balancing the need for data protection with the growth of AI-driven legal analytics. The legal sectors in countries like France and Germany are redefining client confidentiality and data stewardship in the age of AI, forging new paths for privacy law and ethical AI usage. Meanwhile, in Asia, the burgeoning tech industry is accelerating the infusion of AI into legal services. Japan's initiatives to incorporate AI into their judicial system, including AI-assisted court clerks and digital evidence analysis, are particularly noteworthy, signaling a shift toward tech-centric legal proceedings.

Some nations stand out for their pioneering approaches to integrating AI in governance and law. Singapore's Smart Nation initiative and Estonia's e-Residency program are prime examples of government-led strategies that encourage the assimilation of AI into legal processes and broader administrative functions. These programs demonstrate the potential for transformative governance when legal structures and technological advancements evolve in tandem. The legislative agility and digital infrastructure of such countries provide a blueprint for what's possible when innovation is not just adopted but embraced by the legal framework.

This global shift towards AI integration is not without its challenges. The discrepancies in technological infrastructure, the degree of digital literacy among legal professionals, and the variances in regulatory frameworks pose significant hurdles to the uniform adoption of AI in law. However, these challenges are accompanied by substantial opportunities. Legal firms and departments worldwide can derive valuable insights from these diverse international models, discerning best practices for leveraging AI to optimize efficiency, accuracy, and accessibility in legal services.

The spectrum of AI applications in legal innovations across the globe is indeed inspiring. From blockchain's promise to redefine contract law to AI's capacity to streamline dispute resolutions, the transformative potential of AI in law is vast. For instance, the use of AI in predictive justice algorithms in Spain and Brazil points toward a future where legal outcomes may be forecasted with greater certainty, potentially revolutionizing how cases are strategized and argued. The global perspective on these AI-driven innovations not only offers a peek into what is currently technologically achievable but also shapes the collective vision of a legal system enhanced by AI. This is a vision where justice is not only administered more efficiently but is also more accessible and equitable, thanks to the democratizing power of technology.

Engaging with international advancements in AI legal applications broadens the horizon for what legal practice could look like in the coming years. The narratives emerging from various regions underscore the necessity of cross-border dialogue and cooperation in forging an era of legal practice where AI tools are standard. It is within this global context that ethical considerations and standards for AI use in law become paramount. International collaborations are critical in establishing these standards, ensuring that as legal systems become increasingly dependent on AI, they do so with a commitment to upholding the principles of justice, fairness, and human dignity.

This chapter aims not merely to provide a survey of the current state of AI in law globally but to serve as a visionary compendium for future innovation. As AI continues to reshape the legal landscape, it is vital to maintain a global perspective, appreciating the diverse contributions and shared challenges. The synthesis of international experiences and pioneering efforts in AI provides a robust framework for envisaging and realizing the next frontier of legal practice—a practice where AI and law synergize to deliver justice in ways previously unimagined.

In conclusion, the global narrative of AI in law is one of diversity and convergence. From the careful stewardship of personal data in compliance with GDPR to the innovative e-governance models of Estonia and Singapore, each jurisdiction contributes to the collective understanding and implementation of AI in legal practices. As law firms and legal departments worldwide seek to harness the power of AI, they must navigate the intricate balance between innovation and ethical practice. This global perspective serves not only as a reflection of current trends but also as an invitation to participate in shaping the trajectory of AI in law, fostering a future where technology enhances every facet of legal practice for the benefit of all.

Global Diversity in AI-Driven Legal Evolution

As we delve into the proliferation of artificial intelligence (AI) within legal practices globally, it becomes evident that this technological frontier is reshaping the legal landscape in multifaceted ways. The Thomson Reuters Institute, through its incisive research and analysis of legal technology trends, identifies AI as a pivotal force in the modernization of legal frameworks across various jurisdictions. This comprehensive review draws upon the Institute's extensive data to explore the benchmarks and strategic approaches to AI integration, offering a valuable blueprint for law firms and legal departments intent on harnessing AI's transformative potential.

In the competitive crucible of North American legal markets, AI's influence is particularly potent. The United States and Canada demonstrate a robust appetite for AI-driven solutions—evidenced by increasing investments and burgeoning initiatives within law firms and corporate legal departments. These entities are adeptly applying AI to an array of tasks, such as conducting comprehensive legal research, streamlining document review processes, forecasting legal outcomes, and automating routine advisement. The impetus for this rapid embrace of AI technology is driven by a confluence of market pressures that demand enhanced efficiency, elevated accuracy, and superior client service delivery.

The European AI narrative is characteristically heterogenous. The United Kingdom stands out as a bastion of AI adoption within the legal sector. UK-based firms, particularly in London, are pioneering AI's application in contract analysis, litigation forecasting, and a host of other legal services, leveraging these technologies to secure a competitive edge. Conversely, the European Union's rigorous data protection regulations, epitomized by the General Data Protection Regulation (GDPR), pose a dual role as a constraint and a catalyst, compelling the development of AI solutions that are attuned to stringent privacy and security mandates.

Turning our gaze to Asia, nations such as Singapore and China are distinguished by their progressive stance on legal technology. Singapore's government has championed the integration of AI into legal tech through initiatives like the Smart Nation strategy, fostering an ecosystem conducive to innovation and adoption. China's voluminous data reserves and significant AI investments have poised it as an emerging juggernaut in the domain of AI, extending its influence into the legal sector with rapid advancements and applications.

Conversely, many developing economies and less technologically advanced jurisdictions are experiencing AI's infancy within their legal systems. Here, challenges abound: access to robust data sets is often scarce, AI tools can be prohibitively expensive, and there is a dearth of skilled professionals to oversee the development and implementation of AI applications. These hurdles notwithstanding, these regions have the unique prospect of transcending traditional legal operations through strategic AI integration, which holds the promise of radically transforming legal service accessibility and delivery.

The ethical dimension of AI in the legal realm is a subject of intense scrutiny. Across the globe, legal professionals and regulators grapple with the moral quandaries associated with delegating traditionally human-centric legal tasks to algorithms, particularly those that necessitate nuanced judgment. The implications are profound in areas such as legal adjudication prediction and AI-assisted courtroom decision-making, where concerns about inherent biases and the potential for errors are especially heightened.

On the regulatory front, jurisdictions are contending with the complex task of crafting governance frameworks suitable for AI applications within the legal sector. The EU has been at the vanguard, proactively establishing regulatory structures, while others are still evaluating the most effective means to ensure that AI is deployed in a manner that upholds justice, transparency, and fairness.

To leverage the full spectrum of AI's advantages while circumventing its pitfalls, the global legal fraternity could greatly benefit from an ethos of collaborative exchange and mutual learning. By dissecting the success stories and cautionary tales from disparate legal environments, practitioners can glean vital insights into effective AI adoption strategies and the common traps to be avoided.

AI's capacity to democratize access to legal assistance is one of its most stirring prospects. By automating routine tasks, AI has the potential to reduce the cost barrier associated with legal services, thereby expanding its reach to historically underserved demographics. For jurisdictions where legal support is sparse, AI-empowered advisory platforms could represent a vital conduit for legal resources, democratizing the availability of legal knowledge and support.

Education and continual professional development concerning AI technologies are critical for legal professionals to capitalize on AI's advantages and navigate its ethical complexities. As AI's capabilities evolve, staying abreast of the latest developments and understanding the ethical frameworks within which they operate is essential. This necessitates a recalibration of legal education to incorporate AI proficiency and ethical considerations as fundamental elements of the curriculum.

In summary, while the AI adoption narrative varies across jurisdictions, the overarching outlook is one of optimism. Despite the persistent challenges associated with ethical dilemmas, the creation of appropriate regulatory environments, and the disparate rates of adoption, the positive impact of AI on legal practices is irrefutable. As the global legal community continues to integrate AI into its core operations, the primary focus should be on harnessing technology to amplify the effectiveness, accessibility, and equity of legal services.

Thus, legal professionals such as lawyers, paralegals, and legal scholars are positioned at the edge of a profound transformation. Embracing AI technologies means not the obsolescence of human discernment within legal processes, but its enhancement. It empowers legal practitioners to provide greater value to their clientele and contribute more substantially to societal well-being. The journey towards full AI integration into legal practices is ongoing, and maintaining an informed, flexible, and ethical approach to its utilization will pave the way for a future where legal services are more widely accessible, proficient, and just.

The role of institutions like the Thomson Reuters Institute is pivotal in this scenario. As a beacon for dialogue, trend analysis, and insight generation, the Institute assists the legal, corporate, tax, accounting, and government sectors in navigating these transformations. By providing a platform for professionals to discuss and debate these emergent technologies, the Institute is helping to shape the trajectory of AI in the legal field. It encourages a rigorous examination of trends, driving an informed discourse that aids firms and legal departments in making strategic decisions about AI adoption.

A significant offering from the Thomson Reuters Institute is the guidance provided to firms seeking to construct a business case for AI. In an era where AI tools like ChatGPT have the capacity to radically streamline legal workflows, understanding the nuances of AI's benefits against its challenges becomes imperative. Firms must assess not just the direct advantages of automated processes and enhanced analytics but also consider the associated demands on infrastructure, training, ethical compliance, and data security.

To encapsulate, the global legal sector's foray into AI is a tale of great potential shadowed by complex considerations. As firms look to the Thomson Reuters Institute for direction and insight, they are better positioned to navigate the intricacies of AI integration, from logistical implementations to profound ethical inquiries. Legal AI is not just about the automation of tasks but about re-envisioning the delivery of justice and legal services.

AI's adoption across jurisdictions reflects a broader shift in the legal industry, with innovation no longer a luxury but a necessity for survival and growth. As the Thomson Reuters Institute's research indicates, there's a pressing need for a strategic roadmap that law firms and legal departments can follow to achieve the promises of AI. Such a roadmap should encompass the development of AI governance policies, investment in technological infrastructure, and the fostering of a culture that embraces continual learning and ethical vigilance.

Ultimately, the Thomson Reuters Institute's work underscores that the true measure of AI's success within the legal domain lies in its ability to enhance human expertise, not replace it. By fostering a symbiotic relationship between AI and legal professionals, the industry can strive towards a future where the scales of justice balance more evenly for all. The global adoption of AI in law is an evolving narrative, one that continues to unfold with each technological breakthrough and each step taken towards a more informed, ethical, and equitable legal landscape.

In conclusion, AI's journey across various jurisdictions is a testament to the legal profession's adaptability and its enduring commitment to the pursuit of excellence. With entities like the Thomson Reuters Institute illuminating the path forward, legal professionals worldwide are empowered to integrate AI into their practices thoughtfully and strategically. The path is set for a legal future that is more informed, more efficient, and more equitable, with AI serving as a key ally in the quest for justice and legal innovation.

Innovative AI Legal Practices from Around the World

The fusion of Artificial Intelligence (AI) and legal practice is transforming the landscape of the profession, bringing into sharp focus the profound effects of technological advancements on the law. This evolution requires a panoramic view, capturing the mosaic of global contributions that shape the interface between AI and legal systems. The Law Society of England and Wales, along with Stanford Law School's CodeX, stands at the vanguard of this transformation, guiding the legal fraternity towards a future where technology and law coalesce to create unprecedented value.

In the corridors of Europe's legal institutions, AI's integration into legal frameworks is evident, with The Law Society of England and Wales at the helm. As custodians of ethical AI deployment in legal services, The Society champions a vision wherein the legal profession is integral to the advancement of AI. They prescribe a regulatory compass that points towards the recognition of legal expertise, the sanctity of legal professional privilege, a balanced regulatory approach, and the imperative of informed consent from clients when AI is employed. Their work has set a precedent, reflecting the need for a legal industry that is responsible and versed in the nuances of emerging technologies.

Society has not only embraced AI's potential but has initiated a proactive dialogue about its implications. For example, generative AI—a branch of AI specializing in creating novel outputs from extensive data pools—has been recognized for its utility in tasks such as contract analysis, drafting documentation, and even predictive case outcomes. It's noteworthy that a 2020 survey conducted by the University of Oxford in conjunction with The Law Society revealed a tipping point: half of the solicitors surveyed acknowledged the use of AI within their practices. This speaks volumes about the growing acceptance and incorporation of AI in the legal sector.

Parallel to the strides taken by The Law Society, Stanford Law School's CodeX stands as a lighthouse for legal innovation, particularly in the AI domain. CodeX's inter-disciplinary thrusts have galvanized legal tech advancements globally. Its initiatives act as a crucible for progress, merging academic inquiry with pragmatic application, and informing policy-making with the latest technological advancements. Among its many ventures, CodeX has been instrumental in developing novel legal document management methodologies, harnessing AI to streamline information retrieval, outperforming manual processes in both cost and accuracy.

The prowess of Large Language Models (LLMs) such as GPT-4, brought to the forefront by CodeX, exemplifies the seismic shift in legal paradigms. These LLMs have automated routine legal tasks, liberated human lawyers from monotonous duties, and laid down the tracks for return on investment in legal operations. Moreover, CodeX has been proactive in fostering the next wave of legal tech pioneers through its accelerated courses on generative AI. The Stanford Law x LLM hackathon, a crucible of innovation, is one such platform where students are armed with insights on generative AI, designing and pitching AI-driven legal solutions.

CodeX's AI Life Cycle Core Principles represent another significant stride, focusing on the transparency and accountability of AI outputs. This analytical framework scrutinizes the congruence between anticipated and actual outcomes and examines adherence to Data Stewardship Frameworks, ensuring that AI's integration into the legal sphere is both conscientious and in line with professional standards.

The international narrative of AI in law unfolds through a spectrum of innovative applications that both challenge and enrich the profession. Estonia, with its e-Justice initiatives, personifies the drive towards a technologically empowered legal system. AI-driven chatbots in Estonia serve as virtual legal aides, guiding individuals through complex legal terrains with ease. The integration of technology in such a manner exemplifies Estonia's commitment to enhancing legal services and broadening the public's access to justice.

The journey across the globe brings us to Asia, where China's "Internet Courts" emerge as a futuristic fusion of law and technology. These virtual courtrooms, equipped with AI for evidence assessment and judgment drafting, stand as a testament to the efficiency and ingenuity AI brings to judicial proceedings. As a harbinger of what AI-assisted adjudication could look like, China's Internet Courts also present a prototype for future legal systems.

Canada’s legal research tools, powered by AI, illustrate the democratization of legal knowledge. By diminishing the time required to analyze legal documents and case law, these tools enable equitable competition among legal practitioners, regardless of firm size.

In the United Kingdom, AI's predictive capabilities are harnessed to forecast legal outcomes. Such tools guide lawyers in strategizing, albeit with a conscious regard for the ethical dimensions these technologies impinge upon. The predictive prowess of AI elucidates its potential to complement human expertise in legal decision-making processes.

As we navigate further, AI's reach extends to legal drafting and contract management, where, for instance, Australia has seen the development of AI tools that automate the crafting of legal documents, ensuring adherence to legal standards. This not only streamlines workflow but also mitigates the risks associated with human error. The introduction of these tools represents a transformative shift in the legal sector, significantly enhancing the ability of lawyers to produce accurate, reliable legal documents quickly.

These AI-driven tools redefine the norms of legal practice, embodying efficiency and precision. The advancements are not just limited to common law countries but have seen reception in civil law jurisdictions as well. Brazil, grappling with a significant backlog of legal cases, has turned to AI for resolution. Machine learning algorithms sift through cases, drawing from a vast repository of case law and statutes to prioritize and propose resolutions. This is not just a solution to a national issue but a blueprint for legal systems around the world facing similar challenges.

Venturing into Africa, Kenya is pioneering the use of AI to extend legal aid to those traditionally underserved by the legal system. AI chatbots and mobile platforms provide accessible legal advice and documentation support, a clear indication that AI can be a powerful equalizer in the legal field.

Japan's nuanced approach to AI in legal practices emphasizes ethical integration, aiming to develop systems that enhance rather than replace human lawyers. This approach is reflective of a broader cultural mindset that sees technology as an augmentative tool rather than a replacement for human skill and judgment.

In the northern stretches of Europe, Scandinavian nations are experimenting with AI to simplify regulatory compliance and legal advisement, easing the bureaucratic load on businesses and citizens alike. These initiatives are a glimpse into the potential of AI to streamline complex legal and regulatory processes, making them more accessible to the layperson.

The Middle Eastern foray into AI in law is exemplified by the United Arab Emirates' ambitions to interweave AI throughout its judicial system. This initiative is part of a larger vision to establish the UAE as a global nexus for technology and innovation, employing AI for functions such as drafting legal opinions, predicting judicial decisions, and enhancing court administrative functions.

The myriad of AI applications in the legal sector underpins the transformative power of technology across all aspects of legal work. From research and case management to dispute resolution and expanding access to justice, AI is reshaping the legal profession. Yet, with great power comes great responsibility, and these innovations bring to light the critical need for ethical vigilance, privacy protection, and security. As AI becomes a staple in legal systems, it is imperative that its integration adheres to the tenets of fairness, transparency, and human dignity.

The global canvas of AI legal innovations offers a view into a future replete with opportunities yet fraught with complex ethical considerations. The onus is on legal professionals, law firms, and policymakers to capitalize on AI's potential while fostering international collaboration and exchanging knowledge. The insights gained from various jurisdictions can inform the evolution of AI applications in law, ensuring they are ethical, just, and effective.

As AI continues to redraw the contours of the legal profession globally, its ultimate influence will hinge on our collective ability to balance ingenuity with ethical integrity. Drawing on the rich tapestry of international initiatives and maintaining a principled approach will enable the legal community to steer through the AI revolution. This is a journey towards a new epoch where justice and legal services are redefined for the digital age.

In summation, the integration of AI into legal practices is an odyssey that is both exhilarating and formidable. International legal AI innovations not only provide a vision of what the future of law might entail but also act as a clarion call to legal practitioners worldwide. By embracing AI and conscientiously addressing its complexities, the legal profession can vault into a new era marked by enhanced efficiency, wider access, and revolutionary innovation.

Chapter 13:

Artificial Intelligence: Redefining Client Relations and Legal Marketing

In the landscape of the legal industry, the infusion of artificial intelligence (AI) has set the stage for an extraordinary evolution in client relations and marketing. This evolution is characterized by a shift towards personalized client service and a marketing metamorphosis facilitated by AI-driven tools such as HubSpot and Clio Grow. As these tools weave their way into the fabric of law firms' operations, they create a dynamic environment where efficiency and client-centricity become the cornerstones of business strategy.

The potency of AI in reshaping client relationships is undeniable. With HubSpot at the helm, legal professionals can harness the power of marketing automation, enriched content creation, and targeted outreach. HubSpot's AI Content Creation tool, which leverages OpenAI's GPT models, exemplifies this shift. It amplifies the productivity and creativity of content creators, enabling them to produce more impactful and engaging content that resonates with clients and prospects alike. The use of such sophisticated AI tools facilitates a deeper understanding of clients' needs, driving law firms towards more predictive and personalized services.

For instance, Legartis, a legal tech company specializing in SaaS for B2B, experienced firsthand the transformative capabilities of HubSpot. Faced with the challenge of establishing efficient, lean, and agile processes during its growth phases, Legartis integrated HubSpot's marketing automation tools into its strategy. Nicole Schnetzer, the Head of Marketing at Legartis, acknowledged that for a small but expanding company, the choice of HubSpot was pivotal. It provided an intuitive and professional solution that not only aligned with their current needs but also scaled with their growth. By adopting HubSpot, Legartis streamlined its operations and significantly enhanced its marketing strategies, showcasing the profound impact AI can have in scaling a legal firm's presence in a competitive market.

HubSpot's platform extends beyond operational streamlining; it offers legal marketers the ability to dissect large datasets to identify trends and forecast the effectiveness of marketing efforts. By employing these insights, law firms can craft bespoke marketing campaigns that are not just aligned with their audience but are also fine-tuned to the nuances of client behavior and preferences. This strategic marketing is made possible by the predictive power of AI, allowing firms to maximize engagement and optimize their return on investment.

Moreover, the personalization capabilities provided by AI do not stop at analytics. Tools like chatbots and automated content personalization facilitate interactions that are not just consistent but are also tailored to the individual client. This ensures a client experience that is engaging and memorable, thereby elevating a firm's brand in an increasingly digital-dominated landscape. Yet, as firms march forward into the AI-powered future, they must do so with a keen awareness of the ethical boundaries that govern client privacy and data security.

Complementing HubSpot's suite of tools, Clio Grow offers a targeted solution to a perennial challenge in legal practice: client intake and relationship management. Clio Grow's cloud-based software refines the process of converting potential client interactions into successful client relationships. With its advanced features, law firms can efficiently manage and track client intake processes, ensuring that no detail or potential client is overlooked.

Clio Grow's approach to client intake is revolutionary, enabling firms to automate and simplify what was once a labor-intensive process. The platform's customizable Matter Pipeline allows for the organization of potential clients according to the firm's specific intake procedures. Firms can craft personalized journeys for each client, from the initial contact through to the final onboarding, with the click of a button.

The software also provides the Quick Intake feature, which allows for rapid data entry and appointment setting, demonstrating Clio Grow's understanding of the need for speed and efficiency in today's fast-paced legal environment. It's a tool that not only saves time but also enhances the quality of the client's first impression of the firm. Clio Grow recognizes that while a significant portion of new business comes from referrals, the initial client experience is critical in securing repeat business and fostering long-term relationships.

The integration of HubSpot and Clio Grow into law firm operations speaks volumes about the changing tides in the legal industry. These tools stand as testaments to the power of AI in revolutionizing the way law firms interact with clients and market their services. They embody the transition from traditional practice to a more innovative, client-focused approach, heralding a new era where efficiency, personalization, and client satisfaction are paramount.

As legal professionals chart their course in the AI era, they must strike a balance between technological adoption and the preservation of the human touch. While AI offers unparalleled efficiencies and insights, the core of legal practice—personalized, empathetic service—must not be lost in the race to innovate. It is the firms that can harmonize the precision of AI with the relational depth of human interaction that will emerge as leaders in the legal landscape.

In conclusion, the legal industry's embrace of AI, as exemplified by the integration of HubSpot and Clio Grow, represents more than a passing trend. It is a strategic reorientation towards a future where client relations are deeply personalized, and marketing strategies are sharply data-driven. It is a move towards a future where the intersection of technology and legal expertise propels firms into new realms of efficiency and effectiveness.

The benefits of AI-driven tools extend beyond immediate gains in productivity. They lay the groundwork for a deeper, more analytical approach to legal practice—one that embraces the complexities of client behaviors and the subtleties of market dynamics. With platforms like HubSpot, legal professionals can leverage AI to navigate these complexities, using insights to build more robust and enduring client relationships.

In marketing, the agility afforded by AI enables firms to deploy innovative campaigns that cut through the noise of a crowded digital space. The analytics capabilities of these tools allow firms to continually refine their strategies based on real-time feedback and evolving market conditions. This level of adaptability is crucial in today's fast-paced world, where client expectations and market landscapes are in constant flux.

The synergy of AI tools like HubSpot and Clio Grow creates an ecosystem where every client interaction is an opportunity for growth, and every marketing initiative is an exercise in precision. It empowers firms to deliver a level of service that not only meets but exceeds client expectations. In doing so, it sets a new standard for what it means to engage with legal services.

Yet, for all the advancements AI brings, the future of client relations and marketing in the legal sector will not be defined by technology alone. It will be shaped by the ability of firms to integrate these tools with a sense of purpose and a commitment to their core values. It will depend on their ability to use AI responsibly, upholding the highest standards of privacy and ethics.

Firms that successfully merge AI's analytical strength with the human element's emotional intelligence will find themselves at the forefront of the legal industry. They will be the ones to offer their clients a seamless blend of innovation, convenience, and personal touch. As such, the integration of AI into legal practice is not just about adopting new tools; it's about fostering a culture of innovation that respects the past while boldly stepping into the future.

In this context, HubSpot and Clio Grow serve not only as catalysts for change but also as bridges between the old and the new. They embody the potential of AI to enrich the legal profession, offering a glimpse into a future where the lawyer-client relationship is more accessible, transparent, and connected than ever before.

As law firms adopt these tools, they must do so with an eye towards the long term. They must consider how AI will help them build a practice that is not only successful today but also resilient and relevant in the years to come. With these technologies, firms can construct a foundation for growth that is as enduring as it is dynamic.

In sum, the evolution of client relations and marketing in the legal industry, driven by AI and exemplified by tools like HubSpot and Clio Grow, is a multifaceted journey. It is a journey marked by advancements in technology, a deepening understanding of client needs, and a commitment to maintaining the personal touch that lies at the heart of the legal profession. As firms navigate this journey, they will redefine what it means to provide legal services, setting new benchmarks for excellence and innovation in an ever-changing world.

Personalizing Custom Client Experiences with AI

In an era where the impersonal click of a mouse can dictate the nature of professional engagements, the legal industry stands as a bastion where personal connections still hold paramount importance. The contemporary legal professional's repertoire is incomplete without a nuanced understanding of how to foster deeper, more meaningful client relationships. This chapter seeks to dissect the intricate weave of personalization within legal practices, particularly through the sophisticated lens of artificial intelligence (AI) and its offshoots.

At the bedrock of personalizing interactions lies the undeniable necessity to grasp and address the diverse and complex needs of clients. The advent of AI technologies in legal practices ushers in an era of enriched data analysis capabilities. With AI, legal professionals can sift through vast repositories of client interactions, case notes, and preferences. This data, once nebulous and unwieldy, can now be orchestrated into a harmonious symphony of personalized client management strategies. The benefit is twofold: clients bask in the glow of tailored attention, and legal services enjoy an uptick in both efficiency and effectiveness.

The quintessential example of AI's impact on client relations is the emergence of AI chatbots and virtual assistants. These digital sentinels stand guard, offering clients the convenience of 24/7 access to basic legal guidance, document procurement, and case status updates. This ceaseless vigil ensures that clients are neither left in the dark nor feel neglected, fostering a bond that is both strong and secure between client and counsel.

AI further extends its utility in the customization of communication strategies, embracing a variety of client segments with open digital arms. By dissecting and understanding patterns within client communication data, AI can tailor the most effective channels and messages for engaging with diverse client groups. Consequently, information is not merely disseminated but is received in a manner that resonates best with the clients' individual preferences.

The realm of client feedback also falls under AI's transformative gaze. Through machine learning algorithms, feedback from direct surveys, social media, and other interaction points are aggregated and analyzed to distill actionable insights. These insights are then wielded to refine services and interaction strategies, ensuring that the legal practice is not just heard but is responsive and adaptive.

Predictive analytics stands as one of AI's most vaunted offerings in the domain of personalization. Armed with the knowledge of past case data and outcomes, AI algorithms can forecast the legal needs of clients, allowing lawyers to proactively extend pertinent services. This anticipatory approach is invaluable in cementing client trust and underscoring a genuine commitment to their well-being.

In today's digital paradigm, where data breaches headline daily, protecting client privacy and data is more than a legal obligation—it is a cornerstone of trust. Here, AI-driven security measures act as the vanguard, fortifying the confidentiality of client data and, by extension, strengthening the client's faith in their legal advisors.

The personal touch of AI in legal practices is not limited to mere conversation. AI empowers the creation of tailored legal documents and advice, each a reflection of the client's unique circumstances and requirements. Such meticulous customization significantly amplifies the perceived value and impact of the legal services provided.

To navigate the infusion of AI into client interactions, legal professionals must don the cloak of innovation and perpetual learning. As AI technologies evolve, so too should the legal practitioner's knowledge and application of these tools. Mastery of these digital instruments is paramount to maximizing their potential in the service of client relations.

However, this digital integration must be balanced. The allure of AI must not overshadow the intrinsic value of human connection in the lawyer-client relationship. AI should be regarded as a complement to human empathy, not a substitute. Its role is to enhance the natural abilities of legal professionals to understand and connect with their clients on a personal level.

The journey toward the zenith of client relationship management through AI calls for transparency, especially regarding the use of client data. Clients should be apprised of how their information is utilized to personalize the services they receive. Such transparency is a pillar upon which trust and consent are built.

As we stand on the threshold of this new dawn, law firms that embrace AI to personalize and enrich client interactions will likely thrive. The convergence of AI with client care heralds a future where client satisfaction and service excellence are not mere aspirations but tangible realities.

HubSpot's AI-driven Customer Relationship Management (CRM) platform epitomizes the convergence of technology and client care. Through its sophisticated analysis of client interaction data, HubSpot enables law firms to distill vast amounts of information into coherent strategies for personalized communication. The platform's predictive analytics capabilities anticipate client needs before they arise, allowing for a proactive approach to client service. The data-driven personalization that HubSpot offers is not just a convenience—it is a strategic tool for law firms to cultivate and maintain robust client relationships.

Clio Grow enters the fray as another AI-powered ally in optimizing client relationship management. Its capabilities extend from streamlining the client intake process to organizing client information with surgical precision. Clio Grow provides invaluable insights into client behavior, allowing law firms to tailor their services to the unique contours of each client's needs and preferences. The platform's intuitive design ensures that client information is not just collected but translated into a roadmap for building satisfaction and loyalty.

This transformative potential of Clio Grow is particularly evident in the initial stages of the lawyer-client engagement. The intake process, often fraught with paperwork and administrative overhead, is reimagined through Clio Grow's AI mechanisms. By automating and organizing this crucial phase, the platform ensures that clients are ushered in through a gateway that is both welcoming and efficient. The resultant effect is a client, who feels understood from the first interaction, setting the tone for a relationship grounded in efficiency and attentiveness.

The integration of HubSpot and Clio Grow into a law firm's operations is indicative of a broader trend: the shift towards a more client-centric model of legal service delivery. By harnessing these platforms' AI and machine learning capabilities, legal professionals can move beyond traditional reactive service paradigms. Instead, they can adopt a proactive stance, where client needs are not just met but anticipated, where satisfaction is not just achieved but exceeded.

This anticipatory service model is further enhanced by the deep analytics offered by HubSpot and Clio Grow. These platforms provide a granular view of client interactions, identifying patterns and preferences that might otherwise go unnoticed. This intelligence allows law firms to personalize their outreach and services in a manner that feels individualized and thoughtful, rather than generic and perfunctory.

In practice, the personalization afforded by AI and these platforms can manifest in numerous ways. For instance, a client who prefers digital communication and fast responses may receive updates via automated emails or SMS messages that are triggered by specific events in their case progression. Another client who values in-depth discussions could be scheduled for regular video calls that allow for comprehensive briefings and a more personal touch.

The benefits of such tailored interactions are manifold. Clients who feel their preferences are respected and catered to are more likely to engage with their legal representatives actively. They are more inclined to provide the comprehensive information that is crucial for effective legal representation. Moreover, satisfied clients become ambassadors of the firm, sharing their positive experiences with others and, in doing so, becoming a powerful driver of new business.

However, as law firms tread the path of AI-enhanced personalization, they must also be mindful of the ethical implications. The use of AI must be governed by a strict code that prioritizes the client's privacy and the integrity of their data. Law firms must ensure that the AI tools they employ, such as HubSpot and Clio Grow, adhere to the highest standards of data protection and privacy compliance.

This commitment to ethical AI use extends to the transparency of AI processes. Clients should not only be informed about how their data is used but should also understand how AI tools contribute to the services they receive. This understanding can help demystify AI, fostering a relationship where technology is viewed as a beneficial tool rather than an opaque and distant process.

The narrative of AI in legal client relationship management is one of empowerment—of legal professionals and clients alike. For the former, AI platforms like HubSpot and Clio Grow are the heralds of a new era of efficiency and personalized service delivery. For the latter, these technologies promise a level of attention and customization that was once the exclusive domain of the elite but is now increasingly the standard expectation.

In the broader context, the successful implementation of AI in legal practices will likely determine the competitive landscape of the industry. Firms that adeptly wield these digital tools will position themselves at the vanguard of a new order—one where the client is truly king, and the law firm is a trusted, responsive, and insightful advisor.

Concluding this exploration, it is clear that the personalization of client interactions through AI is more than a fleeting trend—it is the future of the legal profession. The judicious application of technologies like HubSpot and Clio Grow promises a revolution in the way legal services are conceived, delivered, and experienced. Legal professionals who recognize and embrace this potential will not only transform their practices but will also play an integral role in defining the very essence of client relationships in the age of AI.

Embracing this transformative potential, however, requires a bold commitment to continuous innovation, ethical practice, and an unwavering focus on the client experience. Those who rise to this challenge will lead the charge in an industry ripe for change, setting new benchmarks in client satisfaction and service excellence. In doing so, they will not just navigate the future; they will shape it.

Data-Driven and Client-Centric: AI's Reinvention of Legal Marketing

The infusion of Artificial Intelligence (AI) into legal marketing strategies marks a revolutionary turn in the way law firms connect with, engage, and retain their clients. This comprehensive examination delves into the nuanced application of AI in legal marketing, underscoring the profound benefits of integrating AI-powered tools such as HubSpot and Clio Grow to not only streamline operations but also to foster deeper, more meaningful client relationships and optimize marketing strategies for an unmatched return on investment.

At the crux of this integration lies the ability to make data-driven decisions. AI's prowess in processing and analyzing vast amounts of data opens up avenues to understand client behaviors, predict their needs, and tailor marketing strategies with exceptional precision. The result? Marketing campaigns that are not only effective but also resonate deeply with the target audience, ensuring that the right message reaches the right client at the right time.

Content creation is one of the primary beneficiaries of AI in marketing. By utilizing AI, law firms can now generate content that is both informative and engaging, answering the pressing legal questions of potential clients and establishing the firm's expertise. This not only strengthens the firm's online presence but also improves its search engine rankings, increasing visibility to those seeking legal assistance. AI's role in crafting such targeted content is integral to a firm's ability to attract and educate potential clients, thereby building a foundation of trust and authority.

Social media, a behemoth in the realm of user engagement, offers fertile ground for AI-enhanced legal marketing. By analyzing data from these platforms, AI can identify current trends and topics of discussion within legal contexts, allowing firms to create content that is not just timely and relevant but also engaging. This application of AI ensures that a firm's marketing efforts on social media are aligned with the interests of potential clients, fostering engagement and, ultimately, leading to client acquisition.

In the realm of email marketing, AI has transformed the game. It allows for the analysis of past interactions and engagements, utilizing this data to craft messages that will likely resonate with different segments of a law firm's audience. The automation of personalized email campaigns, made possible by AI, does not just save time and resources; it creates a pathway for informative and nurturing communication that leads potential clients through the funnel of conversion.

Chatbots epitomize the potential of AI in providing immediate, valuable interactions with potential clients. These AI-driven assistants offer on-the-spot responses to queries on a law firm's website, facilitating appointment scheduling and providing preliminary advice. This level of immediate engagement is crucial for enhancing client satisfaction and increasing the conversion rate of website visitors into clients.

Predictive analytics emerge as another powerful application of AI, particularly in pinpointing potential clients who are most likely in need of legal services. AI algorithms can sift through data points such as web interactions, social media behavior, and public records to identify those who are in the midst of a legal issue or are likely to encounter one soon. This allows law firms to concentrate their marketing efforts on individuals or businesses that are already seeking the type of legal expertise they offer.

When it comes to managing client relationships, AI-enhanced CRM systems like Clio Grow provide a dynamic tool for tracking and analyzing client data. Such systems enable a comprehensive view of client interactions across various platforms, yielding insights that law firms can leverage to tailor their communications and services. This not only enhances the client experience but also ensures that the firm's marketing strategies are responsive to the evolving needs of its clientele.

Adopting AI in legal marketing demands more than just an investment in new technologies; it requires a cultural shift within the firm. This paradigm shift involves moving away from traditional marketing methods to a more sophisticated, data-driven approach that capitalizes on AI's capabilities to understand and engage with the target market on a deeper level.

However, the integration of AI into legal marketing is not devoid of challenges. Issues related to data privacy, the acquisition of large datasets for training AI models, and the necessity for continuous system refinement call for deliberate planning and implementation. Law firms must tackle these challenges with diligence, ensuring that their use of AI adheres to regulations and upholds the highest ethical standards.

The effectiveness of AI in marketing hinges on the quality of data it operates on. Law firms must, therefore, prioritize the gathering and maintenance of high-quality data, implementing stringent data governance policies to safeguard data integrity.

AI's capacity to personalize marketing efforts underscores a new era in client relations. By providing content and communications that are uniquely tailored to individual clients, law firms can signal their commitment and understanding of their clients' needs. This bespoke approach facilitated by AI not only improves client engagement but also promotes loyalty and advocacy, which are instrumental to a firm's prosperity.

Continual education and adaptation to the latest AI advancements in marketing are crucial for legal professionals. Keeping abreast of these developments ensures that law firms can utilize AI to its fullest potential, setting new benchmarks for excellence in client services.

The future landscape of legal marketing, shaped by the integration of AI, suggests not just incremental improvements but a complete redefinition of client acquisition and retention strategies. In this new digital epoch, law firms that adeptly employ AI in their marketing initiatives are poised to gain a significant competitive edge. They stand to achieve levels of efficiency, efficacy, and client satisfaction that were previously unattainable.

In synthesizing the essence of AI in legal marketing, it becomes clear that the technology is not simply an add-on to existing frameworks; it is a transformative force. AI enables firms to leverage data-driven insights to craft highly personalized and effective marketing strategies, tailor communications and automate routine tasks, allowing lawyers to focus on the human aspect of the legal profession – client counseling and courtroom advocacy.

HubSpot embodies the potential of AI in legal marketing through its suite of tools designed to enhance the efficacy of digital campaigns. Its predictive analytics capabilities enable law firms to anticipate the needs and interests of prospects, guiding the creation of content and campaigns that resonate on a personal level. Intelligent automation within HubSpot takes the heavy lifting out of marketing workflows, from lead capture to nurturing strategies, all aligned towards converting prospects into engaged clients.

Similarly, Clio Grow extends the power of AI beyond client relationship management. It serves as a repository of rich client interaction data, which, when analyzed, provides invaluable insights into client behaviors and preferences. This data is the cornerstone of informed marketing strategies, enabling law firms to launch campaigns that are not only targeted but also measured for their impact, ensuring that marketing investments yield the desired outcomes.

The strategic use of platforms like HubSpot and Clio Grow exemplifies the intersection of AI and client-centric marketing. These platforms serve as the pillars of a data-driven marketing strategy that respects the nuances of the legal market. Their capacity to assimilate and interpret client data, to automate personalized client interactions, and to streamline marketing operations, positions law firms to excel in a competitive landscape increasingly defined by technological advancement.

Incorporating AI into legal marketing requires a holistic approach. It is not just about adopting the latest AI tools but also about understanding and aligning them with the firm's strategic objectives. Law firms need to consider the entire client lifecycle, from awareness to retention, and apply AI to enhance every touchpoint. This might mean using AI to generate compelling content, employing chatbots for initial client interactions, or applying predictive analytics to fine-tune client acquisition strategies.

The implications of AI in legal marketing extend to the ethos of the law firm itself. In an era where clients expect tailored experiences and instantaneous responses, AI enables firms to meet these expectations without compromising on the quality of their services. It helps build a brand that is responsive, knowledgeable, and ahead of the curve – qualities that clients value in a world where legal services are increasingly commoditized.

Ultimately, the deployment of AI in legal marketing signals a shift towards a more agile, responsive, and client-focused legal industry. It calls for a commitment to excellence and an openness to innovation. By embracing AI, law firms not only position themselves at the forefront of the current digital revolution but also secure their place in the future of the legal marketplace.

In conclusion, AI is reshaping the landscape of legal marketing by offering unmatched insights, personalization, and efficiency. Law firms that harness the power of AI – through tools like HubSpot and Clio Grow – not only enhance their competitive edge but also redefine what it means to engage and retain clients. As the legal industry continues to evolve, the importance of AI in marketing will only become more pronounced, making it an indispensable asset for any law firm looking to thrive in the digital age. The future belongs to those who recognize the potential of AI to transform the client experience, ensuring that legal services are as accessible and personalized as the clients they serve demand.

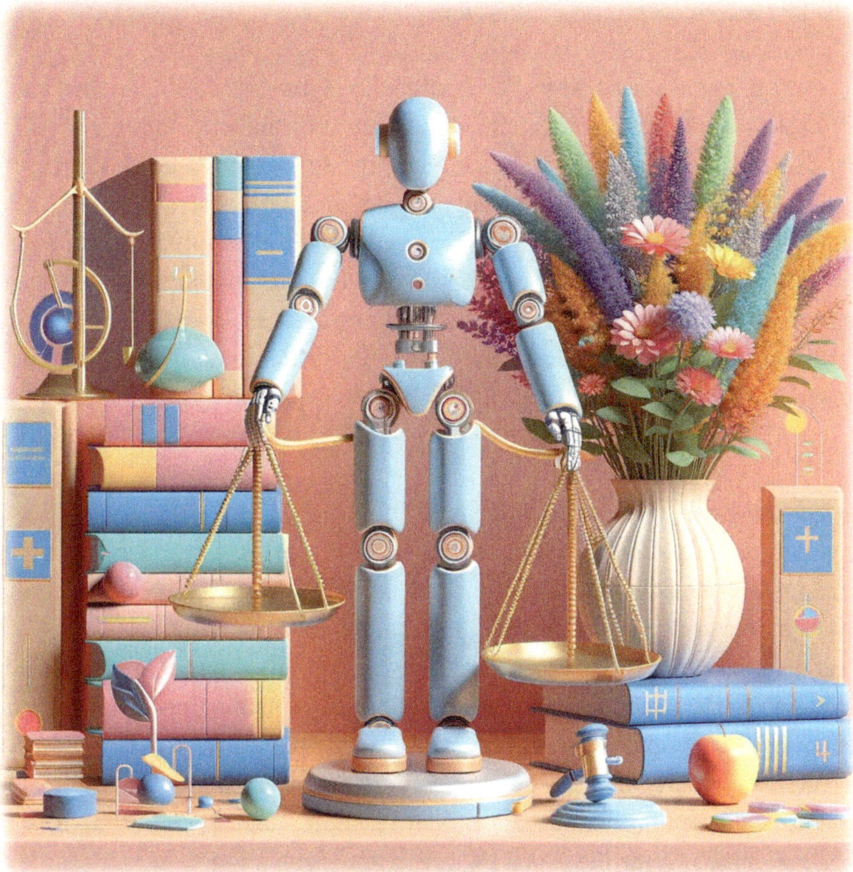

Chapter 14:

The Ethical Integration of AI in Legal Practice

As we stand at the crossroads of technological innovation and professional ethics within the legal sector, the integration of Artificial Intelligence (AI) brings forth a confluence of transformative potential and ethical imperatives. The intersection of AI and legal ethics is not merely a new chapter in the annals of the profession but a complete re-envisioning of its principles under the light of digital dawn. This chapter is an expedition across the ethical terrain sculpted by AI in legal practices, forecasting the paths that are emerging on the horizon.

The ethical landscape of AI in law is rife with complexity, as it touches upon the sanctity of data privacy, the insidious nature of biases within algorithms, and the intricate question of accountability for AI-influenced decisions. Such issues have sparked fervent debate among jurists, scholars, and practitioners alike. There is a pressing need for a robust ethical framework, one capable of harnessing the strengths of AI while upholding the cardinal virtues of the legal profession—justice, fairness, and equity. By critically examining these concerns, we seek not only to illuminate the path towards responsible use of AI in law but also to fortify the profession's dedication to its ethical mandate.

Beyond ethical considerations, AI's impact on the future of legal employment is a subject of acute interest. The prospect of AI supplanting human roles in routine tasks has raised alarms regarding job security, urging the legal fraternity to reconsider the nature of lawyering in the era of intelligent machines. However, this disruption also paves the way for reimagining the role of legal professionals. By channeling their unique capacities for empathy, moral reasoning, and creative problem-solving, lawyers and paralegals can distinguish their irreplaceable value within the legal ecosystem. Embracing AI as a partner rather than a rival offers a vista of opportunities for professional evolution and innovation.

The trajectory of legal services will be indisputably shaped by the interplay of AI, ethics, and the legal profession. The task before us is to steer this journey with prudence and vision, ensuring that the assimilation of AI into the legal sphere augments its efficacy and honor. Promoting continuous discourse, cooperation, and learning about the ethical dimensions of AI and its occupational repercussions is essential. The legal sector must not only join the ranks of this technological march but lead with a commitment to change, ethical adherence, and adaptability to the ever-evolving milieu.

This chapter is a manifesto for legal practitioners to herald an era of transformation, maintain ethical integrity, and acclimate to an evolving environment, thereby cementing a future in which AI revolutionizes legal practice in harmony with the highest ethical standards.

The provided notes underscore the urgency of addressing the ethical boundaries being redrawn by AI within legal practices. Rooted in justice and confidentiality, the legal profession confronts novel ethical challenges presented by AI. The role of legal tech companies in defining these emerging ethical frontiers is significant. As the harbingers of innovation, these organizations play a pivotal role in molding the ethical framework within which future legal technologies will operate.

This thorough examination, coupled with a commitment to ethical fidelity, envisions a future for the legal profession where AI and ethical considerations are interwoven to promote a more efficient, fair, and enlightened practice. Thus, legal professionals of the future must be adept not only in technological prowess but also in ethical vigilance to navigate this new reality. This equilibrium is crucial for a profession on the brink of a revolution, where its conventional paradigms are being redefined by the technologies it seeks to employ.

In navigating these uncharted waters, the ethical principles that have long guided the legal profession face new tests. AI presents scenarios that challenge traditional notions of responsibility, confidentiality, and impartiality. The potential for AI to enhance access to justice is immense, yet so is the risk of exacerbating existing disparities. The algorithms powering legal AI must be designed with an understanding of the legal system's complexities and the ethical standards that must be maintained.

Legal tech companies have become de facto ethicists, shaping the moral contours of legal AI applications. Their products—ranging from predictive analytics to automated document review—carry implicit value judgments and ethical implications. It is incumbent upon these entities to engage in a proactive ethical discourse, ensuring that the tools they create do not inadvertently perpetuate injustice or infringe on privacy.

Education on the ethical aspects of AI in law is crucial, not just for legal professionals but also for those who create the algorithms. A multidisciplinary approach is necessary, one that brings together legal scholars, technologists, ethicists, and practitioners to cultivate a shared understanding of the ethical landscape. As the profession adapts to these new tools, the continuous education of legal professionals on AI's capabilities and limitations becomes indispensable.

The promise of AI in law is not without its perils. Concerns about the opacity of AI decision-making processes and the challenge of ensuring these processes align with legal reasoning are significant. There must be a concerted effort to create transparent AI systems whose reasoning processes can be understood and evaluated by human practitioners. This transparency is essential to maintain trust in the legal system and to ensure that AI aids rather than obscures the pursuit of justice.

The ethical stewardship of AI in the legal realm calls for a new kind of vigilance. It demands legal professionals not only to be knowledgeable about the laws they practice but also about the underlying mechanisms of the technology they use. They must question not just the outcomes but the origins of the AI-generated advice, the data that fed the algorithms, and the values that were prioritized in their creation. In this vein, legal professionals must become interdisciplinary, capable of engaging with technologists on issues such as data integrity, algorithmic transparency, and the prevention of bias.

As legal tasks become increasingly augmented by AI, there is a corresponding need for legal professionals to enhance their skills. Those who can harness AI effectively will find themselves at an advantage, able to navigate large volumes of data with precision and uncover insights that would be unattainable manually. The ethical application of this technology, therefore, hinges not only on the technology itself but on the education and discernment of those who wield it.

This transformation also calls for a reevaluation of the principles of legal education. Law schools must evolve their curricula to include training in data science, statistics, and technology ethics, equipping future lawyers with the tools to thrive in a tech-centric legal landscape. Furthermore, continuing legal education should include modules on AI literacy, ensuring that the current generation of lawyers remains adept and ethically aligned with the new tools at their disposal.

The responsibility for ethical AI in law also extends to the developers and vendors of legal AI technologies. They must work in lockstep with legal practitioners to ensure that their products are not only technologically sound but ethically designed. This means engaging in a dialogue about the goals of the legal system, the nature of the legal reasoning, and the imperatives of justice that the technology must serve.

The rise of AI in the legal profession presents an opportunity to revisit and reinforce the core values that underpin the practice of law. As AI systems handle more tasks, there is a golden chance to amplify the human elements of legal practice—empathy, judgment, and the nuanced understanding of human narratives that machines cannot replicate. This human-AI partnership could lead to a more empathetic, accessible, and efficient legal system.

This future, however, is not preordained. It will be the result of choices made by legal professionals, educators, policymakers, and technologists. It requires a proactive approach to ethics, an openness to innovation, and a commitment to justice that transcends traditional boundaries.

In the end, the ethical integration of AI into the legal profession is not a destination but a journey—one that will be continually navigated as technology evolves. This chapter calls upon the legal profession to lead this journey with courage and foresight, ensuring that AI serves as a tool for enhancing the human pursuit of justice. It challenges legal practitioners to envision a future where AI is a force for good, promoting the ideals of fairness, accessibility, and transparency within the legal system.

It is a call to embrace the potential of AI to transform the legal profession while steadfastly adhering to the ethical commitments that define it. As AI reshapes the landscape of legal services, it is the duty of the profession to ensure that this new territory is charted with a moral compass, guarding against the erosion of the ethical bedrock upon which the law stands. The path forward will require a delicate balance, a synthesis of innovation and timeless principles, ensuring that as the legal profession changes, its core commitment to justice remains unwavering.

The unfolding narrative of AI in law is one of both promise and caution. It is a narrative that demands the legal profession not only to react to technological change but to shape it. By doing so, the legal profession will not only adapt to a new era of legal practice but will lead the way in ensuring that the era of AI in law is marked by an enhanced commitment to justice, fairness, and the ethical practice of law. The opportunity before us is vast, and the ethical imperative is clear. It is an opportunity to redefine the practice of law for the better and to ensure that as the algorithms become more integral to the profession, they do so in a way that upholds the dignity and integrity of the legal system and those it serves.

Navigating Ethical Complexities with AI in Law

The advent of artificial intelligence (AI) in the legal profession marks a transformative era that juxtaposes cutting-edge technology with the time-honored principles of justice. Lawyers, paralegals, judges, and other legal practitioners are increasingly interfacing with AI technologies that promise to redefine the landscape of legal services. This intersection of law and technology brings to light a constellation of ethical considerations that necessitate careful and deliberate navigation to preserve the sanctity of the legal system.

AI's allure lies in its potential to streamline complex processes and enhance the capabilities of legal professionals. However, this technological allure is not without its ethical pitfalls. Bias in AI is a paramount concern. As AI systems are fed with vast datasets, they can inadvertently learn and perpetuate the biases within that data. This can result in outcomes that exacerbate social inequalities, which is antithetical to the notion of justice that legal systems strive to uphold.

The dilemma of accountability is further complicated by the introduction of AI in legal decision-making. As AI systems take on more sophisticated roles, from legal research to predictive analytics, determining who is accountable for the decisions influenced by AI becomes increasingly convoluted. The enigmatic nature of AI's decision-making algorithms—the so-called 'black boxes'—conflicts with the legal sector's foundational demand for clarity and explain ability in reasoning and decision-making processes.

The imperative of confidentiality, a bedrock of legal practice, is also tested in the AI epoch. The integration of AI tools into legal operations requires sharing sensitive information with AI providers and third-party entities, which raises potential issues regarding data security and the preservation of client privacy. The traditional protocols of client confidentiality must, therefore, be reevaluated and restructured to counteract the risks posed by these new technological alliances.

In the face of these challenges, it is essential for the legal profession to establish robust ethical frameworks for AI utilization. These frameworks should not only provide guidance for current ethical dilemmas but also be forward-looking, anticipating the complex ethical situations that future AI developments may present.

Education and training in AI ethics should become an integral part of the legal profession's curriculum and professional development programs. Lawyers and legal practitioners must acquire a deep understanding of both how AI tools function and the ethical considerations surrounding their use. This dual knowledge base will be critical for ensuring that AI is utilized in a manner that is consistent with the legal profession's ethical standards.

The pace at which AI technologies evolve necessitates dynamic regulatory oversight. Static regulatory frameworks are ill-equipped to keep up with rapid technological advancements, calling for regulatory bodies to be both nimble and well-informed. The cooperation between legal professionals, ethicists, and technology experts is crucial in crafting regulations that protect the public while encouraging ethical innovation.

Transparency in AI systems is a requisite for maintaining the trust of clients and the integrity of the legal profession. Legal professionals should prefer AI systems that provide clear insights into their algorithms and decision-making processes, which will facilitate ethical oversight and accountability.

Inclusivity in the development of legal AI tools is paramount in combating inherent biases. Teams composed of individuals with diverse backgrounds and perspectives are better positioned to identify and mitigate bias in AI systems, leading to fairer and more representative outcomes. It is incumbent upon the legal profession to both advocate for and actively engage in the development of AI technologies to ensure their conformity with ethical standards.

Confronting the ethical implications of AI in the legal field presents not only challenges but also an opportunity to reinforce the profession's foundational values. The principles of justice, fairness, and adherence to the rule of law must guide the integration of AI into legal practices. By doing so, we can ensure that the benefits of AI are realized without compromising our commitment to ethical practice.

Collaboration between legal professionals and AI developers is critical in creating technologies that reflect legal ethical standards. Through active dialogue and participation in the design process, legal professionals can influence the creation of AI tools that uphold the principles of confidentiality, fairness, and transparency.

Public engagement and education about the role of AI in the legal system are essential components of fostering trust and acceptance. It is the responsibility of the legal profession to demystify AI's role in legal services, to communicate the safeguards that are in place, and to explain how it influences the delivery of justice.

The establishment of ethical guidelines for the use of AI by professional organizations is a key step in ensuring that the legal profession remains at the forefront of ethical AI utilization. These organizations should lead the effort to define best practices, provide guidance to practitioners, and promote an ethical approach to the adoption of AI within the field.

In summary, the ethical implications of AI in the legal profession demand a thoughtful and comprehensive approach. It requires a concerted effort across education, regulation, and collaboration to navigate the challenges that AI presents. Our collective resolve must be to maintain the highest ethical standards as we incorporate AI into our legal practices. By engaging with AI conscientiously, we have the potential not only to confront these challenges but to thrive, ensuring that justice remains a fundamental and untarnished virtue in an age of technological advancement.

As we embrace AI, we do not simply adopt a set of new tools; we undertake a profound shift in the practice of law. This shift, while filled with complexities, also offers a chance to significantly advance our profession. It is an opportunity to improve our work, serve our clients more effectively, and contribute to a legal system that embodies greater equity and efficiency. By proactively addressing the ethical implications of AI, we ensure that our profession maintains its role as a steadfast guardian of justice in an increasingly digitized world.

Engagement with AI demands a commitment to continuous learning and adaptation. As AI evolves, so must our understanding and strategies for its deployment. We must remain ever vigilant to the nuances of how AI affects the rights and liberties of individuals, particularly those who are most vulnerable to systemic injustices. This means prioritizing the development of AI in ways that serve the public good, resist the perpetuation of bias, and enhance the fairness and accessibility of legal services.

The task of integrating AI into legal practice also calls for introspection about the core identity of the profession. It challenges us to consider what it means to practice law in the 21st century and how we can balance the use of sophisticated technology with the human elements of empathy, moral reasoning, and advocacy that have always been central to the practice.

Moreover, as we harness the power of AI, we must also establish safeguards against the potential erosion of the attorney-client relationship. The trust placed in legal professionals is based on a covenant of confidentiality and personalized counsel; the deployment of AI must enhance rather than diminish this trust. We must ensure that as AI handles more tasks traditionally performed by humans, the fundamental aspects of professional judgment and personalized advice are not lost.

Professional bodies and legal institutions must take a leadership role in this transition. They must provide clear ethical guidelines, promote best practices, and facilitate the development of AI tools that are compatible with the values of the legal profession. This includes advocating for legislation that governs the use of AI in legal contexts, ensuring that these technologies are used in ways that are transparent, equitable, and accountable.

Additionally, we must promote an environment where ethical AI development thrives. This involves supporting research into AI systems that are inherently fair and unbiased, and fostering a culture of ethical AI use among the developers, vendors, and purchasers of these systems.

In conclusion, as AI becomes more deeply integrated into the practice of law, we face a critical period of adaptation. The ethical implications of AI are significant, but they are not insurmountable. With thoughtful engagement, careful planning, and a commitment to our ethical foundations, we can leverage AI to enhance the practice of law, increase access to justice, and maintain the trust of those we serve. The legal profession has always been a bastion of ethical standards and guardianship over the rule of law. As we enter this new technological frontier, it is incumbent upon us to ensure that these standards are not only preserved but are also strengthened for the future.

Through dialogue, education, and collaboration, we can shape the future of AI in law—a future that respects the dignity of all individuals and upholds the highest ideals of justice. By confronting AI's ethical implications with determination, the legal profession will continue to be a beacon of integrity and a pillar of societal trust in our increasingly digital world. As we navigate these changes, we must always remember that at the heart of law is humanity, and our ultimate goal is to serve and protect it.

The Future of Legal Employment in the Age of AI

In the vast expanse of professional sectors feeling the touch of technological evolution, law stands as a prime candidate for transformation under the influence of Artificial Intelligence (AI). This discourse aims to unravel the intricate tapestry of AI's prospective effects on legal employment, scrutinizing the fate that awaits attorneys, paralegals, and legal specialists.

At the forefront of AI-driven change is the streamlining of tasks deemed mundane and repetitive. The AI forte—processing copious amounts of information at breakneck speed—can condense the hours dedicated to legal research, document scrutiny, and aspects of due diligence. Legal experts are not facing the specter of obsolescence but rather an opportunity to pivot towards endeavors that demand greater analytical, strategic, and creative investment.

Contravening the doomsday prophets, AI stands as a harbinger of novel career paths within the legal sphere. The infusion of AI into legal frameworks necessitates a cadre of professionals adept at managing, interpreting, and deploying these systems. These emergent roles, spanning AI legal analysts to tech-fluent legal consultants, represent a synergy between jurisprudence and technology.

Moreover, AI ushers in an era of heightened specialization within legal practice. The digital age spawned domains such as cyber law; similarly, AI paves the way for niches like AI ethics and policy, proprietary rights in AI contexts, and AI-based contract dissection. Legal practitioners can expect to diversify their services and clientele as AI elucidates previously uncharted territories of law.

Yet this progressive perspective is not devoid of ethical conundrums. As AI begins to assert itself in decision-making, the legal sector must counteract biases embedded within AI frameworks. It falls upon legal professionals to wield AI in a manner that reinforces justice, equity, and clarity—mandating a robust grasp of both technological underpinnings and ethical guidelines.

On the client interfacing frontier, AI is poised to effectuate a paradigm shift. AI-powered legal services promise accessibility, expediency, and affordability, broadening the availability of legal aid to marginalized groups. Legal professionals will evolve into stewards of these AI solutions, ensuring congruence with both client expectations and the rigors of legal integrity.

The nurturing of legal minds will need to evolve in tandem to equip the forthcoming generations for an AI-intertwined legal vista. This entails an educational overhaul to include legal theory, data science, AI ethics, and technology stewardship as part of the core curriculum across law schools and continuing education programs.

Notwithstanding the above, anxieties linger over the potential 'depersonalization' of the legal sector in AI's wake. The imperative for legal professionals is to preserve the human touch—empathy, discretion, and ethical reasoning—which remains irreplicable by machines. AI may augment service delivery and operational scope, but the essence of legal practice retains its human core.

Adapting to these changes calls for a proactive and innovative mindset among current and future legal professionals. This is not simply about weathering the storm of change but about guiding AI's integration in ways that bolster the profession, enhance access to justice, and uphold ethical benchmarks. The arc that AI will trace in legal employment is open to influence—it will be defined by the choices and actions of today's legal community.

Furthermore, AI is set to recalibrate the competitive equilibrium within the legal sector. Early adopters who adeptly leverage AI stand to gain significant benefits in terms of efficiency, cost-effectiveness, and innovation. This will ignite a fervor for AI competence within the legal workforce, nurturing a culture predisposed to continuous learning and adaptation.

AI also signals a shift in the global paradigm of legal services outsourcing. With AI's proficiency in tasks conventionally delegated offshore, a reorganization of the legal work geography is imminent, potentially centralizing services in AI-sophisticated regions and reshaping the international legal services marketplace.

Additionally, AI's integration into legal processes prompts critical discussions on data privacy and cybersecurity. Legal professionals must master the intricacies of client data management within AI systems, highlighting the indispensability of solid data governance and cyber fortification protocols.

In synthesis, AI's imprint on legal employment is multifaceted and profound, promising challenges offset by a myriad of opportunities for those willing to embrace and navigate these changes. The legal employment vista in an AI-driven epoch will be characterized by augmented capacities, emergent roles and disciplines, and a renewed dedication to ethical practice.

The march towards AI integration in law is nascent, likely to be punctuated with both strides and stumbles. Yet, with a prescient understanding of AI's potential effects on legal jobs and a readiness for adaptation, the legal field can undertake this transition with assurance and vision. The endgame is to harness AI in manners that elevate legal practitioners' expertise and principles, thereby enriching both the practice of law and justice administration.

Regarding OpenAI's GPT and specifically ChatGPT, its iteration has instigated a renaissance across various sectors, including law. A segment of this discussion could deliberate on the ethical guidelines stipulated by OpenAI for the deployment of its AI, examining the ramifications on legal employment. By advocating for the principled utilization of tools like ChatGPT, the narrative can delve into how such innovations might bolster legal services, while also contemplating the dynamic nature of legal vocations in the AI milieu.

Generative Pre-trained Transformer 4 (GPT-4), a multi-modal language model by OpenAI, represents the zenith of this technological lineage thus far. Released on March 14, 2023, and accessible through various platforms such as the subscription-based ChatGPT Plus and OpenAI's API, it embodies the cutting edge of AI. GPT-4's capabilities in understanding and generating human-like text herald a new epoch for the legal profession.

The ethical guidelines surrounding the use of ChatGPT and similar AI tools are crucial for the legal domain, given the profound implications for employment. The model's design encapsulates ethical use cases, serving as a beacon for legal professionals to navigate the potential quandaries posed by AI. As legal entities integrate such tools, it is incumbent upon them to cultivate an environment where AI complements human expertise, ensuring that justice remains equitable and accessible.

The discourse on AI in law often conjures a binary of trepidation and optimism. On one hand, there is the fear that AI will render certain legal roles redundant, particularly those reliant on tasks susceptible to automation. On the other, there is the anticipation of AI as an enabler—a technological ally that enriches the profession by assuming the yoke of procedural tasks, liberating legal professionals to focus on the quintessentially human elements of their work.

Indeed, as AI like ChatGPT takes on a more prominent role in the legal sector, it is anticipated to augment rather than diminish the value of legal practitioners. The technology is adept at synthesizing vast legal databases, translating legalese into layman's terms, and even drafting preliminary legal documents. Such capabilities can exponentially increase a law firm's productivity, allowing lawyers to scale their services and cater to a broader client base.

This shift, however, will not be without its challenges. As AI systems become more entrenched in legal operations, the potential for machine-borne biases—stemming from flawed data or algorithms—raises significant ethical issues. Legal professionals will need to be vigilant, ensuring that the principles of fairness and impartiality are not compromised by the AI they employ.

There is also the aspect of AI's potential in creating a legal landscape that is more collaborative and diverse. By eliminating geographic and economic barriers, AI-powered platforms can facilitate unprecedented connections between lawyers and clients, as well as among legal professionals globally. This creates a fertile ground for a more inclusive legal ecosystem, enriched by varied perspectives and expertise.

As AI continues to advance, there will be a growing emphasis on roles that blend technological proficiency with legal acumen. The demand for legal technologists, individuals skilled in both the law and AI, will likely surge. These professionals will need to understand not only legal precedents but also the algorithms that underlie AI operations to ensure that these tools are employed judiciously and effectively.

Moreover, GPT-4 and its successors promise to redefine the landscape of legal education. The traditional curriculum, primarily focused on case law and statutory interpretation, will need to expand to include AI literacy. This entails an understanding of AI's capabilities, limitations, and ethical considerations. Legal educators will have the onus to integrate these elements seamlessly into the educational framework, preparing students for a legal market where AI is ubiquitous.

The potential for 'depersonalization' remains a concern. As AI takes on more functions within the legal profession, maintaining the human connection—central to client trust and the delivery of personalized legal advice—becomes crucial. The distinct qualities that define the legal profession, such as advocacy, negotiation, and moral judgment, must continue to be the domain of human practitioners.

To navigate this impending AI epoch, the legal sector will need to foster an adaptive mindset among its members. This involves a proactive approach to embracing AI, not as a competitor, but as a collaborator that can amplify the efficacy and reach of legal services. The trajectory of legal employment in the wake of AI's rise is contingent upon the decisions made by legal practitioners today—decisions that should aim to harness AI's strengths while upholding the fundamental tenets of the legal system.

AI's impact on legal employment portends a reconfiguration of the sector's competitive dynamics. Firms that spearhead the integration of AI stand to benefit from heightened operational efficiency and innovation. This, in turn, necessitates a commitment to continual professional development among legal professionals, with a focus on AI competencies.

AI's implications for the globalization of legal services also merit attention. The capacity for AI to assume functions typically outsourced may catalyze a re-distribution of legal work, potentially consolidating it within regions that exhibit robust AI capabilities. This shift could engender a more uniform global legal services framework, yet it might also necessitate a re-evaluation of jurisdictional laws and regulations.

Chapter 15:

Navigating the AI Evolution in Legal Services

As the tides of change usher in the era of AI in various sectors, the legal profession is not left untouched. The dawn of AI promises an overhaul in the way legal services are delivered and managed. This transformation, however, is not an automatic gateway to improved operational efficiencies; it requires a concerted effort to prepare both the legal practice and the professionals within. "AI Legal Blueprint" is an endeavor to serve as a guide in this preparation, emphasizing the significance of a strategic approach towards the adoption and integration of AI technologies in the legal sector.

The process of preparing a law firm for AI transformation goes beyond mere technical upgrades; it is an intricate process of aligning technological potential with human expertise. Law firms must begin by demystifying AI, understanding its practical applications, and clarifying the misconceptions that surround its capabilities. This understanding is the foundation upon which strategies for integration should be built. It is imperative to recognize that AI is a complement to, not a replacement for, the nuanced judgement and expertise of legal professionals. Its role is to augment the analytical and processing capabilities of legal teams, allowing them to undertake more substantive and strategic work.

An essential step in the preparatory phase is conducting a thorough analysis of the existing workflow within the law firm. This analysis will highlight the processes that can benefit from automation and optimization through AI solutions. Tasks that involve pattern recognition, predictive analysis, or large-scale data processing are particularly well-suited for AI applications. By automating these tasks, law firms can reallocate human resources to areas where personal judgement and interaction are paramount, thereby increasing both efficiency and the quality of client service.

It is also crucial to understand that the AI transformation will require a recalibration of skills within the legal profession. Ongoing education and training will be a cornerstone of this evolution. Legal professionals will need to be trained not only in the use of AI tools but also in the interpretation of their output. In addition, they will need to maintain a critical oversight of these tools to ensure that they are used ethically and effectively.

The strategic plan for AI integration must be comprehensive, covering everything from the selection of the appropriate AI tools to their implementation within the firm’s operational framework. It should also consider the impact on the firm's culture, ethical boundaries, and client relations. A well-designed strategy will provide a blueprint for the firm to follow, detailing the steps and stages of AI adoption. It should be flexible enough to adapt to the rapid pace of technological change while ensuring that the firm remains aligned with its core values and ethical standards.

To create an AI-ready culture within the firm, it is essential to have buy-in from all levels, from the leadership to the support staff. This cultural shift requires a vision that is shared across the organization, promoting an environment of innovation, learning, and adaptability. It is the firm's leadership that must spearhead this cultural transformation, championing the use of AI and encouraging a mindset of continuous improvement and openness to change.

Ethical considerations are of paramount importance when integrating AI into legal practice. Law firms must be diligent in ensuring that the AI tools they use are free from bias, respect client confidentiality, and adhere to the highest standards of data security. The ethical deployment of AI will not only protect the firm from potential liabilities but will also maintain the trust and confidence of its clients.

The AI transformation journey is continuous and requires regular assessment and refinement. Law firms need to establish mechanisms to monitor the performance of AI tools, assess their impact on efficiency and service delivery, and ensure that they are delivering the intended benefits. It is through this ongoing evaluation that firms can make the necessary adjustments to their AI strategies, ensuring that they remain at the cutting edge of legal technology.

In conclusion, the transformation of legal practices through AI is not an event but a process. It requires a deep understanding of AI technologies, a commitment to continuous learning and adaptation, and a strategic approach to the integration of these technologies into the firm's operations. "AI Legal Blueprint" is designed to guide law firms through this transformation, providing them with the knowledge and tools they need to enhance their operational efficiencies and redefine the delivery of legal services in the age of AI.

Building the Foundation: AI Education for Lawyers

The integration of Artificial Intelligence (AI) into legal practice is not just an upgrade to existing tools; it is a fundamental shift in how legal work is conceptualized and executed. As legal practices stand on the threshold of this shift, the role of foundational skills and knowledge in AI becomes unequivocally critical. The promise of AI in the legal profession is not limited to enhancing efficiency and accuracy—it also extends to redefining the nature of many traditional legal tasks. However, realizing the full spectrum of these benefits necessitates that legal professionals not only acquaint themselves with AI technologies but also incorporate them effectively into their daily operations.

To navigate this new paradigm, legal practitioners need to establish a solid understanding of the basics of AI and machine learning (ML). It's an unrealistic expectation for every lawyer to attain the depth of knowledge of an AI expert. Nevertheless, it is essential for them to comprehend the rudimentary aspects of AI—how it functions, its applications within the legal sphere, and the potential impacts on a variety of legal functions. Such knowledge becomes an indispensable tool, aiding in the assessment of which AI solutions are most suitable for their specific needs and understanding how these can be utilized to enhance both efficiency and outcomes.

The journey toward AI competency starts with access to education on the subject. This is where platforms like Coursera and edX become invaluable. They offer a plethora of courses, designed and delivered by world-leading educational institutions, which provide an accessible starting point for legal professionals to commence their AI education. These courses range from the fundamentals of AI and ML to more advanced topics that delve into the nuances of AI's application in legal settings. For example, Coursera offers a course by Austral University on 'Legal Tech and the Digital Transformation of Law', which provides insights into the legal-tech market, the impact of new digital technologies on law firms and justice systems, and key market trends shaping the future of the legal industry.

Similarly, edX offers a course titled 'Artificial Intelligence: Implications for Business Strategy' by MIT Sloan School of Management, which is designed to help professionals integrate AI into their organizational strategy. This course teaches how operations can become more efficient through AI and outlines a strategic approach to implementing AI technologies within a company or law firm. These platforms democratize the learning experience, making foundational knowledge in AI and machine learning accessible to all legal professionals, which is the cornerstone of legal tech proficiency.

For legal teams, training programs should be more than just informational sessions; they need to be comprehensive, covering not only the technical aspects of AI but also addressing the ethical considerations and potential biases inherent in AI systems. Understanding these dimensions is vital for legal professionals to make informed decisions that uphold the integrity of legal processes and guard against unintended consequences.

As AI technologies and their applications are constantly evolving, legal firms must commit to fostering a culture of continuous learning and adaptability. This involves investing in ongoing education and availing resources that empower legal teams to keep abreast of the latest AI developments and understand their implications for legal practice. Encouraging innovation and a willingness to adapt will be crucial for the successful navigation and adoption of AI in the legal sector.

The adoption of AI in legal practices also calls for a reevaluation of hiring practices. Law firms looking toward the future must recognize the importance of AI literacy and technical proficiency when hiring new team members. The inclusion of candidates with a background in technology or data science can bring a diverse and invaluable skill set to legal teams, which is essential for a smooth transition to AI-augmented legal tools.

Interdisciplinary collaboration is the cornerstone of bridging the divide between legal expertise and AI capabilities. Forming teams that combine the insights of legal professionals with the technical acumen of AI experts fosters a collaborative environment conducive to innovative problem-solving. This collaboration should extend beyond the firm's boundaries, involving partnerships with external AI vendors and developers to ensure that the AI tools designed for legal applications are finely tuned to meet the specific needs of legal processes.

The importance of practical experience with AI tools cannot be overstated. Engaging in pilot projects and hands-on experimentation allows legal teams to gain firsthand knowledge of how these tools can be tailored to fit their particular needs and constraints. This experiential learning process is instrumental in demystifying AI and positioning it as a central component of the legal toolkit.

The ethical dimensions of AI utilization in legal contexts also demand special attention. Legal professionals need to be well-versed in navigating the ethical considerations associated with AI, ensuring that its use is consistent with professional standards and societal values. Training in AI ethics is critical, enabling legal teams to leverage AI in a manner that is both effective and ethical, taking into account issues such as data privacy, consent, and the potential for AI systems to perpetuate biases.

Moreover, as AI becomes integrated into legal operations, awareness of regulatory compliance becomes increasingly significant. Legal professionals must have a firm grasp of the regulatory landscape that governs the use of AI, both within their jurisdiction and at the international level. This includes understanding any specific regulations that apply to AI applications in legal contexts to ensure that their practices remain compliant with all relevant legal standards. Knowledge of these regulations not only protects the firm from legal repercussions but also reinforces the commitment to ethical AI use.

Leadership within legal firms plays a critical role in facilitating this transition into AI proficiency. Leaders must set the vision for AI's role within the organization and allocate the necessary resources for training and development. Their example can be powerful, instilling a belief in the potential of AI throughout the firm. By championing the adoption of AI, leaders can cultivate an environment where AI is not seen as a disruptor, but as an enabler of growth and innovation.

The practicality of AI implementation is also significantly enhanced by engaging in peer-to-peer learning networks. These networks allow for the sharing of experiences, challenges, and insights, thus expediting the learning curve for legal teams. By exchanging knowledge and solutions with others in the field, legal professionals can quickly identify best practices and common pitfalls associated with AI in legal contexts. Networks can range from local discussion groups to expansive online communities, all serving as vital resources for collaborative growth.

In the broader context of the legal profession, the integration of AI education into formal legal training cannot be overlooked. Law schools and providers of continuing legal education are instrumental in preparing future legal professionals for an AI-infused practice. By integrating AI and technology-focused courses into their curricula, these institutions ensure that new entrants to the legal field are not just aware of AI, but are proficient in its applications relevant to their profession.

Beyond formal education, informal learning experiences such as workshops, webinars, and online tutorials provide legal professionals with flexible opportunities to enhance their understanding of AI. These learning modules, often available through platforms like Coursera and edX, cater to a range of expertise levels—from novices to those seeking to refine their AI knowledge. They present an alternative pathway for legal professionals to engage with AI concepts and practical applications on their own terms and at their own pace.

Within the firm itself, practical application is key. Legal professionals should be encouraged to initiate or participate in pilot projects that apply AI to specific legal tasks. Through these projects, they can gain a clearer understanding of the practical benefits and limitations of AI. Moreover, these pilot projects serve as a testing ground for refining the use of AI within the firm, providing invaluable insights into how AI can streamline workflows, enhance accuracy, and unlock new capabilities within the practice of law.

The ethical use of AI is a theme that resonates across all aspects of its integration into legal practice. It's not just about using AI responsibly but also about understanding the societal impact of AI-driven decisions. Legal professionals need to be at the forefront of discussions on how AI is changing the landscape of privacy, bias, and equity within the legal system. This requires a proactive approach to learning about and advocating for ethical AI practices, ensuring that the tools they adopt are not only efficient but also just and fair.

Regulatory compliance is another critical area that legal professionals must navigate with confidence. As AI technologies become more prevalent, so do the regulations governing their use. Legal teams must stay informed of the latest developments in AI law and regulation, ensuring their practices do not inadvertently violate these new standards. This requires a commitment to ongoing education and vigilance, ensuring compliance not just with current regulations but also being prepared for future legislative changes.

To support this comprehensive approach to AI integration, firms must prioritize resources for AI education and practical experience. This includes budget allocation for training, the acquisition of AI tools for pilot testing, and time set aside for learning and development activities. By investing in these areas, firms demonstrate their commitment to building an AI-savvy legal team, equipped to leverage AI for improved client outcomes and a competitive edge in the market.

Finally, it's important to acknowledge that the path to AI integration is not without its challenges. It requires a shift in mindset from viewing legal technology as a set of tools to understanding it as a dynamic and integral component of legal practice. Legal professionals need to approach AI with curiosity, openness to change, and a willingness to engage with technology partners to shape the development of AI tools that meet the unique needs of the legal sector.

In conclusion, the journey toward AI proficiency for legal teams is both necessary and complex, demanding an integrated approach that combines education, practical application, collaboration, and a firm ethical grounding. By embracing this multifaceted path, legal practices can unlock the transformative potential of AI, not just to enhance efficiency and outcomes but also to play an active role in shaping the future of law in an increasingly AI-driven world. Through platforms like Coursera and edX, legal professionals have access to the resources they need to embark on this journey, building the skills that will define the next era of legal practice.

Strategizing AI Integration in Legal Operations

The landscape of legal practices is undergoing a profound transformation, spurred by the integration of Artificial Intelligence (AI). For legal professionals, AI is not merely an optional enhancement but an imperative for remaining at the cutting edge of efficiency and competitiveness. Developing a comprehensive AI strategy is essential for those intent on harnessing AI's transformative potential, ensuring it is in lockstep with the firm’s mission, bolsters operational efficiency, and heightens client satisfaction.

In initiating this strategic journey, clarity in objectives is paramount. Legal practitioners must pinpoint what they wish to accomplish through AI — whether its refining document review processes, augmenting legal research, or forecasting litigation outcomes. Such specificity in intent will steer the selection of AI tools and shape the trajectory of AI endeavors within the firm. Equally vital is the establishment of measurable success metrics, which afford a tangible yardstick for assessing AI's impact on legal practices.

Comprehending the technological landscape forms the bedrock of a robust AI strategy. Legal professionals must acquaint themselves with the plethora of AI and machine learning offerings on the market. Understanding their successful implementations within legal contexts will shape a more informed strategy, tailored to the nuances of one's practice.

Yet, technology is not the only facet demanding scrutiny — ethical considerations are paramount. A comprehensive strategy must address the ethical conundrums AI presents, including data privacy, client confidentiality, and the biases that may be intrinsic to AI algorithms. Upholding ethical standards and legal requirements is non-negotiable, for it underpins the trust and integrity vested in AI applications.

Another cornerstone of this strategic framework is team readiness. The seamless assimilation of AI demands a team that is not just versed in legal precedents but is also agile in adopting novel technologies. Investment in AI literacy and ongoing training is crucial, equipping personnel with the prowess to wield AI tools efficaciously.

The selection of AI tools is a critical juncture in the strategic planning process. The legal tech market brims with AI applications, each promising to streamline different facets of legal work. A discerning evaluation of these tools — considering factors like compatibility with existing workflows, precision, cost-effectiveness, and vendor support — is imperative to making choices that resonate with the firm's aims.

The integration of these AI tools within existing systems is a complex yet vital phase. A strategy must delineate a clear plan for meshing AI capabilities with the firm's IT infrastructure, ensuring new tools gel with legacy systems and facilitate an uninterrupted data continuum.

Data stewardship is the linchpin of effective AI deployment. The caliber and structuring of data exert a profound influence on AI performance. Strategies must chart a course for data governance — encompassing collection, storage, and analysis protocols that not only comply with legal benchmarks but also bolster AI tool efficacy.

Ongoing monitoring and evaluation are the lifeblood of a dynamic AI strategy. Consistent appraisal of AI tools against pre-set objectives allows for iterative refinements and progressive enhancement. This evaluative cycle ensures that AI initiatives are scalable, adapting to technological leaps and shifts in legal paradigms.

Client interactions form the face of the firm's AI adoption. The strategy should advocate for client transparency regarding AI utilization, demystifying the technology's role in service delivery, and underscoring its value proposition.

Risk management is an integral aspect of AI strategy. Identifying potential perils — such as technological glitches, data security breaches, and ethical quandaries — and crafting strategies to mitigate them is indispensable. The strategy should enfold a thorough risk assessment and proactive planning to preemptively address such challenges.

Lastly, championing a culture of innovation will act as a catalyst for successful AI strategy realization. Encouraging a mindset of exploration, receptivity to learning, and adaptability can convert skepticism about new technologies into proactive engagement and innovation.

In weaving these threads into a cohesive strategy, the roles of educational institutions and thought leaders like the Practicing Law Institute (PLI) and Thomson Reuters Legal Executive Institute become pivotal. PLI, as an educational and thought leadership bedrock, can be woven into the strategy by emphasizing its offerings in AI literacy and specialized applications in legal settings. PLI’s advanced seminars and workshops on AI in law act as a bridge between theoretical knowledge and practical implementation, equipping legal teams with the necessary acumen to navigate AI's complex landscape.

On the other hand, staying abreast of the rapid developments in AI is critical, and resources like the Thomson Reuters Legal Executive Institute are invaluable in this regard. Their research and analyses furnish legal professionals with cutting-edge insights, giving them a competitive advantage in adopting AI and legal tech innovations.

In the rapidly advancing domain of AI, legal practitioners are standing at the threshold of a significant overhaul in how legal services are rendered. A well-devised AI strategy will not only facilitate navigation through the complexities associated with AI integration but will also propel legal practices to the forefront of innovation. Embracing AI, with a strategy that is educational, ethical, client-centric, and innovation-focused, ensures enhanced services, improved efficiency, and superior client engagement in the digital era.

The Practicing Law Institute (PLI) serves as an invaluable beacon for continuous education and specialization within the realm of AI in legal practices. As part of a comprehensive strategy, PLI is instrumental in offering legal professionals a pathway to deepening their understanding and proficiency in AI applications. Its resources are tailored to facilitate ongoing education, which is vital for staying relevant in a field that's rapidly evolving due to technological advancements. PLI's contributions to a firm's AI strategy can be multifaceted — from providing foundational AI knowledge that supports initial strategy development to offering advanced insights that enable legal teams to refine and expand their AI capabilities.

Moreover, the importance of the Thomson Reuters Legal Executive Institute in the crafting of an AI strategy cannot be overstressed. It is a vital source of up-to-date information and analysis on the intersection of AI, legal technology, and industry trends. As legal professionals endeavor to keep pace with the swift advancements in AI, the institute's resources are indispensable, offering a rich repository of knowledge that empowers law firms to make strategic decisions about AI adoption and optimization. They help ensure that a firm's AI strategy is not only grounded in the latest innovations but is also forward-thinking, positioned to capitalize on emerging opportunities and navigate potential disruptions in the legal tech landscape.

In integrating the insights and educational offerings of PLI and Thomson Reuters Legal Executive Institute, a comprehensive AI strategy should also emphasize the development of internal expertise. This involves nurturing a team of AI-savvy professionals within the firm who can lead AI initiatives, serve as liaisons with technology providers, and foster an environment of tech fluency among their colleagues. By combining external resources with internal talent development, legal firms can create a symbiotic ecosystem that supports sustainable AI integration.

A strategic approach to AI in legal practices also demands a proactive stance on developing proprietary AI applications or customizing existing tools to fit the unique needs of the firm. This can involve collaborating with AI vendors and technology partners to create bespoke solutions that offer a competitive edge. The strategy should outline a roadmap for such collaborations, detailing how to leverage partnerships to achieve the firm's specific AI goals.

Client collaboration is another aspect that enriches a comprehensive AI strategy. Involving clients in the AI journey can not only foster transparency but also encourage feedback that can refine AI tools and processes. This collaborative approach ensures that AI solutions are client-centric, addressing the actual needs and concerns of those the firm serves.

Further, a comprehensive AI strategy should also consider the broader implications of AI on the legal profession and society at large. It should address issues such as access to justice, the potential for AI to democratize legal services, and the social responsibilities of legal professionals in deploying AI. These considerations ensure that the firm's AI initiatives are aligned with broader social values and contribute positively to the evolution of the legal landscape.

In fostering a culture of innovation, it is also vital to recognize and reward creativity and experimentation within the firm. This can be done through incentive programs, innovation labs, or internal think tanks that encourage the exploration of new ideas and applications of AI in legal work. Such initiatives can catalyze the development of groundbreaking AI applications and promote a spirit of continuous innovation.

The strategy should not overlook the importance of cross-disciplinary collaborations that merge legal expertise with insights from data science, behavioral science, and other relevant fields. Such collaborations can enrich the firm's AI applications, ensuring they are not only technically sound but also attuned to the nuances of human behavior and legal reasoning.

Lastly, a comprehensive AI strategy must be flexible and adaptable. It should be reviewed and updated regularly to reflect new developments in AI technology, changes in legal regulations, and shifts in client expectations. This agility ensures that the firm's AI initiatives remain relevant and effective in an environment characterized by constant change.

In conclusion, crafting a comprehensive AI strategy for legal practices is a complex but rewarding endeavor. It necessitates an alignment of technology, ethics, education, and innovation, all directed towards enhancing the quality and delivery of legal services. By leveraging the resources and expertise of institutions like PLI and Thomson Reuters Legal Executive Institute, and by fostering a culture that values continuous learning, ethical practice, and client engagement, legal professionals can ensure that their AI strategy is not just comprehensive but also transformative. As AI reshapes the legal industry, those who embrace a thoughtful, holistic approach to AI strategy will lead the charge into a more efficient, effective, and equitable future.

The Power of Connection: Legal AI Networks and Summits

In an era where the synergy between law and technology is becoming inseparable, networking and knowledge sharing have emerged as cornerstones for transforming legal professionals and law firms. This transformation is geared towards a comprehensive adoption of Legal AI and Legal-Tech strategies, aimed at revolutionizing the practice and business of law. Central to this metamorphosis are seminal conferences like LegalWeek, the American Bar Association TechShow (ABA Techshow), ILTACON, and the British Legal Technology Forum (BLTF). These gatherings provide unparalleled platforms for education, discussion, and the dissemination of the cutting edge in legal technology.

LegalWeek, often celebrated as a grand confluence of legal professionals, serves as a beacon for networking, professional development, and strategy exploration. It's a unique occasion that brings together a diversity of legal professionals, ranging from lawyers to technologists, and from private practice to corporate law departments. The conference offers a fertile ground for attendees to explore new topics, tailor strategies specific to their roles, and acquire the tools necessary to execute legal business effectively. Renowned for being the "one legal event that hits all the marks for information, education, and networking," LegalWeek stands as an essential pivot around which today's law firms and legal departments revolve to restructure, rebuild, and reinvigorate their operations. The actionable insights gleaned from this event empower legal leaders to make informed decisions that shape the future of their practices in the context of Legal AI and Legal-Tech.

The ABA Techshow serves as a nexus for bringing together the worlds of law and technology. With technology now fully integrated into legal practice, lawyers find themselves at a crossroads where embracing technological tools is not just an option but a necessity for efficiency and effectiveness. The Techshow is a catalytic event that forges this integration, teaching lawyers and legal professionals about the most useful and practical technologies available. Through a variety of Continuing Legal Education (CLE) programs, the Techshow compresses a vast wealth of knowledge into an intensive three-day learning experience. This experience is invaluable for legal professionals striving to manage their practices and serve their clients more adeptly through the power of technology.

ILTACON is the premier educational and networking event for the legal sector, distinguished by its genesis and evolution through peer creation and leadership. The conference focuses on legal technology professionals, providing an educational program that spans a broad array of topics and trends relevant to today's legal industry. With co-chairs and experienced ILTA members at the helm, ILTACON's educational content is delivered through various formats, including learning pathways, panel discussions, hands-on training, and attendee-driven sessions. This variety ensures that learning styles across the spectrum are catered to, promoting an inclusive approach to knowledge dissemination. As a multi-day conference, ILTACON offers comprehensive peer-driven programs that facilitate the discovery and evolution of successful legal operation strategies, which are crucial for navigating the rapidly transforming legal industry.

Across the pond, the British Legal Technology Forum (BLTF) occupies a prestigious position as Europe's leading legal technology conference and exhibition. Located in London, the BLTF welcomes an international audience from the spheres of law, legal technology, and IT security. It stands out for its forward-looking approach, bringing together the most respected professionals from the legal and commercial technology sectors. The forum provides an expansive vision that explores the systems, strategies, processes, and platforms poised to drive law firms and legal businesses into the next decade and beyond. BLTF is not just a conference; it is a celebration of the marriage between law and technology, offering a platform for showcasing innovations and learning through discovery. The forum is a testament to the pivotal role that knowledge sharing and networking play in shaping the future of legal technology.

These conferences are more than just annual or biannual events; they are vital ecosystems that facilitate the growth and advancement of Legal AI and Legal-Tech. They represent the collaborative spirit of the legal profession, embodying the principle that the whole is greater than the sum of its parts. In these venues, ideas are exchanged, relationships are forged, and collaborative partnerships are formed, all of which are critical to the collective advancement of the field. These conferences serve as the soil from which the future of legal technology will grow, shaped by the hands of those who attend and contribute to their rich programs.

As legal professionals navigate the complexities of adopting Legal AI and Legal-Tech strategies, it is clear that knowledge sharing and networking are not peripheral activities but essential ones. The sharing of knowledge leads to the empowerment of individuals and organizations alike, ensuring that they are well-equipped to meet the challenges of the present and the opportunities of the future. In essence, conferences like LegalWeek, ABA Techshow, ILTACON, and BLTF are not just events but critical hubs of learning and connection that help to foster a community that is at once deeply knowledgeable and profoundly interconnected. The collective intelligence and camaraderie cultivated in these settings are the engines that will drive the legal profession forward into a new era marked by technological sophistication and strategic innovation.

The significance of these conferences cannot be overstated; they are the crucibles in which the legal profession's future is being forged. LegalWeek, ABA Techshow, ILTACON, and BLTF are much more than platforms for imparting knowledge—they are the catalysts for building a community that is robust, innovative, and ready to face the future. By convening the best minds from across the globe, these events facilitate a cross-pollination of ideas and best practices that are essential for the growth and evolution of Legal AI and Legal-Tech.

Moreover, these conferences act as incubators for legal technology startups and established companies to showcase their latest innovations. From cutting-edge software that streamlines document review to sophisticated algorithms that predict legal outcomes, these gatherings provide a showcase for the technology that will define the next wave of legal practice. For legal professionals, this is a golden opportunity to witness the future of their profession firsthand, to learn about tools that can enhance their practice, and to strategize on how best to integrate these advancements into their work.

Networking opportunities at these events are unparalleled, as they connect practitioners with thought leaders, innovators, and peers who share a common goal of optimizing the practice of law through technology. Relationships forged at these conferences often lead to collaborative ventures, mentorship, and the exchange of services that are mutually beneficial. This interconnectedness is vital, as it creates a support system for legal professionals to navigate the complex and sometimes daunting world of Legal AI and Legal-Tech.

The impact of these conferences extends beyond individual learning and networking; they influence the broader legal industry by setting trends and benchmarks for legal technology. They serve as a barometer for the state of the industry, highlighting successes, pinpointing challenges, and providing a forum for collective problem-solving. The insights gained from these events drive decision-making processes within law firms and legal departments, influencing policy, investment in technology, and ultimately, the delivery of legal services to clients.

Furthermore, the educational value of these conferences cannot be ignored. They provide a multitude of CLE opportunities that help legal professionals meet their educational requirements while simultaneously keeping them abreast of the latest legal technology trends. The range of topics covered is vast, from data privacy and cybersecurity to artificial intelligence and blockchain. This breadth ensures that attendees leave with a comprehensive understanding of current technologies and how they can be harnessed to improve the practice of law.

The future-facing nature of these conferences is particularly critical. As the legal industry grapples with rapid technological advancements, these events provide foresight into the trajectory of legal services. They help legal professionals anticipate changes, prepare for them, and remain agile in an ever-evolving landscape. By highlighting innovations on the horizon, these conferences play a pivotal role in ensuring that the legal profession is not left behind in the wider digital transformation.

In conclusion, LegalWeek, ABA Techshow, ILTACON, and BLTF are much more than annual meetings; they are the heartbeat of the legal technology community. They are where legal professionals go to learn, connect, and grow. They serve as a compass pointing toward the future of the profession—a future that is digital, interconnected, and full of promise. For those legal professionals and law firms seeking to embrace a comprehensive Legal AI and Legal-Tech strategy, these conferences are not just helpful; they are essential. They provide the knowledge, networks, and inspiration needed to transform practices and businesses, ensuring that the legal profession remains at the forefront of innovation and excellence. Through these events, the legal community comes together to share knowledge, learn from one another, and build the future of legal services—a future that is intelligent, efficient, and profoundly human.

Chapter 16:

AI and the Financial Transformation of Legal Practice

As the legal industry stands on the brink of a technological revolution, understanding the economics of artificial intelligence (AI) within legal services becomes paramount. The infusion of AI technology into legal practices isn't just a nod to innovation; it's a strategic economic decision that law firms and legal departments are making to stay competitive, increase efficiency, and deliver superior client outcomes. At its core, the economics of AI revolves around analyzing financial impact and return on investment

(ROI), alongside exploring viable funding options for legal AI initiatives. The potential for cost savings, coupled with the enhancement of service quality, leads to a compelling case for AI adoption in the legal sector. However, it's crucial for legal professionals to approach this transformative journey with a clear-eyed view of the economic implications, balancing initial investments with projected long-term gains.

Investing in AI technologies entails upfront costs, which include acquisition, integration, and training. Yet, the ROI becomes apparent through various avenues, such as significantly reduced man-hours on tasks like document review and legal research, more accurate risk assessment, and predictive analysis. These efficiencies not only transcend cost savings but also free up legal professionals to focus on more strategic, high-value tasks. Furthermore, AI initiatives offer law firms and legal departments a competitive edge by facilitating faster, data-driven decision-making processes and enabling personalized, efficient client service. While funding these innovations may pose a challenge, especially for smaller practices, various models exist to finance AI adoption — from traditional capital investments to innovative subscription models and partnerships with technology providers.

As legal professionals navigate the complex landscape of AI and machine learning, a strategic approach to their economics will be instrumental in achieving optimal outcomes. The focus should extend beyond just the financial aspects, emphasizing how AI can redefine value creation in legal services. By meticulously analyzing AI's financial impact and exploring diverse funding opportunities, legal practices can position themselves to thrive in the AI era. Ultimately, the goal is not only to improve operational efficiency and enhance client satisfaction but also to spearhead a new wave of innovation in the legal industry, paving the way for a future where AI and human expertise collaborate to achieve greater justice and legal precision.

Legal Tech ROI: Assessing Financial Impacts and Returns

In an era where the legal profession is constantly evolving, the integration of Artificial Intelligence (AI) into legal services is not just a forward-thinking strategy, but a necessity for those looking to remain competitive. With the promise to enhance operational efficiency, client satisfaction, and case outcomes, AI comes with its fair share of financial considerations. Assessing its financial impact and Return on Investment (ROI) is crucial for law firms navigating the complexities of this technological transformation.

The integration of AI in legal practice involves a substantive initial investment. Expenses can span from the acquisition of AI software and tools to training legal teams, and potentially, upgrading existing infrastructure to support the new technology. Despite the significant initial costs, the long-term financial benefits of AI often far exceed the initial investments. To understand the ROI of AI, law firms must examine how, when, and where the benefits of AI technology materialize.

Legal research has been dramatically transformed by AI, which can sift through vast quantities of legal documents, case law, and precedents at speeds no human could match. This capability doesn't just cut down on hours of manual labor—it also enhances the accuracy and pertinence of the findings. As a result, productivity increases, freeing legal professionals to concentrate on more strategic tasks and offering significant savings on time and resources.

The enhancement is similarly profound in document review, a fundamental task in legal work. AI-powered tools, such as predictive coding, permit legal teams to process and analyze documents more expediently and accurately. The direct outcome is a cost reduction, particularly in the expenses associated with discovery in litigation.

AI also revolutionizes contract analysis and management, streamlining the entire lifecycle from creation to execution and monitoring. Quick identification of potential risks and compliance issues through AI tools helps avoid costly disputes and fines, contributing to risk mitigation and financial preservation for law firms and their clients.

Furthermore, AI's impact on client relationships and marketing cannot be understated. With AI-powered chatbots and customer service tools providing round-the-clock assistance, firms can enhance client engagement. This leads to increased satisfaction, retention, and referrals—crucial elements for financial growth.

In the realm of litigation, AI tools capable of predicting outcomes give lawyers a strategic edge by offering insights into the potential success of cases. Armed with these insights, law firms can make more informed decisions that can optimize financial results by informing case strategy, settlement discussions, and trial preparations.

Yet, the ROI of AI in legal services extends beyond direct financial gains. Indirect benefits such as bolstered client relationships, an enhanced firm reputation, and the attraction of top legal talent through technological innovation also significantly impact a firm's market positioning and long-term financial success.

Measuring the exact ROI of AI in legal practices, however, is challenging due to the qualitative benefits that aren't easily quantifiable, like brand enhancement and client satisfaction. Additionally, AI's impact varies significantly among firms, based on their size, areas of specialization, and the extent to which AI is integrated into their processes.

Despite these challenges, law firms can use several methods to estimate the ROI of their AI investments. One approach is to compare the time and cost savings facilitated by AI against the costs of acquiring and implementing the technology. Another is to evaluate the increase in successful case outcomes or client acquisitions that can be attributed to AI's predictive analytics and case preparation tools.

To truly maximize AI's ROI, law firms need to be strategic in its implementation, selecting AI solutions that align with their specific goals, investing in training for their staff to ensure AI tools are used efficiently, and continuously monitoring and adjusting AI strategies in response to changing legal technologies and market demands.

The evolving nature of AI promises even greater financial impacts on the legal profession. Law firms that are early adopters of AI and are strategically integrating it into their operations are likely to achieve a competitive advantage, not only in operational efficiency but also in financial performance.

While it's essential to focus on tangible financial returns, it's equally important not to neglect the broader impact AI can have on a law firm's culture, client relationships, and overall market position. The real ROI of AI, therefore, includes both measurable financial gains and qualitative improvements to a firm's practice, both of which are indispensable in today's legal landscape.

In conclusion, analyzing AI's financial impact and ROI presents a compelling case for its integration into legal services. Despite the challenges of quantification, the range of benefits—from direct cost savings to intangible enhancements in service quality—firmly establishes AI as a critical investment for law firms seeking to succeed in a rapidly transforming industry.

Incorporating LawVu's unified legal workspace into this analysis, we observe a case study that exemplifies the economic benefits of AI integration. LawVu's platform, which leverages AI tools, streamlines legal operations and significantly cuts costs while improving service quality for legal teams. Specific metrics and outcomes show the substantial ROI that LawVu's AI tools bring to legal practice management, emphasizing the practicality and profitability of AI in legal processes. LawVu's all-in-one connected platform enables legal teams to manage all matters, contracts, and legal operations efficiently.

LawVu embodies the legal workspace for in-house legal teams, underscoring the impact of a centralized, AI-enhanced legal environment. The platform not only promotes operational effectiveness but also gives legal teams comprehensive visibility and control over their work, which is crucial in maximizing the utility of AI. By empowering legal teams to operate more effectively, LawVu facilitates better work management, all in one place, signifying a pivotal shift from traditional, compartmentalized legal practice to a cohesive, integrated approach.

In practice, LawVu's AI integration manifests in several key areas. First, by automating mundane and repetitive tasks, LawVu allows lawyers to focus on more complex and high-value aspects of legal work. This shift not only enhances job satisfaction among legal professionals but also drives a more strategic allocation of human resources. Second, AI-driven analytics provide in-house legal teams with data-driven insights that support more informed decision-making and strategy development, particularly in risk assessment and management.

Furthermore, LawVu's AI capabilities in contract management epitomize the precision and speed with which legal terms, obligations, and compliance requirements can be identified and managed. This not only expedites the contract review process but also significantly mitigates the risks of costly oversights. By employing AI to perform in-depth analyses of contractual terms, LawVu helps prevent disputes that would otherwise require considerable time and financial resources to resolve.

The economic benefits of LawVu's AI tools extend to predictive analytics in litigation management as well. By analyzing historical data and trends, LawVu provides in-house legal teams with foresights that can influence the outcome of legal actions. These insights enable lawyers to assess the viability of litigation, estimate potential costs, and probabilities of success, thus facilitating more cost-effective litigation management and strategy.

When it comes to ROI, LawVu's platform demonstrates its value through measurable reductions in time spent on tasks, lower operational costs, and decreased risk of non-compliance and associated fines. This translates into direct financial savings for law firms and clients alike. Additionally, by enhancing the quality of legal services, LawVu indirectly contributes to stronger client-lawyer relationships, leading to increased client retention and, potentially, a more significant market share.

The financial analysis of LawVu's AI integration must also consider the time saved across various legal functions. Time, in the context of legal services, is directly correlated with costs. Therefore, the hours saved through LawVu's AI tools can be quantified into financial metrics that contribute to a firm's profitability. These savings are not just in terms of billable hours but also in the opportunity costs of lawyers being able to take on additional work or focus on more strategic initiatives that bring in new business.

In terms of marketing and business development, LawVu's platform enables law firms to showcase their technological prowess and commitment to innovation. This modern image is appealing to prospective clients who are increasingly looking for tech-savvy legal partners to address their complex legal needs in a rapidly digitizing world. As such, LawVu contributes to the branding and positioning of law firms, a qualitative benefit that, while difficult to measure, can significantly impact a firm's growth trajectory.

Lastly, the adoption of AI through platforms like LawVu necessitates an examination of the cultural shift within law firms. AI's integration demands a move towards a more collaborative and interdisciplinary approach to legal work, where technology and legal expertise merge to create more efficient, client-centric services. This cultural evolution can attract top-tier talent, foster a more innovative mindset among staff, and enhance the overall ethos of a law firm.

In closing, the comprehensive analysis of AI's financial impact and ROI, enriched with the integration of LawVu, paints a robust picture of AI as an indispensable asset in the legal industry. The compelling economic advantages, coupled with the qualitative improvements in legal practice brought about by AI, underscore its significance as a transformative investment for any law firm committed to excellence and innovation in the digital age. The strategic implementation of LawVu's AI tools is a testament to the enduring value and potential of AI in enhancing the financial and operational aspects of legal practice management.

Seeding AI Futures: Funding Legal Tech Innovations

In the swiftly advancing domain of legal services, the fusion of Artificial Intelligence (AI) has emerged as a key revolutionizing force, propelling efficiency, precision, and innovation. The integration of AI within legal practices—from mechanizing document analysis to predictive litigation analytics—is gaining pronounced recognition for its transformative capabilities. Yet, the inception of AI endeavors mandates a meticulous evaluation of funding mechanisms to underpin these progressive undertakes.

Internal financing stands as a principal approach for legal entities, necessitating the dedication of a budgetary share to AI ventures. This method underscores a law firm's or legal department's commitment to innovation, necessitating a significant investment of internal assets but affording absolute project oversight. Such reinvestment catalyzes a culture of continuous enhancement and innovative spirit within the legal practice.

Conversely, external funding avenues offer the allure of diversified financial risk and access to broader capital reserves. For legal tech startups infused with technological zeal, Venture Capital (VC) represents an enticing prospect for spawning pioneering AI solutions. Yet, securing such investments demands an enticing value proposition, evidencing the AI project's potential to disrupt or substantially advance the realm of legal services.

An often-overlooked fountain of support is grants and government subsidies, earmarked for initiatives that promise to augment legal services or broaden justice access. These financial injections can infuse vital initial capital for R&D, evading equity dilution or debt accrual. Prospecting and soliciting relevant grants necessitate extensive research and an acute grasp of the synergy between technology and jurisprudence.

Partnerships with legal tech conglomerates or scholarly circles can also act as a strategic funding conduit. Collaborative projects distribute resources, expertise, and financial responsibilities, potentially expediting AI tool development by uniting legal, technological, and research acumen.

Though less conventional in the legal tech sector, crowdfunding presents an enterprising method to garner funds from prospective users or interested parties. This modality validates market demand for the AI offering and necessitates a captivating campaign that lucidly communicates the AI initiative's advantages.

Subscription models and software-as-a-service (SaaS) frameworks can underwrite continuous funding streams for AI developments. By marketing AI utilities on a subscription basis, legal practices access regular revenue to further innovations. This aligns with a trend toward scalable, adaptable legal services responsive to client needs.

Revenue sharing pacts with AI technology purveyors can mitigate initial expenditures and harmonize stakeholder incentives. Such agreements entail sharing the profits from the enhancements AI brings about, spurring collective success investment in the initiative.

Traditional yet still pertinent, debt financing can offer the requisite capital to jump-start or expand AI projects. Legal entities pondering this route must scrutinize borrowing conditions to ensure the AI project's return on investment warrants the borrowing cost.

Angel investors provide not merely funds but also mentorship and expansive networks. Legal tech startups may lure such benefactors intrigued by the intersection of technology and law. This path necessitates an engaging pitch and a lucid expansion and profitability strategy for the AI project.

For self-reliant entities, bootstrapping presents an alternative, relying on internal finances and revenue reinvestment to steer the AI project's course. This requires a disciplined approach centered on swift gains and sustainable expansion to back further AI investments.

Lastly, legal tech-focused incubators and accelerators proffer not only financial backing but also comprehensive mentorship, workspace, and networking access. These entities propel the rapid development of nascent legal tech ventures, including AI-focused entities, highlighting the need for an unparalleled proposition and robust business strategy.

Strategically selecting a funding source must consider the legal entity's fiscal landscape, risk tolerance, and AI integration vision. As AI reshapes the legal sector, those adeptly traversing the funding milieu will not only innovate but also flourish in the emergent legal service paradigm.

The goal, irrespective of the chosen funding route—be it internal investment, alliances, grants, or a blend—is steadfast: to leverage AI's transformative potential in legal practice, magnifying client service quality, and contributing to a more approachable, effective, and equitable judicial system.

PwC's NewLaw enters the fray here as an avant-garde advisory force, steering legal operations towards peak efficiency with AI and tech solutions. NewLaw provides a comprehensive consultancy to legal departments, aiding them in grasping the fiscal intricacies of AI integration, budget planning, and the pursuit of trailblazing funding models. This guidance is pivotal in making knowledgeable financial determinations concerning AI ventures.

NewLaw epitomizes the legal business solution, propelling corporate legal departments to unprecedented heights through the synergistic integration of human expertise, procedural efficiencies, and technological advancements.

The NewLaw team offers profound insights, practical tools, and proficient delivery capabilities, enabling legal teams to uphold their strategic business roles whilst delivering technology-fortified legal services at scale. We conduct assessments and benchmarks for expansive in-house legal teams to pinpoint and act upon opportunities to refine their operational models, adopt automation, harness legal technologies, and manifest substantial cost reductions.

The advantages of legal transformation through NewLaw's guidance are undeniable: enhancing the quality, consistency, efficacy, and velocity of legal support to business operations. With these elements realigned, lawyers are liberated to dedicate their time to more rewarding and high-value tasks, thus elevating the overall performance of the legal team.

NewLaw champions the design of a bespoke LegalTech roadmap, leveraging both prevailing enterprise solutions and cutting-edge technologies tailor-made for legal departments. Our advisory role extends to assisting in the selection, implementation, integration, and ongoing maintenance of LegalTech tools, ensuring these are seamlessly embedded into the broader enterprise technology stack or as a standalone service.

With NewLaw's assistance, legal departments embark on a transformation journey that not only brings their operations in line with contemporary standards but also prepares them for future challenges and opportunities in the digital era. This approach is rooted in a holistic understanding of legal operations, infused with a commitment to innovation and an unwavering focus on delivering value.

NewLaw's expertise is particularly salient when considering funding options for AI initiatives within the legal realm. Our team offers nuanced advice on how to allocate resources smartly, how to evaluate and mitigate financial risks associated with such technology investments, and how to explore and secure diverse funding channels that align with the strategic goals of the organization.

By engaging with NewLaw, legal entities can gain a clearer view of the landscape of potential funding opportunities, from the traditional to the avant-garde. This partnership empowers legal departments to make strategic decisions that are financially sound and technologically forward-thinking, ensuring that investments in AI and legal tech yield the desired returns and drive long-term success.

For law firms and legal departments looking to innovate, NewLaw's holistic approach to funding AI initiatives is of inestimable value. It allows legal entities to navigate the complexity of the current financial landscape for legal tech with confidence, backed by data-driven insights and strategic acumen.

To summarize, exploring funding options for legal AI initiatives is a complex, multifaceted endeavor that requires careful planning, strategic insight, and an understanding of the unique dynamics at the intersection of law and technology. By considering internal financing, external investment, and innovative funding models such as those advanced by NewLaw, legal entities can secure the resources necessary to capitalize on the benefits of AI.

The roadmap to successful AI integration in legal practices is paved with challenges, but also with significant opportunities. The key to unlocking these opportunities lies in harnessing the right mix of funding sources, aligning them with the entity's vision, and leveraging expert guidance to navigate the complexities of the legal and technological worlds. Through a strategic blend of creativity, persistence, and informed decision-making, legal entities can achieve their goals and set new benchmarks for efficiency and innovation in the legal sector.

In navigating the future of legal services, entities that tap into the full spectrum of funding mechanisms—including those offered by NewLaw—will position themselves at the forefront of the industry, ready to lead the charge in the new era of legal tech. This journey requires a bold vision, a collaborative spirit, and a willingness to embrace change. For those willing to take this path, the rewards are a transformed legal landscape that is more accessible, efficient, and just for all.

In partnering with PwC's NewLaw, legal entities are not just investing in AI; they are investing in a strategic partner who will walk with them step-by-step through the transformation process. It is a partnership that promises to redefine what is possible in legal operations, ushering in a new wave of technological empowerment that will benefit not just lawyers and legal departments but the entire fabric of the legal system.

The Strategic Investments in Legal AI Technologies

Investing in artificial intelligence (AI) technologies represents a strategic commitment for legal professionals and law firms to not only stay ahead of the curve but also to redefine the curve itself. The emergence of AI in the legal domain has been transformative, driving efficiencies and enhancing the capabilities of legal practitioners. This is where Elevate enters the narrative as a key protagonist. As an Alternative Legal Services Provider (ALSP), Elevate stands at the intersection of law and technology, proffering both wisdom and tools—expert-led and software-powered—to harness the potential of AI.

Elevate has consistently demonstrated that strategic investment in AI technologies is not merely about adopting new software but reimagining the very essence of legal work. This reimagining transcends traditional boundaries, offering practical ways for law departments and law firms to improve their efficiency, quality, and, ultimately, their business outcomes.

AI in legal services is about harnessing computing power to do more than just streamline workflows—it's about augmenting the intellectual labor of legal professionals. With AI, attorneys and legal staff can transcend mundane tasks to focus on complex legal analysis, strategy, and client engagement. It's a paradigm that Elevate has not only understood but also expertly executed.

One of the key differentiators for Elevate is its Contracts Insights Service, a prescient development that tapped into the power of structured contract data well before the advent of widely known AI platforms like OpenAI's ChatGPT. By leveraging AI to analyze and structure contract data, Elevate has enabled legal departments to extract critical insights from their contracts, insights that inform strategic business decisions, risk management, and operational efficiencies.

The return on investment (ROI) from adopting Elevate's AI solutions is multifaceted. Firstly, there is the direct cost saving from faster contract review and analysis. AI-driven contract management systems can dissect and categorize contract clauses, assess risks, and highlight obligations much faster than a human could, leading to a reduction in hours billed for these tasks. Moreover, the accuracy of AI, which eliminates human error, further ensures that the ROI is not just in terms of time and money saved, but also in the avoidance of potential legal setbacks.

However, the ROI of AI in legal services goes beyond contracts. Litigation support, legal research, and even predictive analytics are areas where AI can and does offer a competitive advantage. Elevate's strategic investments have also led to the development of AI tools that can predict legal outcomes based on historical data, allowing lawyers to better advise their clients and plan their strategies. The use of AI in e-discovery, for instance, has been revolutionary, enabling law firms to sort through vast amounts of data with unprecedented speed and precision.

The transformational effect of AI technologies on legal professionals and law firms cannot be overstated. By automating routine tasks, AI liberates legal minds to engage more deeply with the substantive aspects of their work. This shift is creating a new breed of legal professionals who are not just practitioners of law but strategic advisors who leverage data-driven insights to provide enhanced value to their clients.

Elevate's emphasis on expert-led, software-powered solutions means that technology is not deployed in a vacuum but is integrated into the fabric of legal practice by those who understand the nuances of the profession. This is perhaps the most critical element of their AI strategy—ensuring that the adoption of AI is not just about software implementation but about cultural transformation within the firm.

In the sphere of legal AI and Legal-Tech, investment is not simply a matter of purchasing new tools; it's about nurturing an ecosystem where technology and legal expertise coalesce to drive innovation. The comprehensive Legal AI and Legal-Tech strategy that Elevate advocates for is rooted in this ecosystem approach, underscoring the necessity for a symbiotic relationship between tech and talent.

As legal departments and law firms ponder over the merits of integrating AI into their practices, the narrative crafted by Elevate offers not just compelling evidence but a roadmap. This roadmap begins with an assessment of current practices, identification of areas ripe for innovation, and the strategic deployment of AI solutions that resonate with the specific needs of the legal practice. It's a journey that Elevate has not only embarked on but also invites the wider legal community to join.

Concludingly, investing in AI technologies, as championed by Elevate, is not just about staying competitive. It's about setting new standards for efficiency, efficacy, and strategic acumen in the legal profession. It's a commitment to an evolutionary leap where AI is not an external aid but an intrinsic part of the legal fabric—a thread that weaves through every aspect of practice to create a stronger, more responsive, and more intelligent legal service delivery model. Elevate exemplifies this commitment, ensuring that their partners and clients are not just participants but leaders in this new era of legal services.

Implementing AI: Reframing Legal Operations

In the landscape of modern legal practice, the integration of artificial intelligence (AI) stands as one of the most transformative forces of the 21st century. UnitedLex, a leading provider of data and professional services, has positioned itself at the forefront of this revolution, offering a suite of services that not only streamline legal operations but also engender substantial economic efficiencies and an enhanced return on investment (ROI) for law firms and corporate legal departments.

UnitedLex's inception was fueled by the vision to innovate and push the legal industry beyond its traditional boundaries. By adopting a digital transformation strategy centered around AI, UnitedLex aligns with the needs of legal and business leaders who recognize the necessity of a transformational, data-centric approach. Such a strategy is not merely a pathway to growth; it is a requisite evolution to maintain competitive advantage in a rapidly changing legal landscape.

At the heart of UnitedLex's transformative promise is the ability to reduce the friction from commerce through the mastery of data. This is no small feat in an environment where data proliferation is matched only by the complexity of regulations and the ever-expanding global market. Through the skillful application of AI, UnitedLex turns this burgeoning data from a potential liability into one of an organization's greatest assets.

The strategic approach to AI implementation by UnitedLex is multi-faceted, reflecting the complexity and diversity of the challenges faced by modern legal departments. One critical aspect is the integration of AI into digital litigation. Here, AI tools are used to parse through massive datasets, identify patterns, and predict outcomes, thereby enabling attorneys to formulate strategies based on deep data-driven insights. The efficiency gains in this area alone can significantly reduce hours spent on legal research and review, leading to direct cost savings and faster case resolutions.

In the realm of intellectual property, AI serves as a guardian of innovation. By deploying AI algorithms, UnitedLex helps firms navigate the intricacies of IP law, from patent searches to infringement analysis. AI's ability to rapidly analyze vast arrays of patents and legal documents means that intellectual property can be managed with efficiency and accuracy that human practitioners alone could never achieve.

Contracts, the lifeblood of commerce, are another area ripe for transformation through AI. UnitedLex's AI-driven contract lifecycle management services streamline the drafting, negotiation, and management of contracts. The AI systems can detect risks, ensure compliance, and even predict which negotiations are likely to succeed. This not only saves time but also transforms contracts into dynamic tools for business intelligence.

Compliance, an ever-growing concern in our interconnected world, is another area where UnitedLex's AI implementation proves indispensable. By continuously monitoring and analyzing regulations, policies, and company data, AI can provide real-time insights into compliance risks. This proactive approach can prevent costly legal battles and penalties before they arise, securing peace of mind for clients.

The core of UnitedLex's strategy is its approach to legal operations. By harnessing AI, UnitedLex delivers operational efficiencies that translate into direct financial benefits. Workflow automation, predictive analytics, and data-driven decision-making are just a few examples of how AI can enhance productivity and strategic planning.

The strategic considerations for adopting UnitedLex's services revolve around understanding the symbiotic relationship between technology and the human element. AI is not a replacement for legal expertise; rather, it is a tool that augments the capabilities of legal professionals. By freeing attorneys from routine tasks, they can focus on more strategic, high-value work.

Another key strategic consideration is the integration of AI into the existing legal framework. The transition to AI-enabled practices must be seamless to avoid disruption. UnitedLex excels in customizing its AI implementations to fit within the unique ecosystem of each client, ensuring a smooth transition and immediate operational gains.

UnitedLex's impact on operational efficiency and competitive advantage can be quantified in various ways. From the reduction in time and costs associated with legal processes to the increased accuracy in legal outcomes and the mitigation of risks, the advantages are clear and measurable. The strategic adoption of AI services provided by UnitedLex enables law firms and corporate legal departments to not only keep pace with the current demands of the market but also to position themselves ahead of the curve.

In conclusion, UnitedLex's AI implementation services serve as a comprehensive blueprint for legal departments aiming to achieve economic efficiency and an enhanced ROI. Their commitment to innovation, passion for data mastery, and the strategic integration of AI technologies empower legal entities to elevate their practice to new heights of operational efficiency and maintain a decisive competitive edge in a rapidly evolving industry.

Chapter 17:

Case Studies: Successes and Lessons in Legal Tech

The journey of AI in the legal industry is marked by a number of pioneering efforts and cautionary tales. One of the most remarkable stories is that of IBM Watson. Watson's foray into the legal sector is a narrative that we will explore as a cautionary tale of innovation's fragility within AI legal tech. IBM Watson was an early example of AI's potential to transform legal research, a harbinger of the evolution to come. However, as we delve deeper into the chapter, the tale of IBM Watson unfolds as one of initial groundbreaking strides followed by unexpected stumbles. The insights we gain from IBM Watson's trajectory in the legal tech space shed light on the readiness of the market for such innovations, the criticality of establishing the right market fit, and the imperativeness of early adoption as a catalyst for subsequent technological advancements within the legal framework.

Moving from the instructive lessons of IBM Watson, we shift our focus to a success story within the AI legal tech landscape—Luminance. This segment of the chapter will underscore how Luminance has become a beacon of triumph in employing AI for document analysis and contract review. Luminance's assimilation of cutting-edge AI has revolutionized legal processes, making them more efficient and mitigating the risks associated with human error. As we dissect the roadmap to their successful AI adoption, the narrative will highlight the seamless integration of AI technologies into existing workflows, the strategies for training legal staff to harness these new tools, and the multitude of benefits that ensued. The success of Luminance is not only an inspiring account but also serves as a practical blueprint for law firms that are contemplating similar paths of AI integration. The detailed exploration of Luminance's journey will reveal the profound impact of AI on enhancing accuracy, curtailing processing times, and ultimately elevating client satisfaction.

The final narrative we weave is that of Smith.AI's strategic incorporation of AI into the fabric of a traditional law firm setting. Smith.AI is not just a story but a contemporary illustration of how legal firms can undergo a profound transformation through strategic planning and the adoption of advanced technologies. The chapter will elaborate on the various innovative initiatives undertaken by Smith.AI, the advancements made in their service provision, and the operational efficiencies gained as a result. By focusing on Smith.AI's successful outcomes, this section will serve as a testament to the benefits of AI adoption in the legal sphere. The story of Smith.AI provides a robust template for success, showcasing actionable strategies that other firms might adopt to harness the power of AI and replicate the success that Smith.AI has seen.

Each case study, with its unique experiences and outcomes, offers a distinct perspective on the role of AI in the legal domain. While IBM Watson reminds us of the perils of being ahead of one's time, Luminance and Smith.AI demonstrate the tangible benefits of AI when implemented with careful planning and clear strategic vision. These narratives serve as both a warning and an inspiration, collectively composing a comprehensive overview of AI's diverse potential within legal practices. In the forthcoming sections, we will unpack these stories, extracting the rich details and pivotal moments that have shaped the role of AI in the legal industry.

The IBM Watson Experience: Innovating Legal Processes

Introduction

In the dynamic realm of legal technology, IBM Watson has emerged as a cornerstone of innovation, particularly in the sphere of litigation document preparation and analysis. LegalMation, a groundbreaking entity in legal practice, realized the potential of IBM Watson's AI to revolutionize its litigation processes. This case study examines the synergy between LegalMation's vision and IBM Watson's AI capabilities, showcasing a successful integration that has significantly enhanced the efficiency of legal workflows.

Background: The Drive for Legal Efficiency

LegalMation, operating within the highly competitive legal market, sought to differentiate itself through efficiency and quality service delivery. In the wake of economic pressures, the firm aimed to optimize its litigation processes, which were traditionally manual, time-consuming, and costly. They needed a transformative solution to streamline their operations while maintaining high-quality outputs.

Selecting IBM Watson: Embracing Advanced Legal AI

After assessing various AI technologies, LegalMation was drawn to IBM Watson's sophisticated capabilities. The firm required a seamless transition to modernize its processes without hindering its existing operations. IBM Watson's AI offered an intuitive platform that could be quickly integrated into LegalMation's workflow, promising a significant reduction in document preparation time.

IBM Watson in Action: Automating Litigation Documents

IBM Watson's AI was utilized by LegalMation to address one of the most arduous tasks in legal practice: drafting responses in the early phases of litigation. IBM Watson's technology allowed LegalMation to automate this process, enabling the generation of initial litigation documents in minutes instead of hours, fundamentally transforming their practice.

The Impact of AI on Legal Efficiency

Prior to integrating IBM Watson, LegalMation's legal team managed tasks manually, a method that was not only inefficient but also prone to human error. IBM Watson's AI revolutionized this process, handling the analysis swiftly and accurately for each document, thus enabling significant cost savings and reallocating time towards more complex legal tasks.

Post-Implementation Review and Analysis

Following the deployment of IBM Watson's AI, LegalMation was able to analyze vast quantities of legal documents rapidly, gaining insights into key legal concepts and the overall document landscape. This capability allowed the legal team to offer more proactive and strategic advice to their clients.

IBM Watson: A Holistic Business Solution

IBM Watson's influence extended beyond the confines of LegalMation's legal department. It proved invaluable in assisting cross-departmental functions such as finance, where the insights provided by Watson's AI informed financial reporting and compliance across multiple office locations and teams.

The Outcome: Measurable Achievements

The adoption of IBM Watson's AI by LegalMation resulted in:

- Dramatic reductions in time spent on document preparation
- Decreased dependency on external counsel, leading to cost savings
- Enhanced ability to provide strategic legal advice in real-time

Conclusion: IBM Watson as a Paradigm for AI Integration in Legal Practice

LegalMation's experience with IBM Watson stands as a robust model of the power of AI within the legal industry. IBM Watson not only streamlined litigation document preparation but became an integral part of the business, fostering a more agile, responsive, and strategic legal function. For legal practices contemplating AI adoption, IBM Watson and LegalMation present a compelling case for the transformative impact of AI when implemented with clarity and foresight in the legal domain.

The Luminance Success Story: AI in Contract Analysis

Introduction

In the cutting-edge domain of legal technology, Luminance has emerged as a beacon of innovation, particularly in the sphere of document analysis and contract review. Feature Space, a global pioneer in the fight against financial crime, recognized the potential of Luminance's AI technology to revolutionize their in-house legal operations. This case study explores the symbiosis between Feature Space's needs and Luminance's AI capabilities, painting a comprehensive picture of a successful integration that has dramatically reshaped legal process efficiency.

Background: The Quest for Enhanced Legal Efficiency

Feature Space stands at the forefront of enterprise financial crime prevention, and its expertise is sought after by over seventy major financial institutions worldwide. The critical nature of their work demands legal precision and rapid contract turnaround times. However, the traditional manual methods of contract negotiation and analysis were increasingly misaligned with the speed and scalability of their operations. In their quest for a transformative solution, Feature Space turned to Luminance.

Selecting Luminance: A Leap into Next-Generation Legal Tech

Conducting an exhaustive market analysis, Feature Space was captivated by the sophistication of Luminance's AI technology. The team required a swift transition that would not disrupt their existing workflow, and Luminance's plug-and-play offering promised a stark contrast to the three-to-six-month implementation period other providers proposed. Within a mere week, Feature Space was operational with Luminance, setting a new standard for agility in legal tech adoption.

Luminance at Work: The Traffic Light Analysis

A standout feature that impressed Feature Space was Luminance's Traffic Light Analysis. This innovative tool integrates seamlessly into Microsoft Word, providing instant assessments of contract clauses using color coding—green for acceptable, amber for reviewable, and red for unacceptable, based on the organization's predefined standards. The AI extends its utility by suggesting alternative phrasings from internal precedent banks when terms do not align with the company’s criteria.

Don Riddick, Chief Legal Officer at Feature Space, attests to the transformative effect: "With Luminance, our lawyers have become proactive business enablers rather than mere repositories of information."

The Impact of AI Automation

Before Luminance, Feature Space's legal team relied on manual checklists—a laborious and fallible process. Luminance's AI now conducts this scrutiny for every incoming contract within seconds. This shift to automation, particularly for repetitive tasks like NDA and employment contract reviews, has enabled Feature Space to halve its expenditure on outside counsel, reserving external advisors for only the most intricate IP and local law-related projects.

Post-Execution Contract Landscape Analysis

Luminance's AI doesn't stop at contract review. It extends into the post-execution phase, offering Feature Space a panoramic view of its contractual ecosystem, encapsulating approximately 15,000 pages of documentation. The AI highlights key details like document types, governing laws, and anomalies and identifies over 1,000 key legal concepts. This granular insight empowers the legal team to guide the business on policy, commitments, and strategic direction proactively.

Beyond Legal: A Business Integration Tool

Luminance's influence radiates beyond the legal department at Feature Space, streamlining functions across the enterprise. Lawyers, equipped with Luminance's insights, guide finance teams on revenue recognition, ensuring fidelity in finance and accounting. Furthermore, with offices spanning from London to the Asia-Pacific and the US, Luminance's cloud deployment and project management tools are indispensable for a distributed team, fostering collaboration and ensuring transparent and efficient work progression.

The Outcome: Quantifiable Success

The integration of Luminance's AI into Feature Space's legal operations has yielded tangible benefits:

- A 50% reduction in outside counsel spend

- Expedited analysis of over 15,000 pages of contracts

- Real-time provision of business-critical legal advice during board meetings

Conclusion: Luminance as a Blueprint for AI Adoption in Legal Practice

Feature Space's experience with Luminance serves as a testament to the power of AI in legal practice. Luminance not only optimized contract negotiation and analysis but also ingrained itself as a vital cog in the broader business mechanism, enabling a more dynamic, responsive, and strategic legal function. For law firms and corporate legal departments seeking to harness similar efficiencies, Luminance offers a compelling blueprint for success—demonstrating the transformational potential of AI when applied with precision and vision to the world of legal operations.

The Smith.AI Revolution: AI-Enhanced Family Law Practice

Introduction

The integration of artificial intelligence (AI) into the legal industry has been transformative, and the alliance between Smith.AI and Sibus Law Group is a testament to the power of this technology in enhancing legal firm operations. This case study will detail how Smith.AI's virtual receptionist services and AI technologies have been successfully integrated into the traditional law firm setting of Sibus Law Group, bringing about significant improvements in client intake, communication, and overall efficiency.

The Challenge at Sibus Law Group

Based in San Diego, California, Sibus Law Group is a law firm specializing in family law. Sara Kelley, who spearheads the firm, found that their previous answering service was inadequate, lacking detailed call summaries and data integration with their Customer Relationship Management (CRM) system. This shortcoming was a significant impediment to the firm's responsiveness and client service quality.

Smith.AI's Innovative Solution

Addressing Sibus Law Group's challenges required a tailored approach. Smith.AI stepped in with a suite of services designed to enhance client interaction. Their solution included comprehensive call summaries and seamless CRM data integration, which proved to be a game-changer for the firm.

Smith.AI's virtual receptionists specialize in capturing crucial information from potential clients, managing new client intake, and scheduling appointments promptly. This level of detail and efficiency provided by Smith.AI's services differentiated them from the plethora of virtual receptionist services available, ensuring that Sibus Law Group could provide personalized attention around the clock.

24/7 Coverage and Improved Responsiveness

One of the critical changes Sibus Law Group experienced was the ability to update their business hours to 24/7 on their Google Business profile, thanks to Smith.AI's round-the-clock coverage. This change was pivotal in increasing the firm's lead generation, as it assured potential clients that they could reach out to the firm at any time.

Smith.AI's commitment to delivering detailed call summaries meant that the information was readily transferable to Sibus Law Group's case management platform. This process significantly reduced the time spent on relaying information to other staff members and allowed for prioritizing urgent calls over less critical communications.

The Results: Enhanced Efficiency and Client Service

Sibus Law Group witnessed a tangible improvement in operational efficiency. With Smith.AI handling initial client interactions, the firm could connect with new clients and schedule meetings in as little as one or two days—half the time it previously took.

Sara Kelley notes the positive impact of Smith.AI's services, stating, "We are very pleased that Smith.AI has increased our business hours by answering calls 24/7. That's a huge plus for us because we know people don't always think to call an attorney during the typical 9-5 hours of a Monday - Friday workweek."

The Broader Impact of Smith.AI

Beyond Sibus Law Group, Smith.AI's approach serves as a beacon for small businesses seeking to improve their client relations and operational efficiency. Their 24/7 virtual receptionist service ensures clients receive the personalized attention they deserve, capturing more qualified leads and filtering out spam calls. This level of service enables business owners to dedicate their time to more productive tasks, fostering sustainable growth.

Conclusion

In conclusion, the collaboration between Smith.AI and Sibus Law Group illustrates the profound impact AI can have on a legal firm's operations. By enhancing the quality and responsiveness of client services, Smith.AI has not only propelled Sibus Law Group to new heights of efficiency but has also set a standard for the integration of AI in legal practice. Smith.AI's success story with Sibus Law Group can inspire other firms to embrace AI and technology, underscoring the potential for such tools to revolutionize the business of law.

Embracing AI as the Future of Legal Services

As we reach the culmination of our exploration into the transformative world of AI in legal firms, it is clear that the journey is just beginning. The case studies of IBM Watson, Luminance, and Smith.AI serve not just as milestones but as beacons guiding the path forward. Each story, with its lessons and successes, heralds a future where AI and Legal Tech are intrinsic to the legal landscape.

The Cautionary Tale of IBM Watson

IBM Watson's venture into the legal industry began with great promise, offering a glimpse into a future where legal research and document analysis were to be revolutionized. The partnership with LegalMation was a bold step into automating litigation documents, showcasing the potential for AI to enhance efficiency dramatically. Yet, the story also served as a cautionary tale, highlighting the challenges of pioneering in an evolving market. IBM

Watson's experience underlines the importance of market readiness and the need for technologies to align with the specific demands of legal practitioners.

Luminance: A Model of Success

In contrast, Luminance stands as a testament to the power of AI when it fits snugly into the existing fabric of a legal firm. The partnership with Feature Space revealed that with careful planning and execution, AI could elevate the practice of law to unprecedented levels of efficiency and accuracy. The Traffic Light Analysis, a standout feature, exemplified innovation, seamlessly fitting into the day-to-day work of legal professionals and providing real-time insights that empowered decision-making.

Smith.AI's Paradigm Shift in Client Services

Smith.AI's integration into Sibus Law Group's operations narrates a story of exceptional client service, driven by AI. By providing a virtual receptionist service and integrating seamlessly with CRM systems, Smith.AI demonstrated how AI could transform client intake and communication processes, enhancing responsiveness and operational efficiency.

Envisioning a Tech-Driven Legal Ecosystem

Looking ahead, the narrative of Legal AI and Legal Tech is one of inevitable growth and increasing adoption. The legal industry stands on the cusp of a digital revolution, where AI's role will become increasingly central to every aspect of legal work. The future portends a legal ecosystem where AI-driven analytics inform strategic decision-making, machine learning algorithms predict legal outcomes, and virtual assistants handle routine tasks, allowing lawyers to focus on complex legal reasoning and client relations.

Predicting the Rise of Personalized Legal Services

The legal industry is poised to experience a paradigm shift towards more personalized and client-centered services. Legal firms will leverage AI to understand client needs better and tailor services accordingly. Predictive AI will enable law firms to anticipate client issues and address them proactively, fostering deeper trust and client loyalty.

The Democratization of Legal Services through AI

AI will also play a pivotal role in democratizing legal services. Automated self-help solutions and AI-based legal advice platforms will become more sophisticated, making legal assistance more accessible to underserved communities. This increased access to justice will transform the societal perception of legal services, turning them from a luxury for the few into a standard provision for the many.

AI as a Catalyst for Collaboration and Innovation

The integration of AI will encourage collaboration across various sectors. Law firms will increasingly partner with tech companies to co-create solutions tailored to the specific nuances of the legal domain. This collaboration will spark innovation, leading to the development of new legal tech tools that can adapt to the changing landscape of the legal industry.

Strengthening Ethical and Regulatory Frameworks

As AI becomes more prevalent, the legal industry will need to reinforce ethical and regulatory frameworks to govern the use of AI. There will be a concerted effort to ensure that AI respects privacy, maintains confidentiality, and upholds the highest standards of legal ethics. Regulation will evolve to keep pace with technological advancements, ensuring that AI acts as an ally rather than a disruptor of legal integrity.

Final Thoughts: The Legal Industry's AI-Enabled Horizon

The case studies of IBM Watson, Luminance, and Smith.AI are early chapters in the ongoing saga of AI in legal practices. They symbolize a broader movement towards an AI-enabled horizon, where the legal industry embraces technology not just as a tool, but as an indispensable ally. As we stand on the brink of this new era, the legal profession is poised to transform itself in ways we are only beginning to imagine. The promise of AI and

Legal Tech is not just in the efficiencies they bring but in the potential they unlock for a more equitable, agile, and innovative legal future. The dawn of AI in law is not just approaching; it is already upon us, and its full glory remains to be written by the collective efforts of visionary practitioners and technologists. As we look forward, the legal industry awaits a renaissance of its own, a renaissance powered by AI and Legal Tech.

Crafting the Blueprint for Legal Tech Advancements

The adoption of Legal AI is akin to assembling a complex puzzle. Each piece, from IBM Watson's analysis tools to Luminance's contract automation and Smith.AI's client service enhancements, contributes to a holistic vision for the sector. The future will likely hold a legal profession where AI tools are not only commonplace but also deeply integrated into the fabric of daily practice, enhancing the counsel provided by attorneys and enabling firms to deliver services with unprecedented precision and insight.

Harnessing AI for Global Legal Integration

The global nature of business today demands that legal services transcend borders, and AI is poised to facilitate this integration. Multilingual document review, cross-jurisdictional compliance checks, and international case law analysis are areas where AI will make significant strides, making global legal practice more seamless and interconnected.

Driving Legal Education and Continuous Learning

As AI reshapes legal practices, it will also redefine legal education. The next generation of lawyers will not only need to be adept at legal reasoning but also proficient in the use of AI and technology. Legal education providers will incorporate AI training into their curricula, ensuring that graduates are equipped with the skills necessary to thrive in a tech-driven legal landscape.

AI's Role in Fostering Inclusivity and Diversity

The neutral nature of AI, when properly designed and monitored, can be leveraged to promote inclusivity and diversity within the legal profession.

AI-based hiring and promotion tools, blind review of legal documents, and unbiased legal decision-making aids will help minimize unconscious bias, ensuring a level playing field for all legal professionals and clients.

Enhancing Legal Research and Development

AI will revolutionize legal research, enabling lawyers to sift through vast amounts of case law, statutes, and legal writings in a fraction of the current time. This will not only make legal research more efficient but also more comprehensive, potentially uncovering insights and precedents that might have otherwise been overlooked.

The Role of AI in Sustainable Legal Practices

Sustainability is becoming a guiding principle across industries, and the legal sector is no exception. AI will aid in developing sustainable legal practices by optimizing resource allocation, reducing paper usage through digital workflows, and minimizing the carbon footprint associated with traditional legal work.

Optimistic Projections for Legal Tech Growth

The trajectory for Legal AI and Legal Tech is overwhelmingly positive. As technology continues to mature, its adoption will become more widespread, and its benefits more pronounced. Legal Tech startups will flourish, fueled by a growing recognition of the value these innovations bring to the legal profession. These enterprises will become the crucible for the next wave of legal tools, continually expanding the capabilities and reach of AI within the industry.

A Call to Action for Legal Professionals

For current and aspiring legal professionals, the message is clear: embrace AI and technology as partners in practice. By adopting a mindset of continuous learning and adaptability, legal practitioners can leverage these tools to amplify their expertise, deliver superior service to their clients, and contribute to the creation of a more just and efficient legal system.

Envisioning a Future Powered by Legal AI

In closing, the future of the legal industry is not one of AI versus human practitioners, but of AI in concert with legal expertise. This harmonious blend will unlock new dimensions of legal practice, driving innovation, and excellence. The narrative that began with pioneering firms like IBM Watson, Luminance, and Smith.AI will continue to unfold, shaped by the collective endeavors of the legal community and the promise of technology. As we stand at the threshold of this new era, it is with a sense of optimism that we envision a future where Legal AI and Legal Tech are not just auxiliary tools but the very bedrock upon which the edifice of modern legal practice is built.

In this dynamic and rapidly evolving landscape, the legal industry is primed to reach unprecedented heights, guided by the stars of AI and Legal Tech. The chapters we've explored in this book provide the groundwork for what is sure to be an exhilarating journey forward. It is a future that beckons with the promise of greater efficiency, deeper insights, and a more profound commitment to justice. A future where the law is not only practiced with technology but is also perfected by it.

Chapter 18:

The AI Revolution in Legal Services: Transformative Technologies and Future Directions

Artificial Intelligence (AI) is reinventing legal operations, heralding a new age of service delivery and expanded access to legal resources. The integration of AI technologies like Neota Logic's no-code platform into legal practices is not merely a step towards modernization; it's a stride towards fundamentally transforming the accessibility and delivery of legal services.

Neota Logic emerges at the forefront of this evolution, offering a beacon for the democratization of AI tools within the legal industry. With its innovative no-code platform, Neota Logic is shattering the traditional barriers that have confined the development of custom legal applications to those with extensive programming knowledge. This shift towards accessible legal tech is empowering law firms of all sizes, enabling them to harness powerful AI tools to automate processes, streamline workflows, and elevate the quality of their services without writing a single line of code.

The implications of Neota Logic's entry into the market are profound. Law firms, once reliant on manual processes and traditional legal service models, now have the opportunity to revolutionize these paradigms. Automation through Neota's platform facilitates the scaling of knowledge across the firm, ensuring that valuable insights and expertise are not confined to the minds of a few but are embedded into the very processes that drive legal operations. This capability extends from automating mundane document workflows to managing complex legal requests, all while ensuring accuracy and compliance with prevailing laws.

The impact of such technology on law firm innovation cannot be overstated. Legal professionals are now equipped to supercharge their innovation roadmaps, forging new pathways for revenue generation, efficiency, and productivity. By crafting digital service models that strengthen client relationships, law firms can achieve superior growth and profitability at an unprecedented pace. Neota Logic's solutions are particularly adept at transforming the perception of legal departments from cost centers to value drivers, liberating legal teams from the repetitive, low-value work that has traditionally consumed their resources.

Increased productivity is another cornerstone benefit of Neota Logic's offerings. The platform enables the creation of self-service tools that not only streamline business operations but also allow legal professionals to focus their talents where they can make a significant impact. As legal teams are freed from the mundane, they can dedicate more time to nuanced legal analysis, strategy, and client engagement—activities that truly capitalize on their expertise and training.

Efficiency gains are similarly transformative. Neota Logic's platform facilitates the rapid deployment of solutions, effectively eradicating the bottlenecks often associated with legal service delivery. This agility allows for quicker deal progressions and ensures that compliance is seamlessly integrated into business operations. With the ability to rapidly iterate and deploy solutions, legal teams can respond more dynamically to the evolving needs of their clients and the market.

Accuracy in legal advice is a paramount concern, and AI tools like those developed on Neota Logic's platform ensure that the advice provided to businesses is both accurate and consistent. The platform's expert logic and reasoning capabilities are designed to mimic the thought processes of seasoned legal professionals, thereby minimizing the risk of error that can arise from manual oversight or the misinterpretation of complex legal frameworks.

Compliance is another area where Neota Logic's tools make a significant difference. The platform's intelligent systems are adept at guiding users through the intricate maze of legal regulations, ensuring that businesses remain on the right side of the law. This is particularly valuable in industries where regulatory compliance is both complex and dynamic, requiring constant vigilance and adaptation.

The breadth of solutions that can be created on Neota Logic's platform is staggering. From Master Service Agreements, which allow for the online creation, negotiation, and execution of contracts, to the management of Legal Service Requests, the platform handles a vast array of legal operations with ease. Even complex Non-Disclosure Agreements can be generated, negotiated, and executed within a framework that ensures legal precision and efficiency.

Furthermore, Neota Logic's no-code platform excels in process automation, which is pivotal in transforming chaotic and inefficient workflows into streamlined and effective ones. Tasks and emails are managed with a drag-and-drop interface, simplifying interactions and coordination between parties. Knowledge automation is another area where the platform shines, with intelligent web-based solutions guiding users through fact-gathering and decision-making processes to produce contextually appropriate outputs. This technology is ideal for applications such as risk assessments and immigration advisories, where the correct application of legal knowledge is crucial.

Document automation is where Neota Logic's capacity for innovation becomes particularly evident. The platform enables the creation of bespoke documents tailored to clients' needs, automating not just the generation of such documents but also the negotiation and execution processes. This extends to the development of digital policies, guidelines, and training programs, all designed to deliver significant value and an enhanced user experience.

In the legal AI landscape, where the prediction of legal outcomes is increasingly driven by predictive analytics and where AI-powered legal assistants are expected to offer nuanced interactions, the capabilities of Neota Logic's platform are particularly relevant. These technologies are laying the groundwork for a future where the delivery of legal services is not only more efficient and accurate but also significantly more accessible. The potential for AI to process and analyze large datasets in real-time augments legal research, contract analysis, and risk management, which in turn allows lawyers to deliver prompt and well-informed counsel.

As legal AI tools like Neota Logic become more integrated into everyday legal practice, they are also beginning to address broader social implications, such as improving access to legal assistance for underserved populations. With platforms capable of interpreting complex legal documents and providing preliminary advice, the doors to legal advice and representation swing wide open for those who may have been previously excluded due to cost or complexity barriers.

The role of AI-powered tools in law extends beyond process efficiencies and enters the realm of strategic business insights. The data harvested from these AI solutions offers law firms unprecedented insights into their operations, allowing them to make data-driven decisions to improve processes and direct their focus where it's most needed. In essence, these tools are not only changing how work is done but also how it is evaluated and enhanced over time.

Moreover, Neota Logic's impact on the legal industry underlines the vital nature of adaptability within the profession. As AI technologies evolve, so too must the legal frameworks, ethical standards, and professional skills adapt to meet the changing landscape. It's not just about adopting new technologies but also about developing a culture of continuous learning and evolution, where legal professionals remain at the cutting edge of their field.

While AI's role in law heralds many positive changes, it also brings challenges and ethical considerations to the fore. The legal profession must navigate issues related to privacy, data security, and the implications of automated decision-making. Building AI systems that respect individual privacy rights and ensuring the integrity of automated legal advice is essential to maintaining trust in the profession. Similarly, the legal community must continue to engage in conversations about the balance between leveraging AI for efficiency and preserving the indispensable human elements of judgment and empathy in legal services.

AI technologies like Neota Logic are, therefore, not just tools but catalysts for a broader conversation about the future of law. They compel us to consider how legal services are structured, priced, and delivered, and they challenge us to reimagine the lawyer's role in society. As AI becomes increasingly capable, lawyers will likely transition from being mere providers of information to becoming strategic advisors who interpret and add value to AI-derived insights.

The future of AI in law promises a landscape where legal professionals are not displaced by technology but are enhanced by it. Lawyers will harness AI to handle routine tasks while dedicating more of their time to complex reasoning, client counseling, and strategic decision-making. In this regard, platforms like Neota Logic serve as a bridge between the traditional practice of law and a future where legal services are more effective, equitable, and aligned with the digital realities of the 21st century.

In conclusion, as we anticipate the advancements that AI will bring to the legal sector, it is clear that platforms like Neota Logic will play a pivotal role. They will not only refine the tools and processes available to legal professionals but will also reshape the very nature of legal service delivery. As we move forward, it is incumbent upon the legal profession to harness these technologies ethically and responsibly, ensuring that the evolution of legal services continues to uphold the highest standards of justice and fairness. The integration of AI in law is not an endpoint but a journey—a journey that will require foresight, adaptability, and an unwavering commitment to the principles that have long defined the legal profession.

The Leading-Edge of Legal AI: IBM Watson's Transformative Impact

The legal industry stands at the threshold of a revolutionary shift, powered by the advent of artificial intelligence (AI). This transition from theoretical discourse to tangible application signifies a future where legal practice is deeply entwined with technological innovation. Among the vanguard of this movement is IBM Watson, whose capabilities in natural language processing (NLP), predictive analytics, and cognitive computing are setting new benchmarks in legal tech. This section explores the transformative potential of emerging AI technologies, with a particular focus on IBM Watson's role in reshaping legal services.

IBM Watson's foray into legal AI marks a significant leap forward. Utilizing advanced NLP, Watson accelerates legal and contract management, enabling professionals to sift through vast repositories of documents and data swiftly, thereby providing relevant insights and answers expeditiously. This not only enhances productivity but also augments the quality of client services, catalyzing revenue growth. IBM Watson Discovery epitomizes this advancement, automating the extraction of pertinent information from extensive document collections, which traditionally consumed considerable time and resources.

Predictive analytics in legal practice is undergoing a metamorphosis through IBM Watson. By harnessing historical data and leveraging its cognitive capabilities, Watson can forecast legal outcomes with remarkable precision. This not only aids legal practitioners in formulating robust strategies but also optimizes resource allocation, ensuring that efforts are directed toward the most impactful areas. The integration of IBM Watson into legal workflows epitomizes the shift towards data-driven decision-making, enabling firms to anticipate trends and adapt strategies accordingly.

The automation of routine legal tasks is another frontier in which IBM Watson is making significant inroads. Document automation and the management of smart contracts are becoming increasingly streamlined, with Watson's AI-driven systems ensuring the execution of contractual terms efficiently and accurately. This shift not only expedites legal processes but also minimizes the margin for error, thus enhancing the reliability of legal transactions.

As legal AI progresses, the potential for AI-powered virtual legal assistants, like those developed with IBM Watson's technology, is becoming more tangible. These systems are designed to provide comprehensive legal support and advice, broadening the accessibility of legal guidance. By democratizing legal assistance, technologies like IBM Watson contribute significantly to enhancing access to justice, particularly for underserved populations.

The application of emotion recognition technology in legal settings is an area of burgeoning interest. IBM Watson's ability to analyze emotional nuances in testimonies and interrogations could offer critical insights, adding a layer of depth to legal analysis that traditional methods may overlook. This technology, although still in its early stages, holds the promise of transforming aspects of legal practice by providing a more nuanced understanding of human behavior and interactions.

In compliance and risk management, IBM Watson's AI capabilities are set to revolutionize the way firms navigate complex regulatory environments. Continuous monitoring and analysis allow for the early detection of potential compliance issues, facilitating a proactive approach to risk management. This not only ensures adherence to regulatory standards but also mitigates the risks associated with non-compliance.

The role of AI, particularly through tools like IBM Watson, in enhancing cybersecurity within the legal sector cannot be overstated. Given the sensitive nature of data handled by legal firms, the application of AI in detecting and preventing cyber threats is crucial. IBM Watson's ability to learn from and adapt to evolving cybersecurity threats ensures robust protection for legal entities, safeguarding their digital infrastructure against potential breaches.

The integration of AI technologies like IBM Watson into legal education is preparing the next generation of legal professionals to navigate a tech-centric landscape. By incorporating AI tools and methodologies into their curricula, educational institutions are equipping students with the skills and knowledge necessary to excel in a future where law and technology are inextricably linked.

AI's impact on alternative dispute resolution (ADR) mechanisms illustrates another dimension of its potential. By streamlining negotiation and mediation processes, technologies like IBM Watson can make ADR practices more efficient and accessible, reducing the time and resources typically required for these processes.

However, the integration of AI into legal practice necessitates a balanced approach. Ethical considerations, data privacy concerns, the digital divide, and the potential for algorithmic bias are critical issues that need careful deliberation to ensure that the deployment of AI in legal contexts is responsible and inclusive.

The legal profession is on the cusp of a technological renaissance, with AI, specifically IBM Watson, playing a pivotal role in this transformation. The capabilities of AI to enhance legal research, improve predictive accuracy, automate routine tasks, and make legal assistance more accessible are reshaping the future of law. For legal professionals, adapting to and embracing these technological advancements is imperative for maintaining relevance and competitiveness in an evolving legal landscape. By investing in AI technology and training, legal firms can foster a culture of innovation and adaptability, ensuring that they not only keep pace with technological progress but also leverage it to improve the delivery of legal services.

The trajectory of law is set towards a more efficient, accessible, and equitable system, with AI and IBM Watson at the helm. By spotlighting these upcoming technologies, we pave the way for a legal profession that is not only prepared for the future but also actively shaping it. The integration of IBM Watson and other AI technologies in legal practice heralds a new era of efficiency and innovation. As legal practitioners and firms navigate this evolving landscape, the focus must be on harnessing the full potential of AI to enhance service delivery and ensure justice is accessible to all.

IBM Watson's expanding AI capabilities, particularly in legal research, are transforming the way legal professionals approach case preparation and strategy. By automating the analysis of large volumes of legal documents and data, Watson enables legal experts to focus more on nuanced legal reasoning and less on time-consuming research tasks. This shift not only improves the speed and efficiency of legal services but also enhances the accuracy of legal outcomes, as cognitive computing allows for a more comprehensive analysis of case law and precedents.

The democratization of legal services through AI, spearheaded by technologies like IBM Watson, is a significant development. By making legal research and assistance more accessible, AI is breaking down barriers that traditionally hindered access to legal support, especially for individuals and communities with limited resources. This increased accessibility is likely to have a profound impact on the legal landscape, ensuring that more people can obtain the legal assistance they need, thereby promoting fairness and equity in the judicial system.

IBM Watson's role in advancing legal AI extends to predictive capabilities, which are set to revolutionize how legal outcomes are anticipated and strategies formulated. With the ability to analyze past case data and legal decisions, Watson can predict potential outcomes with a high degree of accuracy, providing legal professionals with invaluable insights that can inform case strategy and decision-making. This predictive power not only enhances the strategic planning process but also allows for a more targeted and efficient use of legal resources.

Cognitive computing, a cornerstone of IBM Watson's technology, is reshaping the decision-making process in legal practice. By mimicking human thought processes and reasoning, cognitive computing facilitates a more nuanced and sophisticated analysis of legal issues and cases. This capability not only streamlines the decision-making process but also ensures that legal judgments are based on a thorough and comprehensive analysis of all relevant factors and precedents.

As legal AI continues to evolve, the integration of IBM Watson and similar technologies into legal practice is expected to grow, offering unprecedented opportunities for innovation and improvement in legal services. The continued development and adoption of these technologies will likely lead to more predictive and analytical tools that can further enhance the accuracy and efficiency of legal processes.

The role of AI in legal practice, particularly through the application of IBM Watson, is not just about technological advancement but also about fostering a more equitable and just legal system. By improving access to legal resources, streamlining case analysis, and enhancing the predictability of legal outcomes, AI is playing a crucial role in shaping a legal landscape that is more accessible, efficient, and fair.

The future of legal practice, with AI and IBM Watson at the forefront, promises a landscape where legal services are more aligned with the needs and expectations of society. As we continue to explore and integrate these technologies into the legal domain, the potential for innovation and improvement is vast. The journey towards a more technologically advanced and equitable legal system is ongoing, and with AI and IBM Watson leading the way, the possibilities are as broad as they are promising.

Envisioning a New Era: The Intersection of OpenAI and Legal Practice

As we navigate through the early years of the 21st century, the legal profession stands at the cusp of a technological renaissance, spearheaded by rapid advancements in artificial intelligence (AI). OpenAI, a vanguard in this transformative era, postulates a future where the synthesis of machine learning, natural language processing (NLP), and predictive analytics redefines the very edifice of legal practice. This section delves into a prospective vision, unraveling the intricate tapestry of AI's prospective trajectory within the legal domain, orchestrated by the pioneering developments at OpenAI.

The forthcoming decade heralds an unprecedented convergence of AI with legal operations, a blend that promises to recalibrate the scales of justice. At the forefront are machine learning algorithms, growing ever more intricate, which are poised to decipher the complex language of the law with startling acuity. OpenAI's endeavors in refining models such as GPT-3 signal the dawn of a new epoch in legal document drafting, case predictions, and interactive legal assistance. This AI-driven evolution beckons a future where the meticulous and labor-intensive tasks once borne by legal professionals are transmuted into a streamlined digital workflow, enhancing accuracy, expediency, and, most vitally, the democratization of legal assistance.

The interplay of OpenAI's breakthroughs with legal research tools will see lawyers armed with deep learning-powered instruments, capable of sifting through the quagmire of case law, statutes, and precedents with a deft ease once thought impossible. The implementation of such tools is anticipated to cut down the laborious research hours, presenting a two-fold boon: the elevation of legal services' efficiency and the extension of their reach to a wider populace.

In the document review and contract analysis arena, AI's prowess will undergo a significant evolution. Future AI systems will not only identify crucial clauses and flag potential legal risks but will also proactively suggest alterations aligned with prevailing legal standards and best practices. Such advancements will particularly revolutionize corporate law, offering a sturdy bulwark against risk and ensuring unwavering regulatory compliance.

Litigation strategies, too, will be reshaped by the predictive might of AI. By analyzing the wealth of historical legal data, AI will render litigation predictions with a finesse bordering on prescience. Lawyers will harness this foresight to sculpt litigation strategies with a precision that mirrors a chess grandmaster's forethought. This element of predictive analytics, as it garners widespread acceptance, could also begin to percolate through to judicial decision-making, adding a layer of statistical probability to the art of judgment.

The rise of AI legal assistants and chatbots stands as a testament to AI's potential in enhancing public access to legal guidance. These digital assistants, growing ever more sophisticated in understanding and generating natural language, will offer laypersons readily accessible legal advice and automate the drafting of rudimentary legal documents. Such strides in AI democratize legal knowledge, eroding the barriers that have traditionally kept legal expertise ensconced within the ivory towers of the profession.

OpenAI's trajectory in legal AI also contemplates the ethical and regulatory tapestry that must envelop AI's integration into legal services. The responsibility to address AI biases, the opacity of algorithmic decision-making, and the sanctity of client data privacy will demand rigorous ethical discourse and policy-making. This foresight ensures that as AI solutions proliferate, they remain both innovative and grounded in the principles of justice and equity.

The education of legal professionals will undergo a metamorphosis to encompass an AI-literate curriculum, ensuring that emerging lawyers are not merely versed in the letter of the law but also in the digital tools that are increasingly underpinning its practice. This evolution in legal education is imperative, equipping the custodians of law with the dual expertise of jurisprudence and technology.

The utility of AI will breach the traditional confines of legal practice, infusing vigor into non-conventional legal services such as alternative dispute resolution and legal marketing. In these realms, AI's pattern recognition and insight generation capabilities will foster innovative approaches that streamline service delivery and augment client relations.

As AI's legal applications attain a global footprint, international law firms and corporations will harness AI to tackle the labyrinthine challenges of cross-border legal complexities. AI will emerge as a crucial navigator through the mazes of international compliance, intellectual property rights, and multifaceted global disputes.

Data security, always of paramount concern within the legal sector, will catalyze the development of AI tools architected with formidable security features. In adherence to global data protection mandates, these tools will enshrine client confidentiality at their core, thus setting new benchmarks for data stewardship in the legal field.

The next decade may well witness the consolidation of legal AI technologies as startups amalgamate and partnerships between tech behemoths and law firms solidify. This consolidation will birth comprehensive legal AI platforms that proffer holistic, end-to-end solutions, thereby streamlining the legal practice continuum.

Lastly, the nascent field of quantum computing beckons with the promise of multiplying AI's prowess within the legal sector. The amalgamation of quantum computing with AI has the potential to redefine the paradigms of data processing and analysis, catapulting legal AI into realms of efficiency and capability currently unimaginable. The ramifications for legal data analysis, encryption, and complex problem-solving are profound, positing a future where the speed and depth of legal insights evolve beyond today's standards.

In synthesis, the horizon of legal AI is radiant with the light of impending innovation. The next decade is set to unfurl a tapestry of advancements where AI becomes not just an adjunct but a central pillar of legal practice. OpenAI's role in this transformation is critical, providing the foundational technologies that will enable these shifts. Legal professionals will need to be at the vanguard of adopting these changes, retooling their skills to marry the precision of AI with the nuanced discernment of human judgment.

This new era will likely redefine the traditional lawyer's role, shifting from scribe and orator to strategic advisor and technologist. As AI handles the routine and procedural, lawyers will have the bandwidth to focus on the creative, interpersonal, and complex aspects of legal work that require human empathy and ethical considerations—elements that are yet to be replicated by machines.

Moreover, AI's advancements could spawn new legal specialties and services, such as AI ethics consultation and AI systems auditing, offering lawyers fresh avenues to apply their expertise. Firms that integrate AI into their service offerings will likely see a competitive advantage, attracting clients who expect cutting-edge, efficient, and cost-effective legal solutions.

One can envisage a future where AI, through OpenAI's continuous advancements, offers a level of customization in legal services previously unattainable. Legal solutions will be tailored to the individual needs of clients with precision, and the scalability of services will be dramatically enhanced. Such personalized attention, coupled with the economies of scale afforded by AI, could lead to a more just legal system where quality legal advice and representation are accessible to a broader segment of society.

Yet, amidst this technological ascent, the legal profession must be mindful of the digital divide. The unequal access to advanced AI technologies can lead to a bifurcated legal system where only those who can afford the latest AI services receive the best legal advice. It is imperative, therefore, for the legal community, policymakers, and AI developers to collaborate on ensuring equitable access to these transformative tools.

The intersection of AI and law will also extend beyond practice to reshape legal education and thought. Future legal scholars will engage with AI as a subject of study, examining its impact on legal theory, the philosophy of law, and jurisprudential principles. Legal education will have to be reimagined to prepare lawyers not just to use AI tools but to understand the underlying principles and engage critically with the technology's implications for law and society.

The next decade in legal AI, driven by the pioneering work of entities like OpenAI, will be marked by rapid change, growth, and the continuous rebalancing of the relationship between technology and human expertise. As these changes unfold, the legal profession must stay vigilant and adaptive, ensuring that the benefits of AI are harnessed ethically and equitably, and that justice remains the paramount virtue of the law.

In summary, the next decade of legal AI, shaped by OpenAI's developments, stands to be transformative, potentially disruptive, but ultimately beneficial for the legal field. It is an invitation to legal professionals to anticipate change, embrace innovation, and contribute to a future where the law is upheld not just by the human intellect but also by the profound capabilities of artificial intelligence. This era of legal AI will not only redefine the tools and methods of the profession but also enhance the delivery of justice, ensuring that it remains fair, accessible, and in step with the digital age.

Please note that the actual transformation of legal practice by AI, including the developments from OpenAI and the adoption of quantum computing, may unfold differently than forecasted. These predictions are based on current trajectories and are subject to change with new technological breakthroughs and societal developments.

Chapter 19:

Strategic Integration of AI in Law: Challenges and Pathways

A pivotal juncture emerges in legal practice as Artificial Intelligence (AI) steers the profession towards uncharted challenges and strategic innovations. This transformation, while promising to revolutionize how legal services are provided, introduces an array of challenges that must be acknowledged and addressed. As this chapter unfolds, it is crucial to delve deep into the multifaceted barriers to AI adoption and explore strategic solutions that legal firms can employ. The journey of integrating AI into the legal realm is complex, filled with cultural resistance, a skills gap, concerns about data security, and ethical questions that demand a sophisticated, nuanced approach.

At the forefront of these challenges is the issue of cultural resistance within law firms. The traditional practice of law is deeply rooted in the human intellect, with a strong reliance on the nuanced judgment and expertise that seasoned legal professionals provide. The mere suggestion that AI could replicate or replace the functions of legal practitioners often sparks a defensive stance. This resistance stems from a fear of the unknown and a misunderstanding of AI's role in the legal field. However, these fears are often based on misconceptions about the capabilities and purpose of AI. Rather than replacing human intelligence, AI is designed to augment it, enhancing the efficiency and accuracy of legal tasks. To overcome this barrier, firms must initiate a cultural shift by emphasizing the collaborative relationship between AI and legal professionals. Highlighting the synergies between human and machine intelligence can dispel fears and build a foundation of acceptance and curiosity.

To achieve this cultural shift, education and hands-on experience with AI are vital. Law firms must invest in comprehensive education programs that extend beyond simple software training. By fostering an environment where legal professionals of all levels engage with AI tools and understand their practical applications, firms can demystify AI and demonstrate its value as a complement to human expertise. Educational initiatives could include workshops, seminars, and collaboration with technologists, allowing lawyers to see firsthand how AI can enhance their work rather than diminish their role.

Alongside the cultural and educational aspects, the technical challenges associated with AI adoption cannot be ignored. Data security is a paramount concern in the legal industry, where the confidentiality and integrity of client information are of utmost importance. The integration of AI necessitates robust cybersecurity measures that ensure the protection of sensitive data. Law firms must develop and implement comprehensive data security protocols that align with both national and international standards, addressing the unique vulnerabilities introduced by AI technologies.

Furthermore, ethical considerations must be at the heart of AI adoption strategies. The potential for bias in AI algorithms, the implications of automated decision-making, and the accountability of AI systems are ethical issues that require careful consideration and regulation. Law firms must establish clear ethical guidelines for AI use, ensuring that these tools are employed in a manner that is fair, unbiased, and transparent.

The skills gap presents another significant barrier to AI adoption in the legal sector. The effective use of AI requires a blend of legal knowledge and technical expertise—a combination that may be lacking among current legal professionals. Bridging this gap is a complex task that involves rethinking legal education and professional development to include AI literacy. Law firms must become proactive in training their staff, either through the development of in-house educational programs or through partnerships with academic institutions and technology firms. By equipping their lawyers with the necessary skills to utilize AI effectively, firms can unlock the full potential of these technologies.

Drawing on the experiences of companies that have successfully integrated AI can provide valuable insights into overcoming these barriers. These companies often advocate for an incremental approach to AI adoption, tailoring the integration process to the specific needs and capabilities of the firm. They highlight the importance of strong leadership to guide and inspire the organization through the transformation, articulating a clear vision and commitment to navigating the complexities of the new technological landscape.

By adopting a holistic approach that addresses the cultural, technical, and strategic dimensions of AI integration, law firms can not only surmount the barriers but also fully harness the power of AI to enhance their efficiency and productivity. This approach is in line with the broader themes of the book, reinforcing the notion that AI adoption is a dynamic, iterative process. It's a journey that requires thoughtful navigation and strategic planning, with the promise of redefining the practice of law in an increasingly digital world. Firms that embrace this journey will find themselves at the forefront of the legal industry, offering innovative, ethical, and efficient legal services.

In summary, the integration of Artificial Intelligence (AI) in legal practice presents a multifaceted transformation, requiring law firms to navigate a series of substantial challenges. This chapter details how overcoming these obstacles not only enhances the efficiency and capabilities of legal services but also demands a reevaluation of traditional practices to stay relevant in the digital era. The experiences of companies that have successfully managed these changes offer invaluable lessons and affirm the transformative potential of AI. By adopting AI with strategic foresight, resilience, and a commitment to ethical standards, law firms can position themselves as leaders in a new landscape of the legal profession.

Cultural Transformation: Overcoming Technical Barriers

The infusion of artificial intelligence (AI) in legal practices marks not a transient inclination but a fundamental paradigm shift in the delivery of legal services. Yet, the trajectory towards embracing AI is not without its impediments. In this extensive discourse, we draw upon the profound insights and methodologies of McKinsey & Company to navigate through the labyrinthine challenges of AI adoption. Through strategic frameworks and a focus on the culturally transformative role of AI, McKinsey delineates the roadmap for seamlessly integrating AI with the pinnacle of legal expertise.

QuantumBlack's hybrid intelligence model is particularly pertinent in dispelling the aura of complexity that often shrouds AI technologies. McKinsey's tailored communication strategies emphasize the practical and manageable aspects of AI. In lieu of portraying AI as a distant, arcane technology, McKinsey positions it as a suite of tools that legal professionals can harness to augment their existing competencies. This approach demystifies AI and invites law firms to envision a future where AI is intricately woven into the fabric of legal services, offering support, enhancement, and strategic insight.

The concern over costs and investments in AI is another hurdle that McKinsey tackles head-on. By analyzing the long-term trajectory of AI implementations, McKinsey articulates a value proposition that transcends initial expenditures. It is not just about the immediate reduction of operational expenses but the comprehensive re-envisioning of service models that can burgeon into new sources of revenue. McKinsey's financial models and case studies furnish tangible evidence of AI's ROI, making the case for AI as a prudent long-term strategic investment.

In the legal profession, where confidentiality is sacrosanct, McKinsey underscores the importance of adopting AI systems that are not only robust and advanced but also staunchly secure. QuantumBlack's pedigree of blending the precision of data and technology with human creativity ensures that AI solutions meet the highest standards of data protection. By aligning with AI providers who are unequivocally committed to upholding security and privacy standards, McKinsey instills confidence that AI adoption can be both innovative and secure.

Tackling the fear of job displacement requires a nuanced approach. McKinsey elucidates that AI's role is not to replace but to release legal professionals from the tedium of routine tasks. This liberates them to invest their expertise in more intellectually stimulating and higher-value work. McKinsey's focus on upskilling and reskilling initiatives is a testament to its belief in the symbiosis between AI and human intelligence, ensuring that the workforce evolves in tandem with technological advancements.

The inertia of tradition within law firms often impedes the adoption of new technologies like AI. McKinsey's strategy involves engaging firm leadership to act as champions of change, driving the cultural shift necessary for technological integration. Success stories and case studies of AI in action become a powerful narrative in this journey, offering compelling evidence to dismantle resistance and foster an organizational mindset that is receptive to innovation.

Knowledge gaps pose a significant barrier to AI adoption. McKinsey addresses this by promoting comprehensive education and literacy in AI. Through their learning initiatives and knowledge dissemination, they equip legal professionals with the tools to understand and leverage AI in their practices. It is through such enlightenment that the latent capabilities of AI can be fully realized within the legal domain.

Accuracy and reliability are cornerstones of legal work, and any AI system employed must meet this threshold. McKinsey's strategy includes a meticulous evaluation of AI solutions, ensuring that only those with a proven track record are recommended. Pilot programs, which are a McKinsey hallmark, provide a controlled environment to test the waters with AI solutions, verifying their precision and efficacy.

The regulatory landscape for AI in the legal sector is in flux, and navigating this evolving terrain requires both legal acumen and technological savvy. McKinsey leverages its interdisciplinary expertise to provide guidance on regulatory compliance, ensuring that firms adopting AI remain on the right side of the law. By maintaining a finger on the pulse of regulatory changes and fostering dialogue with policymakers, McKinsey helps law firms anticipate and adapt to the regulatory shifts that accompany AI innovations.

Lastly, the technical challenge of integrating AI with existing legal systems and workflows is an area where McKinsey's expertise shines. With an understanding that each law firm operates within a unique ecosystem, McKinsey and QuantumBlack offer bespoke AI solutions. These are not off-the-shelf products, but tailored tools designed to mesh with the firm's existing practices, enhancing their functionality and efficiency.

Forging the Path Forward with AI

McKinsey's multifaceted approach to overcoming obstacles to AI adoption begins with a steadfast commitment from leadership. It requires building an innovative culture within law firms and fostering a deep understanding of AI benefits and applications. Forming strategic partnerships with AI providers who not only understand the legal domain but also offer robust support and training is essential to this process.

The role of pilot projects is instrumental in McKinsey's approach to AI adoption. By enabling law firms to engage with AI on a trial basis, firms are afforded the opportunity to see firsthand the practical benefits and to calibrate the technology to their specific needs. These pilots act as microcosms of the potential AI has to revolutionize practices, allowing firms to iteratively adjust their approach and fully understand the technology's impact before a wider rollout. McKinsey's emphasis on this step-by-step exploration mitigates the risk and allows for a smoother transition to AI-enabled operations.

Continual adaptation and learning stand at the heart of McKinsey's philosophy. The legal profession, like AI technology itself, is ever-evolving. As such, McKinsey advocates for ongoing education and training programs that ensure legal professionals remain at the cutting edge of AI advancements. This proactive stance is pivotal in ensuring that law firms do not merely adopt AI but become vanguards of AI innovation, harnessing new tools and methodologies that refine and redefine their service offerings.

Achieving Sustainable Growth with AI

The narrative of AI adoption in legal practices is multifaceted, requiring a strategic and informed approach to unlock its full potential. McKinsey & Company, leveraging QuantumBlack's expertise, offers more than a blueprint for integration; they provide a comprehensive partnership for navigating the journey toward innovation. This journey is characterized by a fusion of strategic foresight, technical excellence, and a deep understanding of the legal domain. It involves not only implementing cutting-edge AI solutions but also reshaping the culture within law firms to one that is agile, forward-thinking, and continuously evolving.

Moreover, the focus extends beyond the immediate transformation to envisage a future where AI continually drives growth, efficiency, and the capability to handle an increasingly complex legal landscape. McKinsey's holistic approach looks to the creation of sustainable, inclusive growth, ensuring that the benefits of AI are distributed widely across all levels of the firm, enhancing the quality and reach of legal services.

Leveraging Hybrid Intelligence

QuantumBlack's concept of hybrid intelligence, where the foresight of data and technology blends with the creativity and expertise of people, is particularly compelling. It presents a vision where AI is not a siloed entity but an integrated aspect of the legal process that complements and amplifies human intelligence. The result is a legal practice that is more dynamic, more precise, and more attuned to the nuances of both the law and the needs of clients.

QuantumBlack Labs: Pioneering AI Solutions

The pioneering spirit of QuantumBlack is embodied in QuantumBlack Labs, McKinsey's incubator for AI innovation. Here, interdisciplinary teams of technologists, designers, and product managers collaborate to develop the next generation of AI tools. These tools are designed not just for impact but for risk mitigation, ensuring that each stage of the AI transformation journey is grounded in stability and reliability.

In conclusion, the path to AI adoption in legal practices, while fraught with challenges, is navigable with the right partner. McKinsey & Company, through QuantumBlack, provides the depth of knowledge, strategic frameworks, and a proven track record of success to guide law firms through this transformative process. By embracing AI, law firms can position themselves at the forefront of the industry, equipped to deliver superior service and achieve sustainable, long-term success. With AI, the legal profession can transcend traditional boundaries, innovate smarter, and anticipate the future with greater confidence.

Leadership and Innovation: How Law Firms Can Leverage AI for Better Outcomes

The legal sector's exploration of artificial intelligence (AI) is a testament to the industry's commitment to innovation and excellence. Deloitte Insights and Accenture are at the forefront of this transformation, offering cutting-edge solutions and strategic consulting to help legal practices harness the power of AI. This extended narrative will unpack how these two global giants advocate for and facilitate the integration of AI, providing legal professionals with the tools they need to navigate the changes and capitalize on the opportunities that AI presents.

AI's potential in the legal field has sparked considerable interest, signaling a transformative shift in how legal work is conceptualized and delivered. Deloitte Insights, with its comprehensive guide for corporate legal departments, addresses the pressing needs of Chief Legal Officers (CLOs) who play a pivotal role in business outcomes, risk management, and cost control. CLOs are increasingly involved in company decisions from the onset, shaping legal strategies that are integrally linked to business success. Deloitte's approach is to view the law as a business enabler rather than a constraint, offering solutions that encapsulate this philosophy.

Generative AI, a particularly potent form of AI, stands at the heart of Deloitte's offerings. With large language models reaching a tipping point, the capability of Generative AI to process natural language and generate predictive models is proving to be a game-changer. Deloitte blends business acumen with legal expertise, providing comprehensive solutions that address the evolving demands of the legal department. This strategy involves a partnership model where Deloitte's legal professionals, who are also well-versed in technology, work alongside legal consultants and technologists. They bring a wealth of industry knowledge to the table, helping legal departments navigate the complex legal and business landscape.

Deloitte's global reach and proven approach enable CLOs to journey through unprecedented change and transformation. As legal departments grapple with the implications of emerging technologies, Deloitte stands ready with its tech-forward reputation and deep expertise in Generative AI. Their strategy is to partner with clients, providing support every step of the way, ensuring that AI adoption is not only strategic but also in line with the firm's broader objectives and culture.

Accenture, on the other hand, brings its own suite of AI capabilities to the table. They offer AI consulting services and solutions aimed at accelerating business objectives and promoting sustainable growth. Accenture's focus on data-led transformation and scalable AI solutions places it in a unique position to offer responsible and trustworthy AI services. According to Vishal Gupta of Everest Group, Accenture's offerings are infused with data and AI, enabling clients to derive higher value from their digital transformation initiatives.

Accenture has launched specialized services to help companies customize and manage foundation models. This includes a proprietary "switchboard" that can adapt to various business contexts based on factors such as cost, accuracy, or other specific needs. Services to customize large-language models (LLMs) for particular use cases are also part of Accenture's portfolio, ensuring that the AI is tailored to the unique needs of each legal practice. Alongside these offerings, Accenture provides training and skills to help clients effectively use and manage LLMs throughout their value chain.

Julie Sweet, chair and CEO of Accenture, emphasizes that Generative AI is accelerating reinvention and delivering business value at an unprecedented pace. The transition from experimentation to scaled implementation requires a robust data and AI foundation. Accenture's services help clients contextualize enterprise data, driving tangible value from Generative AI with their new foundation model customization services. The proprietary gen AI model "switchboard" and customization techniques are part of Accenture's strategy to ensure that AI services are not only sophisticated but also adaptable and client centric.

The specialized training programs offered by Accenture are designed to equip clients with the necessary skills to navigate the AI landscape effectively. Their collaboration with leading academic institutions, including the Stanford Institute for Human-Centered AI, has led to the creation of the Foundation Model Scholar Program. This initiative provides certifications related to large-language model skills, ensuring that legal professionals are prepared for the emerging demands of AI-driven legal practice.

In both Deloitte's and Accenture's strategies, the management of change is central. As law firms embark on the AI journey, the need for strong leadership is emphasized. Leaders must articulate a clear vision for AI adoption and foster a culture that values innovation and adaptability. They must also ensure that their teams are equipped with the necessary resources, training, and support to adapt to new methodologies and technologies.

The integration of AI into legal practice requires a balance of education, ethical conduct, and collaboration across various domains of expertise. As AI technologies are iteratively implemented, law firms can refine processes, minimize disruption, and align AI initiatives with their strategic objectives. Communication is pivotal, with transparent dialogues ensuring that everyone in the firm is aligned with common goals. Engaging clients in these conversations helps to set realistic expectations and build trust as the nature of legal services evolves with AI.

The ethical implications of AI are not to be overlooked. Law firms must adhere to the highest standards of conduct, assuring that AI applications within legal practices are transparent, unbiased, and respect clients' privacy and confidentiality. It is essential that as AI takes on a more significant role in decision-making processes, legal professionals are equipped with skills in AI ethics, further ensuring the responsible use of technology.

In managing the risks associated with AI adoption, it is crucial for law firms to develop strategies that address challenges such as data security, compliance with regulations, and biases in AI algorithms. Proactive risk management is key to maintaining the trust of clients and the integrity of the legal profession. This means implementing safeguards that prevent pitfalls that might otherwise undermine the benefits of AI.

To gauge the effectiveness of AI initiatives, law firms should establish performance metrics and benchmarks. These tools are indispensable for demonstrating the return on investment (ROI), refining AI strategies, and maintaining momentum towards more advanced deployments. By systematically evaluating the impact of AI, law firms can make informed decisions that optimize their practices and ensure continued growth and innovation.

Deloitte and Accenture both showcase how AI can be tailored to the unique contexts of different legal practices, emphasizing the importance of customization in the process. The bespoke approach to AI integration is vital in addressing the specific needs and challenges of each law firm. Whether it's through the use of Deloitte's business solutions or Accenture's model switchboard, the aim is to ensure that AI tools enhance the practice rather than disrupt it.

Accenture's model, in particular, highlights the importance of flexibility. By allowing firms to select from a range of AI models, they can find the perfect fit for their business context, balancing cost, accuracy, and the specific demands of their practice. This kind of customization is invaluable in a field as diverse and complex as the law, where the nuances of each case and each client relationship can greatly affect the desired outcomes.

The emphasis on training and skills development is another critical aspect shared by both companies. Accenture's collaboration with academic institutions to create certification programs is a forward-thinking approach to building AI competency within the legal field. These programs aim to equip legal professionals with the deep understanding of AI necessary to leverage its full potential while maintaining ethical standards and client trust.

In conclusion, managing the change that comes with AI integration into legal practices is a multifaceted challenge that requires a holistic approach. It involves leadership vision, education, collaboration, communication, ethics, and risk management strategies. By adopting best practices from industry leaders like Deloitte and Accenture, law firms can not only navigate the complexities of AI adoption but also position themselves to harness AI's full potential. The result is a legal practice that is more efficient, innovative, and aligned with the evolving expectations of the 21st-century client.

As AI continues to permeate the legal sector, it is clear that the firms that will excel are those that can adapt and innovate. By embracing AI as a critical component of their service delivery, these firms will not just keep pace with technological advancements—they will be at the forefront, shaping the future of the legal industry. The integration of AI into legal practices is an ongoing process, one that will continually evolve as the capabilities of AI grow and the needs of clients change. Firms that strategically manage this change, drawing on insights and services from companies like Deloitte and Accenture, will be well-equipped to meet the demands of the digital age and provide top-tier legal services that reflect the dynamism and innovation inherent in today's world.

Future-Proofing Law Firms: A Strategic Approach to AI Empowerment

As we encapsulate the voyage through AI's landscape in legal practice, it's evident that the path toward technological empowerment requires a dedicated balance of innovation and prudence. The final chapter in our journey underscores not merely the adoption of new tools but the embracing of a mindset that welcomes digital transformation with strategic action.

In this closing section, "Strategic Pathways to AI Empowerment in Law," our narrative weaves the lessons of foresight and collaboration into a tapestry of opportunity. The legal sector's venture into AI is not about the cold adoption of algorithms but about fostering a harmonious partnership between human wisdom and machine intelligence. It's about setting the stage for a future that values enhanced legal service, client engagement, and the relentless pursuit of knowledge.

As we turn the page from understanding the intricacies of AI integration to putting these strategies into practice, we stand ready to address the final and perhaps most crucial steps in this transformative process. The following paragraphs will guide you through the tangible actions, the rigorous standards, and the dynamic leadership necessary to not only navigate but also thrive in the era of AI. Let us move forward with confidence, propelled by the knowledge that the strategies laid out here are not just theoretical musings but the groundwork for a legacy of innovation in law.

Education and Skills Development

The metamorphosis of legal practices through AI is inextricably linked to the education and enhancement of skills within the legal workforce. Confronting the skills gap that exists between the current legal expertise and the technical knowledge required for AI, law firms must pursue vigorous educational strategies. This involves not only formal training but also a cultural shift towards continuous learning and adaptability. It is through such strategic educational initiatives that law firms can transcend the initial trepidations of AI and uncover its true potential as a collaborator to human intellect, rather than a competitor.

In this vein, we witness the initiatives undertaken by firms like QuantumBlack, which illustrates how hands-on workshops and partnerships with tech innovators can demystify AI. By aligning themselves with academic powerhouses and incorporating AI tools into legal curriculums, these firms blaze a trail for a new breed of legal professionals—ones who are as comfortable with algorithms and data analytics as they are with case law and statutes.

Technical Challenges and Data Security

As with any technological adoption, the introduction of AI into legal practices presents a myriad of technical challenges, with data security standing at the forefront. Law firms handle a wealth of sensitive information, making robust cybersecurity a non-negotiable aspect of AI integration. This requires a meticulous approach, from the initial encryption of data to the constant vigilance against ever-evolving cyber threats.

Companies like QuantumBlack, which operates at the intersection of strategy, technology, and design, provide compelling examples of how AI can be implemented securely. They showcase the development of bespoke cybersecurity protocols that not only safeguard against data breaches but also engender trust among clients and stakeholders. By implementing rigorous security measures and adhering to the highest standards, these firms not only protect their clients' data but also reinforce their reputation for integrity and reliability in the digital age.

Ethical Considerations and Bias

In an industry governed by the principles of justice and fairness, the ethical considerations surrounding AI cannot be overstated. Concerns over inherent biases in algorithms and the need for transparency and accountability are paramount. Law firms are, therefore, tasked with the critical job of crafting ethical frameworks around the use of AI—guidelines that ensure every AI application is vetted for fairness, devoid of prejudice, and transparent in its decision-making processes.

The insights from McKinsey & Company are instructive here as they highlight the importance of an ethical AI framework. They advocate for the creation of multi-disciplinary ethics committees, which include legal experts, technologists, and ethicists, to oversee AI implementations. Such committees evaluate AI applications not just for their technical performance but also for their adherence to ethical standards. They are tasked with ensuring that AI decision-making aligns with societal values and legal principles, safeguarding against biases that could erode the trust in legal processes.

Overcoming the Skills Gap

The disconnect between legal acumen and technical know-how represents a significant barrier to AI adoption. To bridge this divide, forward-thinking law firms are reimagining their approach to professional development. Internal training programs are tailored to enhance AI literacy among legal professionals, ensuring that they possess the skills necessary to leverage AI in their practices effectively.

Case studies from leading legal firms demonstrate the effectiveness of such training initiatives. For example, a global firm's partnership with an AI technology provider led to the development of an AI certification program for its attorneys. This program enabled lawyers to gain hands-on experience with AI tools, fostering a deeper understanding of their practical application in legal work and allowing the firm to capitalize on AI's efficiencies.

Incremental Approach and Strong Leadership

Adopting AI in a legal setting is not a switch to be flipped but a path to be traversed with intentionality and foresight. The experiences of pioneering firms reveal that an incremental approach, championed by strong and visionary leadership, is crucial for success. Firm leaders must not only articulate a clear AI vision but also embody the cultural change, inspiring their teams to embrace AI through their actions and commitment.

A law firm that exemplifies this principle began its AI journey with a small, dedicated innovation team. This team worked on specific, manageable projects, demonstrating quick wins and tangible benefits that gradually won over AI skeptics within the firm. By leading from the front and showcasing the value of AI in real terms, firm leaders were able to facilitate a firm-wide cultural shift towards embracing AI technologies.

The Role of Pilot Projects

Pilot projects serve as a microcosm for AI's potential in legal practice. They provide a sandbox environment for testing, learning, and adjusting AI applications before scaling them across the firm. Successful pilot projects not only prove the concept but also help refine the AI tools to the specific nuances of the firm's operations.

For instance, a leading law firm's pilot project with a contract analysis AI tool resulted in a significant reduction in time spent on due diligence, without compromising accuracy. This initial success led to broader adoption and a strategic rethinking of the firm's due diligence processes, transforming a historically labor-intensive task into an efficient, AI-driven operation.

Sustainable and Inclusive AI Growth

The ultimate goal of AI adoption in law firms is to foster sustainable and inclusive growth. The idea is to integrate AI in such a way that it benefits all stakeholders—lawyers, staff, and clients alike. By leveraging AI, firms can enhance service delivery, increase efficiency, and free up their legal experts to focus on the complex, value-added aspects of legal work.

A case in point is a firm that implemented an AI-driven legal research tool. This tool not only improved the speed and comprehensiveness of research but also allowed junior associates to contribute more substantively to cases, thereby fostering a more inclusive and collaborative environment.

Conclusion: Embracing the AI Journey in Law Firms

In concluding Chapter 19, we synthesize the myriad lessons gleaned from firms that have successfully navigated the AI landscape. It is clear that the journey towards AI adoption in legal practices is multifaceted and ongoing. By considering the insights from industry leaders and embracing a holistic strategy that includes education, ethics, incremental integration, and strong leadership, law firms can surmount the challenges of AI.

This chapter serves as a blueprint for law firms to not only implement AI but to do so in a way that aligns with the broader themes of efficiency, productivity, and innovation that are woven throughout this book. As firms continue on their AI journey, they must remain agile, continuously learning and adapting to new developments in AI technology and the evolving needs of their clients.

Ultimately, the transformative potential of AI in legal practices is immense, offering a new horizon for the industry. The law firms that will thrive are those that can look beyond the challenges, recognize the opportunities, and navigate the complexities of AI adoption with strategic foresight and a commitment to excellence.

Chapter 20:

Solo Legal Practices and AI: A Path to Enhanced Efficiency with Legal Tech

In an evolving legal landscape where the adoption of artificial intelligence (AI) has become a linchpin for operational efficiency and competitive advantage, solo practitioners might feel daunted by the perceived complexity and cost associated with these technologies. However, the integration of AI into the workflow of the solo legal practitioner not only demystifies this groundbreaking technology but also presents an invaluable opportunity to streamline operations, enhance the accuracy of legal research, and foster stronger client relationships. As we pivot towards a more technologically savvy legal arena, the imperative for solo practitioners to harness AI's potential grows increasingly critical, promising to level the playing field with larger firms by enabling more focused, effective, and personalized legal services.

The toolkit for the modern solo practitioner extends far beyond traditional legal research databases and practice management software. AI-driven tools now offer capabilities ranging from predicting legal outcomes to automating routine document analysis and management tasks, freeing up invaluable time for attorneys to concentrate on strategy and client care. Such tools not only augment the efficiency and effectiveness of legal research but also open the door to predictive analytics, where AI's ability to process vast datasets can provide insights into case outcomes, judge biases, and more. This revolution in legal technology empowers solo practitioners to deliver services with a degree of sophistication and insight that was once the exclusive domain of their larger counterparts.

Accessibility and affordability are central to the adoption of AI technologies among solo practitioners. Recognizing the budget constraints and resource limitations typical of smaller practices, developers and service providers are increasingly offering scalable AI solutions. These range from subscription-based models to 'pay-as-you-go' options, allowing for the customization of services to the specific needs and capacities of the solo practitioner. Furthermore, the current trajectory of legal technology innovation indicates a future where AI tools become ubiquitous and indispensable across the legal profession, ensuring that investments made today will yield long-term benefits. Through strategic adoption and utilization of AI, solo practitioners can not only enhance their practice's efficiency and service quality but can also position themselves at the forefront of legal service innovation, ready to meet the challenges and opportunities of a rapidly evolving professional landscape.

Transforming Solo Legal Practices with Smokeball's AI Tools

In the era of technological transformation, solo legal practitioners stand at a pivotal juncture where embracing artificial intelligence (AI) is not just advantageous but essential for the sustenance and growth of their practices. AI emerges not only as a trailblazer of innovation but as a partner that redefines productivity and service delivery in the solo practice arena.

The advent of AI in the legal industry marks a significant shift from traditional practice methods to an automated, efficient, and client-centered approach. For solo lawyers, this means an unprecedented opportunity to enhance the caliber of their services without proportionately increasing operational costs. Smokeball's AI-enhanced case management software is at the forefront of this revolution, serving as an indispensable tool that automates daily activities and document management, boosting the solo lawyer's productivity substantially.

Smokeball's first-to-market features, including Intake.ai, Ask.ai, and AutoTime.ai, signify a turning point in legal case management. Intake.ai revolutionizes client onboarding, transforming even paper forms into digital assets with ease and efficiency. This pivotal feature enables lawyers to handle intake faster than ever, circumventing the hassle of manual data entry and the errors that come with it.

At the core of Smokeball's AI capabilities lies Ask.ai. This powerful amalgamation of three AI engines permits lawyers to delve into their own matters and documents to extract concise answers, comprehensive summaries, and facilitate the drafting of legal documents within seconds, all within the ecosystem of their Smokeball matters. This feature symbolizes the zenith of legal research and writing efficiency, offering instant access to relevant matter information and summaries of extensive documents.

Smokeball's AutoTime.ai represents a paradigm shift in time tracking and billing. No longer is timekeeping a dreary task; it's an automated process that not only records tasks and duration but also crafts beautifully tailored timesheet descriptions, blending accuracy with eloquence.

Smokeball's AI tools are constructed with the solo practitioner in mind. They harness AI to elevate existing intake data, transforming it into customizable digital forms expeditiously. The elimination of paper forms and PDFs is not a mere convenience; it is a substantial leap towards quality control over client data and service excellence.

Moreover, Smokeball Ask.ai is not just a feature; it's akin to a research assistant that works with unerring accuracy and tireless zeal. By providing the ability to draft communications directly connected with Outlook, it ensures secure access and ease of use that seamlessly integrates into the lawyer's workflow.

As for AutoTime.ai, this feature is a testament to the sophistication that AI brings to timekeeping. It's not just about recording what was done and how long it took; it's about creating a narrative that reflects the value of the work undertaken. The software's precision in generating the most accurate and automatic timesheet globally is a promise of time saved and an enhanced client experience.

Hunter Steele, Smokeball's CEO, aptly remarks, "AI won't replace you, but those who use it will soon eclipse those who don't." This statement encapsulates the essence of AI in legal practice — a tool not of replacement, but of empowerment and distinction.

Smokeball's AI tools forge a path to better law firm productivity. With Intake.Ai, the drudgery of dealing with paper forms is obliterated, replaced by swift, digital solutions that save hours for both lawyer and client. The Ask.Ai feature redefines efficiency, allowing for rapid answers, summarization of legal matter documents, and drafting of correspondence. AutoTime.Ai, with its unrivaled accuracy, provides automated, tailored descriptions that dovetail with the work performed, encapsulating the nuances of every case in eloquent narratives.

These tools epitomize the synergy of AI and legal expertise, where technology accelerates the pace of legal practice without compromising the personal touch that is the hallmark of a solo practitioner's service.

LawDroid: Elevating Solo Practices through Smart Automation

Moving to the other side of the AI spectrum, LawDroid presents itself as the quintessential partner for solo practitioner keen on scaling their practice in a digital age where immediacy and efficiency are non-negotiable. It is a testament to the transformative power of AI legal assistants that not only augment productivity but redefine client engagement and service scalability.

LawDroid offers a no-code AI creation platform that empowers lawyers to automate client interactions and routine inquiries through legal chatbots. This level of automation means more than just efficient client service; it is about liberating the lawyer's time to concentrate on complex, nuanced tasks that necessitate deep human expertise — a rebalancing of workload that aligns with the evolving needs of the legal marketplace.

Supercharging a law firm with AI legal assistants from LawDroid is akin to a productivity multiplier, effectively enabling lawyers to undertake more work without incurring additional overhead. These AI assistants can be utilized for a plethora of tasks that range from researching case law to summarizing documents and drafting communications, providing invaluable support to solo practitioners who often juggle multiple roles simultaneously.

LawDroid's key product, LawDroid Copilot, functions as an always-on aid, assisting in tasks such as legal issue research, document summarization, and correspondence drafting. This AI assistant can dramatically increase a lawyer's productivity, ensuring they can stay on top of their workload and maintain a healthy work-life balance.

LawDroid Builder, on the other hand, offers a no-code platform allowing solo practitioners to encapsulate their expertise into scalable legal services. This innovative tool lets lawyers build chatbots to handle phone calls, capture leads, and even facilitate legal document self-service, thus extending their service offerings to clients even outside office hours.

Intake of new clients is streamlined through engaging and interactive chatbots that capture vital information, automatically creating new leads and contacts in the case management system. This approach not only saves time but also enhances the client experience by making the intake process more engaging.

Moreover, LawDroid's analytics provide in-depth insights into client behaviors and preferences, enabling data-driven decisions that can refine and personalize the client experience further. The transformation of Word documents into dynamic templates imbued with conditional logic exemplifies how AI can tailor document creation to the specific needs of clients.

In instances where a human touch is necessary, LawDroid ensures a seamless transition with its 'Human in the Loop' system, enabling practitioners to take over automated conversations and ensuring that the personal aspect of legal service is never lost.

Building scalable tools through LawDroid also means that solo practitioners can monetize their expertise by offering self-serve legal documents and other services directly through their websites or the LawDroid platform. Such scalability was previously unattainable for many solo practitioners, but with AI, it's a tangible reality.

The power of natural language processing enables LawDroid to provide immediate, accurate answers to frequently asked legal questions, freeing up valuable time for the solo practitioner. This level of automation means clients get the information they need without delay, enhancing satisfaction and the likelihood of engagement with the practice.

Testimonials from industry professionals further solidify LawDroid's impact. Frances Wipf, an Immigration Consultant at The Way Immigration, speaks to the trust and client conversion facilitated by LawDroid: "When we provide value to people up front, instantly, at no cost, it builds trust, and they are more likely to turn into paying clients."

Patrick Palace, Principal Attorney at Palace Law, encapsulates the value proposition of LawDroid Copilot: "I was going to hire a paralegal, but after trying out LawDroid Copilot, I now have the help I need." This sentiment resonates with the solo legal practitioner's desire to maximize efficiency without the additional cost of hiring more staff.

Solo practitioners harnessing the power of AI through tools like Smokeball and LawDroid are not only optimizing their work processes but are also carving a new paradigm in legal service delivery. These technologies level the playing field, allowing solo lawyers to provide services that once required the backing of a full-fledged firm.

Embracing such AI tools involves a willingness to engage with new technologies and embark on a continuous learning journey. This investment into understanding and utilizing AI within legal practices promises a future rich with efficiency, precision, and growth potential.

While cost considerations and the learning curve associated with AI adoption may seem daunting, the return on investment is compelling. Many AI tools are offered on scalable subscription models, fitting various budget sizes, and the benefits of efficiency and client satisfaction far outweigh the initial costs.

The commitment to client privacy and data security remains at the heart of AI tool selection for legal practices. Ensuring adherence to stringent data protection standards is not only a legal requirement but a core element of maintaining the integrity and trust inherent in lawyer-client relationships.

Navigating ethical considerations around AI usage in legal practice demands conscientious decision-making, ensuring justice and fairness are upheld. Transparent communication with clients about AI integration aids in setting proper expectations and building a foundation of trust.

To seamlessly integrate AI into their practices, solo legal professionals must remain abreast of the latest advancements in legal technology. Networking with peers, engaging in professional development, and staying connected with industry trends are all crucial steps in harnessing AI effectively.

The journey toward AI empowerment in solo legal practices is not merely about adopting new technology; it's a transformative process that leverages innovation to provide exemplary legal services. In an increasingly competitive landscape, AI offers solo practitioners the tools to excel, delivering quality, efficiency, and accessibility that redefine what is possible for a solo legal practice.

As AI continues to weave itself into the fabric of legal services, its integration within solo practices is set to expand further, catalyzing a transformative effect on the profession. For the forward-thinking solo practitioner, the embrace of AI is not just a strategic move but an imperative one, keeping them ahead in the legal services domain. With the correct application of AI tools like Smokeball and LawDroid, solo practitioners are poised to revolutionize their practices, enhance client relations, and achieve new heights of success.

The future of solo legal practice through the lens of AI is not only bright but also revolutionary. The integration of AI tools is paving the way for a new breed of legal professionals who are efficient, scalable, and technologically empowered. As solo practitioners navigate this landscape, they find that AI does not replace the lawyer but rather, enhances the lawyer's capabilities, allowing them to offer more to their clients and to the legal profession as a whole.

The capabilities of Smokeball, with its advanced case management and productivity tools, and LawDroid, with its client interaction automation and scalable services, are integral to this journey. They are not merely tools but catalysts for change, enabling solo practitioners to reimagine how legal services are delivered and how clients are engaged.

In summary, the empowerment of solo practices through AI tools such as Smokeball and LawDroid reflects a profound shift in legal services. It is a movement from solo practitioners working harder to working smarter, with technology as a cornerstone of their strategy. These tools embody the essence of modern legal practices: a harmonious blend of technology and human expertise where each complements the other, leading to a service that is greater than the sum of its parts.

Solo practitioners equipped with AI are setting new standards in legal service delivery, standards that resonate with efficiency, client engagement, and excellence. As AI becomes more sophisticated and integrated into everyday practice, the solo practitioner will not only keep pace with larger firms but may also set the pace, leading innovation and redefining what it means to practice law in the 21st century.

For solo legal practices, the AI revolution is not coming; it is already here. It's a matter of embracing the change, harnessing the potential, and leading the charge towards a future where technology and law converge to create a better, more accessible, and more efficient legal landscape for all.

Clio: Democratizing AI for Small Law Firms and Solo Practitioners

In the current technological renaissance, small law firms and solo practitioners are increasingly looking to leverage Artificial Intelligence (AI) to augment their practices, aiming for higher efficiency, precision, and client satisfaction. The key lies in demystifying AI and ensuring its benefits are within reach, a challenge Clio has risen to meet with exceptional prowess.

At its core, Clio serves as a beacon for small firms, illuminating a path to AI-driven insights that were once the exclusive domain of larger counterparts. Clio's comprehensive cloud-based platform revolutionizes the day-to-day operation of legal professionals, encapsulating client intake, case management, and intricate financial matters into a centralized system that is as intuitive as it is powerful.

For the solo legal practitioner, Clio is a game-changer. The platform streamlines case and document management with remarkable fluidity, allowing for the automation of time tracking, billing, and scheduling—all tasks that, when done manually, drain precious hours that could be devoted to legal analysis and client engagement. Clio's infrastructure is robust yet flexible, catering to the multifaceted nature of legal work with a suite of tools designed to interface seamlessly with each other.

With Clio Manage, lawyers find themselves at the helm of a sophisticated law firm management software, commanding features that address every facet of practice management. From the meticulous organization of case files to a client communication portal that connects lawyers and clients with transparency and ease, Clio Manage addresses the quintessential needs of the modern attorney.

Clio Grow, on the other hand, extends the functionality to the very beginning of the client journey. As a client intake and legal CRM tool, it refines the process of lead generation, capturing prospective clients through tailored intake forms and automated follow-ups. This preemptive component of Clio ensures that no opportunity for new business slips through the cracks.

The financial realm is addressed with equal finesse by Clio Payments, an integrated payment processing solution that embodies convenience and compliance. By offering means to accept payments via credit card and ACH, and automating the reconciliation process, Clio Payments not only expedites cash flow but also instills confidence in clients regarding the security of their transactions.

Collectively, the offerings under the banner of Clio Suite exemplify the synergy of Manage and Grow. They create a unified experience where the intricacies of legal practice management are distilled into a dashboard that delivers insights and analytics—insights that inform decision-making and enable small firms to punch above their weight class in a competitive legal market.

Clio's benefits aren't purely functional; they resonate with the foundational goals of any small firm—efficiency and scalability. By automating daily tasks, Clio grants legal professionals the gift of time—time that can be redirected to sophisticated legal reasoning and personalized client care that define the soul of a successful practice. Furthermore, its cloud-based nature obliterates geographical constraints, empowering attorneys to operate and thrive in a remote work landscape that has become the new normal.

Security, a paramount concern within the legal profession, is addressed with rigor. Clio adheres to stringent data protection standards, ensuring client information is guarded with the highest levels of encryption and security protocols. In doing so, it nurtures an environment where client confidentiality is sacrosanct.

For small firms intent on embarking on an AI journey, Clio is not merely a tool but a strategic partner that aligns with their aspirations and scales with their growth. It provides a technological infrastructure that is not only advanced but also accessible, without the prohibitive costs or complexity that often preclude small practices from harnessing the power of AI.

Empowering Clients through AI: Hello Divorce's Approach

As the legal industry evolves, small law firms and solo practitioners specializing in family law are presented with a unique opportunity to redefine the client experience. Hello, Divorce emerges as a paragon in this space, exemplifying how AI and technology can streamline the divorce process, making it more comprehensible and less intimidating for those navigating one of life's most challenging passages.

Hello, Divorce's approach is decidedly human-centered, leveraging AI to facilitate a smoother divorce process. The platform's suite of self-help tools democratizes legal information, offering a plethora of DIY resources that empower individuals. From downloadable forms and guides to educational content, Hello Divorce demystifies the divorce process, enabling users to take control of their legal journey with confidence.

Legal professionals can draw inspiration from Hello Divorce's model, integrating similar tools into their practice to offer transparent, straightforward guidance. This not only enhances the firm's value proposition but also distinguishes its services in a saturated market. For clients, the clarity and support provided by such tools can be the differentiating factor in their choice of legal representation.

Document preparation, often a source of confusion and stress for clients, is handled with precision by Hello Divorce. With their expertise, they ensure that every form is meticulously completed, reviewed, and ready for proceedings. Small firms can incorporate these principles into their workflow, utilizing AI for document automation to increase accuracy and reduce turnaround times, thereby enhancing client satisfaction.

The personal touch of legal advice and representation is not lost in Hello Divorce's tech-centric approach. By offering various levels of legal support, from on-demand advice to full representation, Hello Divorce creates a scalable model of service that can be mirrored by small firms. This modular approach allows firms to provide personalized attention where it is most needed, ensuring that clients feel heard and understood.

Membership plans are another innovative feature of Hello Divorce. These plans offer continuous support and resources, a concept that small firms can adopt to maintain client engagement and loyalty. Through memberships, firms can offer exclusive content, tools, and discounted services, fostering a long-term relationship that extends beyond the resolution of a single case.

Settlement agreements and mediation services offered by Hello Divorce underscore the potential for technology to facilitate amicable resolutions. Small firms can emulate this by adopting AI tools that aid in drafting agreements and conducting mediations. This not only expedites the resolution process but also positions the firm as a peacemaker, aligning with the growing preference for alternative dispute resolution.

The transparency of flat-fee pricing, a standout feature of Hello Divorce, introduces predictability into the often uncertain financial aspect of legal services. Small firms can take a cue from this approach, providing clear and upfront cost structures that align with clients' expectations for financial clarity.

The impact of Hello Divorce on the legal industry extends beyond its offerings. It serves as a testament to the profound difference AI and thoughtful tech implementation can make in the lives of individuals. By simplifying complex legal procedures and fostering a supportive environment, Hello Divorce not only changes how clients experience divorce but also challenges small firms to reimagine their role and service delivery in the context of family law.

The adoption of AI and technology solutions from pioneers like Hello Divorce marks a progressive step for small firms. It allows them to offer nuanced, efficient, and client-focused legal services that resonate with the modern consumer. In a field fraught with emotional and logistical complexities, small firms that harness such technologies stand to deliver a caliber of service that is compassionate, clear, and streamlined.

In the grander scheme, the integration of Hello Divorce's philosophies and tools into small legal practices signifies a shift towards a more accessible and humane legal system. It encourages small firms to engage with legal tech to not only enhance operational efficiency but also elevate the standard of care provided to clients. The technologies espoused by Hello Divorce can serve as a blueprint for small firms looking to invest in AI, ensuring their offerings are as sophisticated as they are sympathetic to the needs of those they serve.

In a landscape where technological advancement is relentless, the example set by Hello Divorce offers invaluable insights for small firms. It represents a convergence of innovation, empathy, and legal expertise, pointing towards a future where technology is not a barrier but a bridge—connecting clients to clearer, kinder, and more equitable legal solutions.

Chapter 21:

Legal AI: Architecting Data Privacy and Trustworthiness

OneTrust: Redefining Legal AI with Trust Intelligence

As we delve into the intricacies of data privacy and security in legal AI applications, it is imperative to acknowledge the revolutionary contributions of OneTrust. This section unfolds the story of how OneTrust has been a frontrunner in equipping AI applications with the tools necessary to comply with a kaleidoscope of global privacy laws, thus fortifying the bedrock of compliance and trust.

OneTrust emerges as a paradigm of innovation, offering a Trust Intelligence Platform that transcends conventional compliance measures. By fostering an environment where trust catalyzes innovation, OneTrust has dismantled the barriers that once hindered technological advancement within the legal sector. Their platform is not a mere tool; it is a harbinger of transformation, infusing trust into the DNA of businesses, thereby propelling them towards a horizon of boundless potential.

The platform's architecture is an alchemy of four foundational clouds integrated seamlessly into one comprehensive ecosystem. This integration is meticulously designed to simplify the metamorphosis of a firm's trust posture across pivotal domains: privacy and data discovery, Governance, Risk, and Compliance (GRC), ethics, and Environmental, Social, and Governance (ESG). With this multi-faceted approach, OneTrust safeguards the sanctity of legal AI applications, ensuring they are robust enough to withstand the tempest of cyber threats and global crises. The platform's GRC and Security Assurance cloud, in particular, scales risk and security functions to maintain resilience, instilling confidence in firms to operate without trepidation.

The journey towards trust transformation is laden with challenges, with the establishing and nurturing of trust at its core. OneTrust, through its innovative technology, embeds trust by design across businesses. Supported by a bedrock of regulatory intelligence and a wealth of over 300 patents, their platform empowers firms to action data responsibly, transcend compliance, identify and mitigate risk, and erect a trust program tailored to their unique ethos.

With OneTrust, legal practitioners can now navigate the labyrinth of data privacy with a guide that is not only adept but also anticipatory of the evolving landscape. Backed by this pioneering platform, legal professionals can usher in an era where client data is not just protected but its integrity unassailable. This evolution marks a pivotal shift in the paradigm of legal AI applications, where the dual sentinels of privacy and security are no longer mere regulatory checkboxes but are enshrined as the guardians of trust.

The role of OneTrust in transforming legal AI applications is thus not just functional but foundational. By prioritizing the trustworthiness of AI systems, OneTrust ensures that these systems are crafted to meet the highest echelons of privacy standards. It is through this dedication to excellence that legal AI applications can achieve the zenith of both innovation and integrity.

Adaptive AI Compliance: TrustArc's Global Perspective

Continuing our exploration into the realm of data privacy and security within legal AI applications, the spotlight now shifts to TrustArc. Their role is pivotal when dissecting the geographical quilt that makes up the legislative frameworks, illustrating their expertise in sculpting AI solutions that are agile enough to comply with the ever-changing global legal standards such as the GDPR and CCPA.

TrustArc stands as a beacon of adaptability and compliance, offering a suite of solutions that provide not only navigation but automation and certification of compliance in a complex legal landscape. Their pioneering Responsible AI Certification signifies a commitment to data protection and governance, which is becoming increasingly crucial as organizations seek to demonstrate their dedication to ethical data practices.

At the heart of TrustArc's offerings is the enhancement of brand trust through AI. By pursuing TrustArc's Responsible AI Certification, organizations can publicly affirm their commitment to principled AI deployment. This certification integrates the foundational principles from industry standards such as NIST and OECD, as well as emerging AI regulations like the EU AI Act. TrustArc recognizes that the wave of AI privacy regulations is formidable but believes that with a robust AI governance framework, organizations can navigate these waters fearlessly, showcasing a commitment to data protection and privacy in their AI applications.

The process instituted by TrustArc is methodical and reflective of their vast experience in the field. It begins with a phase of discovery and evaluation, where the organization's AI data governance practices are thoroughly assessed by an adept member of TrustArc's Global Privacy Solutions team. This meticulous assessment, grounded in a proven methodology and bolstered by powerful technology, lays the groundwork for a comprehensive understanding of an organization's privacy posture.

Following this, a gap analysis is conducted, offering organizations a detailed report with actionable recommendations, ensuring that compliance is not just a destination but a journey of continuous improvement. The insights for remediation that TrustArc provides are rooted in years of privacy expertise, complementing the gap analysis with practical steps for enhancing privacy and security postures.

The path to certification is further smoothed by the provision of essential operational templates and an accessible audit trail through TrustArc's platform. This level of transparency and organization is invaluable for responding to inquiries and maintaining audit compliance with finesse. Once the rigorous certification process is completed, the organization is awarded the coveted TRUSTe Seal along with a Letter of Attestation, symbols of their steadfast commitment to privacy and data governance.

TrustArc's support does not culminate with certification. They offer continuous guidance, including annual reviews and third-party dispute resolution services, to mediate privacy issues that may arise. This ongoing support encapsulates their belief in the importance of not just achieving but maintaining and evolving privacy standards in legal AI applications.

The presence of TrustArc in the legal AI landscape is a testament to the possibility of seamless adaptation to diverse legal standards while maintaining the integrity of client relationships. Their solutions provide a compass for legal professionals to navigate the seas of global privacy laws, ensuring that the trust placed in them by clients remains unshaken.

In summation, the harmonious integration of AI into legal practices is contingent upon the existence of robust privacy and security frameworks. Platforms like OneTrust and TrustArc serve as the architects of these frameworks, ensuring that as legal professionals reach for the innovative stars, they remain firmly grounded in the ethical soil of data privacy and security. This dual allegiance to advancement and responsibility is what will ultimately chart the course for the successful and trustworthy incorporation of AI in the legal realm.

Defending Data Integrity in Legal AI Practices

In the rapidly evolving landscape of legal services, artificial intelligence (AI) has emerged as a beacon of innovation, driving efficiency and redefining the traditional methodologies of legal practice. Yet, the shimmering promise of AI is not without its shadows, particularly in the critical domain of data privacy and security. For legal professionals, the sanctity of client data is sacrosanct, a principle that holds paramount significance as we venture into the integration of AI within our operations. The imperative to protect our clients' confidential information is inexorable, extending with undiminished force into the realm of AI projects, where every byte of data must be shielded with meticulous care.

As we harness the capabilities of AI in legal practice, we enter a world abundant with data, an expanse that stretches across the horizon with its repository of personal details, corporate secrets, and sensitive legal communications. The journey to safeguard this data begins with a profound understanding of the distinct risks posed by AI technologies, a spectrum that spans inadvertent exposures to orchestrated attacks by nefarious cyber entities.

Central to the edifice of data protection is the establishment of ironclad data governance policies. Within the framework of AI, data governance transcends into a meticulous choreography of access control and data stewardship. Legal practitioners must delineate the bounds of data interaction within AI systems, discerning the authority to access, the conditions for manipulation, and the methods of engagement with the precision of a master strategist.

In this digital armory, encryption stands as a stalwart guard. As data traverses the cyber terrains and nestles in digital repositories, encryption serves as an impermeable veil, ensuring that, in the event of a breach, the information remains an enigmatic puzzle to unauthorized entities. The selection of encryption algorithms is a delicate balance between the fortress of security and the agility of AI systems, a decision that demands both wisdom and foresight.

Vigilance in the form of auditing and monitoring forms the sinew of our defense strategy. The dynamic landscape of AI projects necessitates a watchful eye for anomalous patterns, signaling potential breaches. Audits, conducted with relentless regularity and incisive scrutiny by internal and external cybersecurity experts, illuminate vulnerabilities within the AI infrastructure, allowing for pre-emptive fortification.

Adherence to data protection laws, such as GDPR and CCPA, is a cornerstone of our ethical and legal mandate. The legal professional's pilgrimage through the complex terrain of regulatory compliance is arduous but fundamental to upholding the covenant of trust with our clientele. The legal frameworks may be intricate, and their reach extensive, but deviation is not an option in our quest to fortify client confidence.

The selection of technology vendors for AI collaboration is a pursuit marked by meticulous assessment. Legal entities must engage with partners whose allegiance to data privacy and security mirrors their own. This requires a probing evaluation of the vendors' security pedigree, a history of their fortitude against breaches, and their policies on data management that endure beyond the terminus of partnership.

Transparency and client consent are the bedrock upon which the edifice of attorney-client trust is constructed. As AI becomes an integral part of legal service, clients must be enlightened about the stewardship of their data—the protective measures in place, their rights, and the consent they grant for AI to process and analyze their information. This candor is the foundation of unshakeable trust.

Preparation for the worst-case scenario—a data breach—is an acknowledgment of the fallibility of even the most secure systems. Incident response protocols, meticulously crafted and routinely refined, articulate the tactical and communicative maneuvers to be executed in the face of a breach, aiming to preserve the integrity of client trust.

A culture of security within the legal practice is not an optional adornment but an essential fabric of our professional ethos. From the paralegals to the partners, every member of the firm contributes to the bastion of client data protection. Education and awareness initiatives reinforce this culture, ensuring that the principles of data security are not just understood but are woven into the daily tapestry of legal practice.

The complexities of protecting client data in AI projects are as formidable as they are critical. By comprehending the risks, erecting robust security measures, and nurturing a culture of transparency and trust, legal professionals can navigate this intricate landscape. Our objective transcends mere protection; it is to cultivate an environment where AI enhances the very essence of legal integrity and trust.

The commitment to data privacy and security is the fulcrum upon which the credibility of AI in legal practice pivots. It is not simply a matter of technical implementation but embodies a profound duty—to our clients, our profession, and the societal principles of justice we serve. The symbiosis of AI within legal operations is a testament to our unwavering dedication to upholding these values.

In this context, Cybereason emerges as an exemplary sentinel in the cybersecurity landscape, offering a bastion of defense with its AI-driven threat detection systems that resonate profoundly with the needs of legal practitioners. Cybereason's dedication to safeguarding the digital perimeter against sophisticated cyber threats aligns with the legal sector's imperative for uncompromising data protection.

The Cybereason Defense Platform epitomizes the fusion of advanced AI and machine learning algorithms to detect, prevent, and respond to cyber threats in real-time. Legal firms dealing in the currency of sensitive information stand to benefit immensely from the comprehensive protection afforded by Cybereason's platform, which is adept at parrying a myriad of cyberattacks, from ransomware to sophisticated persistent threats that lurk in the cyber shadows.

What distinguishes Cybereason is its planetary-scale capability to ingest and normalize data from vast IT environments. The platform's operational-centric approach provides not just alerts but a panoramic view of malicious operations, offering the clarity needed to nullify threats effectively. For legal practices where the stakes are perpetually high, this level of detail is not a luxury but a necessity.

Additionally, Cybereason's predictive response mechanism is designed to automatically predict and respond to attacks, often without the need for human intervention. In a sphere where time is of the essence, and the pace is set by the swift turn of legal events, such automation is invaluable. It ensures that legal professionals can focus on their core responsibilities while Cybereason's platform silently fortifies their cyber defenses.

When integrated into Legal Tech solutions, Cybereason's technology amplifies the security of legal practice management software and e-discovery platforms. Such synergy ensures that client data is not merely protected by conventional means but is secured through cutting-edge cybersecurity technology that anticipates and neutralizes threats before they can breach the digital walls of legal databases.

Cybereason's ethos of cyber vigilance and its unwavering commitment to data integrity provide a model framework for legal firms. By embracing such AI-driven cybersecurity solutions, legal entities do not just defend against external threats; they proactively contribute to shaping a cybersecurity-conscious legal landscape.

This preemptive stance is crucial when considering the sensitivity of the data handled within the legal industry. Client information, case details, and privileged communications must be protected not only from common cyber threats but also from the more insidious and advanced tactics used by adept cyber adversaries.

Cybereason's role extends beyond mere defense; it offers a strategic advantage. The platform's ability to handle petabytes of data ensures that legal professionals are not overwhelmed by the volume of information but are empowered by it. The operation-centric perspective grants legal professionals the insight to discern patterns and anomalies that could suggest a breach or a potential security risk.

The significance of Cybereason's predictive response feature cannot be overstated. In an environment where reactive measures may not suffice, the capacity to predict and counteract an attack preemptively is indispensable. This not only enhances the resilience of legal practices to cyber threats but also serves as a testament to the proactive culture that the legal industry aspires to embody.

As legal professionals continue to weave AI into the fabric of their practice, the incorporation of platforms like Cybereason is not a mere addition but a transformation of their cybersecurity posture. It is a decisive step towards a future where legal tech is synonymous with security, efficiency, and unassailable trust.

In summary, the safeguarding of client data in AI projects within the legal sector is a multifaceted endeavor that requires continuous vigilance, an unwavering commitment to legal and ethical standards, and a proactive approach to risk management. Cybereason stands as an ally in this endeavor, offering tools and expertise that fortify the legal profession's defense against the ever-evolving cyber threat landscape. By embracing such partnerships, legal practitioners can ensure that their use of AI delivers not only operational efficiencies but also upholds and advances the highest standards of data privacy and security.

Navigating AI in the Sea of Data Privacy Laws

In an era where compliance with data privacy laws has become as imperative as the air we breathe in the realm of legal technologies, OneTrust stands out as an exemplary embodiment of innovation and precision. As organizations worldwide grapple with the complexities of integrating artificial intelligence (AI) within their operations, they are simultaneously required to navigate the rigorous demands of data protection laws such as the General Data Protection Regulation (GDPR) and the California Consumer Privacy Act (CCPA). It is within this intricate dance of compliance and technological advancement that OneTrust's AI-powered tools have become indispensable. These tools are not merely functional; they represent the synthesis of legal acumen and cutting-edge technology, providing a suite of solutions that empower organizations to manage consent, conduct data discovery, map data flows, and respond to consumer rights requests with unparalleled efficiency.

For GDPR compliance, the essence of the AI's utility in OneTrust's offering is most vividly observed in its consent management functionality. This facet of their suite ensures that consent governance is not merely an exercise in checkbox compliance but a robust, dynamic process that organizations can administer across various channels. It is AI that enables the seamless alignment of data processing activities with the granular requirements of GDPR, which has planted the seeds for a new era of customer data processing that is built on the bedrock of explicit consent. Furthermore, OneTrust's AI-driven tools automate the Data Subject Access Request (DSAR) process from initiation to completion. This includes the delicate tasks of data identification, extraction, and the redaction or erasure of sensitive information, thereby bolstering compliance and mitigating risk with a level of precision and speed that is humanly unattainable.

Moreover, OneTrust extends its AI capabilities to navigate the complex waters of vendor risk management—a significant aspect of GDPR compliance that underscores the responsibility of data controllers for their processors. Here, the AI framework performs an in-depth risk assessment, enabling organizations to execute due diligence with sophistication and clarity. Not stopping at compliance facilitation, OneTrust also provides a route to GDPR compliance certification, including Euro privacy certifications recognized by the European Data Protection Board, using AI to demystify and streamline the journey towards achieving these critical compliance benchmarks.

Parallel in importance and complexity to the GDPR, the CCPA imposes its distinct compliance challenges, wherein OneTrust again deploys its AI solutions to offer a unified platform. The platform, enriched with AI, automates the entire spectrum of consumer rights mandated by the CCPA. The AI's role in OneTrust's toolkit allows for the precise mapping and classification of data, thus furnishing a prompt and accurate response to consumer inquiries, ranging from information access to the opt-out of data sales—a revolutionary step in consumer data rights management.

Additionally, OneTrust's AI-infused solutions offer a proactive approach to managing the CCPA's stringent 30-day cure period for data breaches. The AI streamlines incident analysis and orchestrates a response mechanism, thus equipping organizations to respond swiftly and effectively to any breaches that might occur, minimizing repercussions and aligning with CCPA's rigorous standards.

In the grand tapestry of data privacy compliance, OneTrust's AI-driven solutions emerge as both a shield and a sword—protecting organizations from the pitfalls of non-compliance while empowering them to harness the full potential of their data in a manner that is ethical, secure, and aligned with the global privacy landscape. By integrating OneTrust's offerings into their operations, businesses are not only safeguarding their interests but are also demonstrating a commitment to the sanctity of personal data—an endeavor that has become the hallmark of reputable and trustworthy organizations in the digital age.

The advent of artificial intelligence (AI) in the field of legal technology has brought about a transformative shift in the way organizations approach data privacy laws. With the stringent requirements of regulations like the General Data Protection Regulation (GDPR) and the California Consumer Privacy Act (CCPA), the need for comprehensive and efficient compliance tools has never been more pronounced. In this landscape, OneTrust, with its suite of AI-powered solutions, emerges as a beacon for organizations striving to navigate the complexities of data privacy regulations. The integration of OneTrust's AI tools within an organization's compliance framework signifies a commitment to upholding the highest standards of data protection, ensuring that every facet of personal data handling is in strict adherence to the legal mandates set forth by the GDPR and CCPA.

The GDPR, with its expansive reach and stringent requirements, has set a new precedent for data privacy regulation, and OneTrust's AI solutions are impeccably designed to meet these challenges head-on. Central to GDPR compliance is the principle of consent management, which OneTrust's AI tools handle with finesse. The AI-enabled consent management platform facilitates organizations in not just collecting and managing consent across various touch points, but also in maintaining a living document of consent records that can withstand the scrutiny of regulatory oversight. This level of meticulousness ensures that the cornerstone of GDPR — the lawful basis for processing personal data — is firmly in place, thereby enabling organizations to process customer data with confidence and legal compliance.

The Data Subject Access Request (DSAR) process represents a significant obligation under GDPR, and here too, OneTrust's AI shines by automating this intricate process. The AI-driven DSAR automation tool expedites the handling of requests, ensuring that individuals' rights to access, rectify, delete, or port their data are respected and responded to promptly. The AI's ability to trawl through vast datasets to identify, categorize, and act on personal data in accordance with data subject requests is not only a testament to its efficiency but also to its precision in safeguarding personal data against unauthorized access or processing.

Vendor risk management is yet another domain where OneTrust's AI tools demonstrate their worth. The GDPR places significant emphasis on the accountability of data controllers, extending to their relationships with data processors or vendors. OneTrust's AI-powered vendor risk management tools automate the due diligence and ongoing risk assessment processes, enabling organizations to monitor and manage the privacy risks associated with third-party vendors effectively. These tools become an essential part of the compliance ecosystem, ensuring that every entity involved in the processing of personal data is in lockstep with GDPR's requirements.

In addition to facilitating compliance, OneTrust's AI solutions also assist organizations in achieving GDPR compliance certification. The journey towards certification is made less daunting as the AI navigates the complex terrain of data protection principles, guiding organizations towards recognized certification standards such as Euro privacy. This certification serves as a tangible affirmation of an organization's dedication to data privacy, enhancing its reputation among stakeholders and regulators alike.

Moving across the Atlantic, the CCPA presents its unique set of challenges and compliance mandates, particularly in granting consumers unprecedented control over their personal data. OneTrust's AI-enriched platform provides organizations with a comprehensive toolset to manage all aspects of CCPA compliance. From automated responses to consumer rights requests to the management of data sale opt-outs, the AI-driven platform ensures that organizations can efficiently meet their obligations under the CCPA. The platform's intelligent mapping and classification of data ensure that CCPA requirements are met with precision, enabling organizations to maintain compliance amidst the evolving landscape of state privacy laws.

Moreover, the CCPA's breach notification requirement is a critical compliance aspect that OneTrust's AI platform adeptly manages. The AI's ability to centralize and streamline incident analysis, coupled with an automated response mechanism, ensures that organizations can respond to data breaches within the CCPA's stringent timelines. This proactive approach to incident management not only mitigates the risk of penalties but also reinforces an organization's commitment to protecting consumer data, a crucial factor in maintaining consumer trust.

OneTrust's AI-driven solutions for GDPR and CCPA compliance are a testament to the power of technology in fortifying data privacy. The role of AI in streamlining compliance processes cannot be understated; it is the linchpin that ensures accuracy in data handling, reduces the administrative burden of compliance, and minimizes the risk of human error. By leveraging OneTrust's tools, organizations are empowered to navigate the treacherous waters of data privacy regulations with agility and assuredness. In a world where data breaches and privacy violations can tarnish reputations and incur substantial financial penalties, OneTrust's AI solutions stand as a vigilant guardian, enabling organizations to not only comply with the law but also demonstrate a principled approach to data stewardship.

The integration of OneTrust's AI tools into an organization's data privacy framework is not just a strategic move; it is an essential one. The solutions provided by OneTrust bridge the gap between legal compliance and operational efficiency, ensuring that organizations can meet their data privacy obligations without compromising on the speed and quality of their service. As AI continues to redefine the contours of data management, OneTrust's pioneering suite of tools serves as the catalyst for this evolution, allowing organizations to embrace the future of privacy law compliance with confidence.

As regulatory landscapes continue to evolve, the need for agile and intelligent compliance solutions becomes more critical. OneTrust's AI-powered suite rises to this challenge, providing an adaptable and robust framework for navigating the shifting sands of data privacy laws. In the domain of GDPR compliance, the platform's capacity to streamline and automate data governance processes stands unparalleled. It delivers a comprehensive approach to managing the lifecycle of personal data, from initial consent to the eventual fulfillment of a data subject's right to erasure. This end-to-end management is made possible by AI's advanced capabilities in data identification, classification, and automated workflow management, ensuring that compliance is maintained throughout the data processing lifecycle.

Furthermore, the deployment of OneTrust's AI tools for vendor risk management reflects an acute understanding of GDPR's broader implications. It underscores the interconnected nature of data processing ecosystems and the need for thorough risk assessments of vendors and third parties. The AI-driven process ensures that due diligence is not merely a perfunctory check but a continuous and dynamic process, reflecting the ongoing nature of risk management in an environment where data flows across borders and systems with ease.

As for GDPR compliance certification, OneTrust's AI-enabled guidance provides organizations with a clear pathway to achieving and maintaining recognized standards of data protection. This facilitates not only adherence to the regulation but also a competitive advantage in the marketplace. Certification serves as a mark of excellence and trust, reassuring clients and partners of an organization's commitment to data privacy and security.

In the context of the CCPA, OneTrust's AI tools offer organizations the capability to respond to consumer requests with unprecedented efficiency. These requests, which range from accessing personal data to opting out of its sale, are managed by AI processes that ensure timely and accurate fulfillment. This automation extends to the handling of data breaches, a critical aspect of CCPA compliance. AI's role in incident response is to provide a rapid, systematic approach to managing and mitigating breaches, ensuring that organizations not only respond within the required timeframe but also take proactive steps to prevent future incidents.

OneTrust's AI solutions go beyond mere compliance; they enable organizations to elevate their data privacy posture, weaving the principles of transparency, accountability, and user control into the fabric of their operations. By integrating AI into the heart of their data privacy frameworks, organizations can transcend the traditional approach to compliance. They move towards a future where compliance is not seen as a regulatory burden but as a strategic advantage that promotes trust and innovation.

The intricate relationship between AI and data privacy is one of mutual reinforcement. AI augments the capability of organizations to adhere to data privacy laws, while these laws, in turn, drive the development of more sophisticated AI tools. In this symbiotic relationship, OneTrust's suite stands as a paragon of what is achievable when technology is harnessed to serve the interests of privacy and protection. The AI-infused tools within the OneTrust suite ensure that every touchpoint of data interaction is monitored, governed, and managed with the utmost regard for the principles enshrined in GDPR and CCPA.

In embracing OneTrust's AI solutions, organizations can effectively respond to the twin demands of innovation and compliance. They can leverage the full potential of AI to enhance their data privacy efforts, transforming compliance from a static, box-checking exercise into a dynamic, strategic function that drives business success. This transformation is not merely about staying ahead of regulatory curves; it is about setting a standard for privacy excellence that resonates with customers, regulators, and the broader industry.

To conclude, the importance of OneTrust's AI-driven compliance solutions in today's data-centric world cannot be overstated. As guardians of data privacy, these tools are instrumental in establishing and maintaining the trust that is critical to any organization's long-term success. They serve as the cornerstone of a robust compliance strategy, ensuring that as laws evolve and new regulations emerge, organizations can remain steadfast in their commitment to data protection. In the final analysis, OneTrust's AI-powered tools are not just a means to achieve compliance; they are a declaration of an organization's unwavering dedication to the principles of data privacy and the rights of individuals in our increasingly digital world.

AI-Enhanced Transparency: TrustArc's Compliance Mastery

The pursuit of data privacy compliance, in the wake of escalating global data protection regulations, necessitates an intricate understanding of the multifaceted nature of personal data handling. In this realm, TrustArc positions itself as a pivotal figure, providing dynamic solutions that meld the agility of AI with the stringent demands of privacy laws. As organizations seek transparency and control within their AI systems, TrustArc emerges as the steward of an automated compliance management system that embodies the adaptability required to thrive within the ever-evolving landscape of data privacy.

The essence of TrustArc's approach lies in its sophisticated suite of AI-enabled tools, each designed to streamline privacy program compliance and risk management. TrustArc's commitment to privacy by design is manifested in PrivacyCentral, a central hub for privacy tasks that not only automates alignment with laws and regulations but also fosters a culture of privacy across the organization. It is AI that empowers PrivacyCentral to provide real-time insights into compliance statuses, making privacy management a proactive rather than reactive endeavor.

Delving into the granularity of data governance, TrustArc's AI solutions offer an extensive overview of an organization's data through tools like the Data Inventory Hub and Risk Profile. These tools employ AI to illuminate the darkest corners of an organization's data landscape, ensuring full visibility and control over personal data. The Data Inventory Hub, with its AI-augmented capability, crafts a comprehensive inventory of data processing activities, offering organizations a clear view of their data terrain.

Meanwhile, the Risk Profile applies AI algorithms to evaluate the risk levels associated with different data sets and processing activities, allowing for informed decision-making regarding AI risks.

The Assessment Manager, another pivotal tool in TrustArc's arsenal, leverages AI to automate the privacy assessment process, including AI risk assessments. This automation translates into efficiency gains, reducing the time and effort required to conduct comprehensive assessments while enhancing the accuracy of the outcomes. The AI's analytical prowess enables it to score privacy assessments, prioritize risks, and recommend actionable steps, thereby simplifying the complex landscape of compliance obligations.

Moreover, TrustArc offers a certification that speaks volumes about an organization's commitment to responsible AI practices — the Responsible AI Certification. This certification is more than a badge of honor; it is a declaration of an organization's allegiance to transparency and ethical data handling. The certification process, supported by TrustArc's AI tools, assesses an organization's AI applications against a set of established criteria, ensuring that these applications uphold the principles of transparency, fairness, and accountability.

In tandem with its suite of tools, TrustArc provides a plethora of educational resources that bolster an organization's ability to navigate the privacy landscape. Nymity Research, part of the TrustArc platform, serves as a treasure trove of the latest privacy regulations, legal summaries, and operational templates. This repository, enhanced by AI, grants privacy professionals access to a constantly updated source of privacy knowledge, enabling them to stay abreast of the latest developments in the field. Complementing this, TrustArc offers a comprehensive range of resources, including webinars, articles, and an Academy that disseminates the latest insights on global privacy regulations and best practices.

The crux of TrustArc's value proposition lies not merely in its solutions but in its role as a partner in the journey through the labyrinth of AI privacy challenges. Their services represent a blend of compliance automation, expert guidance, and regulatory knowledge. This triad ensures that organizations can maintain a dynamic posture in response to privacy challenges posed by AI technologies.

TrustArc's AI-driven compliance solutions provide a tapestry of resources that enable organizations to not only meet but exceed the requirements of privacy laws. These solutions transform the nebulous and often intimidating process of compliance into a structured, manageable, and even strategic advantage. They endow organizations with the tools to build a foundation of trust with customers, regulators, and the public at large, by demonstrating a steadfast commitment to privacy and data protection. As privacy regulations continue to expand and mature, TrustArc's AI-enhanced offerings will undoubtedly become even more integral to organizations worldwide, ensuring that privacy remains at the forefront of technological innovation and business operations.

E-Discovery Reimagined: Zapproved's AI Solutions

As organizations increasingly operate in the digital domain, the management of legal compliance risks, particularly those associated with e-discovery, has become paramount. Zapproved enters the stage as a pioneering force, providing AI-enhanced e-discovery solutions that exemplify the practical application of AI in minimizing data privacy risks during the discovery phase of legal proceedings. Zapproved's platform, ZDiscovery, is designed specifically for corporate legal teams who face the challenge of navigating complex legal landscapes while protecting sensitive data.

ZDiscovery is a testament to the harmonious integration of AI with legal processes. The platform's suite of e-discovery tools addresses the full spectrum of the discovery phase — from legal holds and data preservation to collection and review. AI's role in this suite is to streamline these processes, ensuring that legal teams can efficiently manage the lifecycle of data involved in litigation. This efficiency is critical in an environment where the volume of electronic data is growing exponentially, and the risks of data mismanagement are ever-present.

Zapproved's platform shines brightest in its ability to simplify and secure the e-discovery process. In today's legal environment, where vast quantities of data can be called into question during litigation, ZDiscovery's AI-driven approach mitigates the risk of exposing sensitive information. Its AI algorithms are adept at identifying relevant data, distinguishing between privileged and non-privileged information, and facilitating the review process to ensure that only pertinent data is presented in legal proceedings. This targeted approach to data management aligns with privacy regulations such as the CCPA and GDPR, which demand meticulous handling of personal data across jurisdictions.

The challenges of e-discovery are multifaceted, encompassing not only the identification and collection of data but also the security and privacy implications inherent in handling such data. Zapproved's solutions are constructed with these challenges in mind, integrating tools for early case assessment, internal investigations, and responding to information requests. AI's role extends to improving the efficiency and effectiveness of these tools, managing costs, and optimizing resource allocation amidst the financial constraints that legal departments often face.

Data security is the bedrock upon which Zapproved has built its offerings. The platform enforces stringent security protocols to safeguard sensitive data involved in legal processes. Compliance with high data security standards, such as the SOC 2 Type 2 certification, demonstrates Zapproved's commitment to maintaining the security, confidentiality, and privacy of data under its purview. This focus on security is complemented by a commitment to training and education, equipping employees with the knowledge and tools needed to navigate the cybersecurity landscape and establish robust data retention and deletion policies.

Zapproved's offerings exemplify the proactive stance that corporate legal teams must take towards legal operations in the current era. By employing the ZDiscovery platform, teams are empowered to ensure compliance and minimize risks associated with data privacy throughout the e-discovery process. The AI-enhanced tools provide a safeguard against the complexities of litigation, enabling legal teams to focus on the substantive aspects of their cases rather than being bogged down by the intricacies of data management.

In conclusion, Zapproved's AI-powered e-discovery tools represent a critical innovation in the intersection of law and technology. They provide legal teams with the capabilities needed to tackle the dual challenges of data volume and privacy regulation, ensuring that the e-discovery process is conducted with precision, efficiency, and compliance. As the legal industry continues to evolve, and as privacy regulations become increasingly stringent, the role of AI in managing legal compliance risks will only grow in importance. Zapproved's suite of tools will undoubtedly play a central role in this evolution, guiding corporate legal teams through the complexities of e-discovery with confidence and assurance.

Chapter 22:

Forging the Future Lawyer: AI in the Classroom

Imagine stepping into a classroom where the familiar rustle of pages and the solemn recitations of precedent are infused with the hum of computers processing complex algorithms. Here, at the crossroads of tradition and innovation, legal education is undergoing a radical transformation. This is not the distant future; this is the now. This chapter embarks on an exploration of the emerging realm where artificial intelligence and legal education converge—a dynamic and compelling frontier reshaping the very core of how legal minds are cultivated.

In the heart of this frontier, we find a landscape where law students no longer solely rely on thick tomes of cases and statutes. Instead, they are equipped with sophisticated AI tools that can analyze legal patterns, predict judicial outcomes, and present insights with a speed and accuracy that rival the most experienced of legal minds. But this new territory is not without its trials and tribulations. The fusion of AI into legal education presents formidable challenges that demand innovative solutions and an unyielding commitment to ethical practice.

The journey through this chapter is a deep dive into how AI is not merely a subject to be taught, but a profound partner in teaching. Legal professionals are navigating an increasingly complex world—one where algorithms dictate outcomes, data privacy concerns are at the forefront, and where digital disruption is the norm. As a consequence, the marriage of AI with legal education is no longer a choice but an imperative. It is a decisive response to a world where the immutable laws written in books must now be interpreted and applied within the mutable, digital context that AI brings.

This intersection beckons us to reimagine a curriculum that is not only reactive to the changes but also anticipatory of the evolution of law in the context of technology. Future lawyers are not just expected to understand the law but also to harness AI with precision and ethical consideration. To meet this end, legal education must metamorphose, embracing an interdisciplinary approach that weaves AI literacy into its very essence from the ground up.

Legal education's pivot towards AI is essential for preparing future lawyers for a digitized industry. This preparation extends beyond mere familiarity with AI applications. It's about cultivating a profound comprehension of how machine learning models are constructed and employed, the ethical implications of AI in litigation and data analysis, and the far-reaching ramifications of AI-driven decisions. The objective is not to replace the lawyer's intuition with impersonal algorithms but to enhance their reasoning, expand their analytical prowess, and ensure they can deftly navigate the moral complexities that these powerful tools may present.

The re-imagination of legal education to integrate AI knowledge is, thus, not merely a scholarly endeavor but a pragmatic necessity. It's an exercise that bridges the gap between conventional legal training and the requirements of a technologically driven world. As we stand on the precipice of this new era, agility in adapting to technological advancements becomes a cornerstone of professional competence. The next generation of legal practitioners will need to be as adept at leveraging AI for predictive analytics and automating routine tasks as they are at interpreting legislation.

Therefore, legal education is facing a paradigm shift. It's an invigorating challenge that calls for a holistic overhaul of pedagogical strategies, course offerings, and the very skill sets that legal education aims to impart. We're looking at a future where legal professionals are expected to be as comfortable with coding as they are with case law. This chapter aims to dissect the elements of this transformation, to understand the synergies and tensions at play, and to propose frameworks that can effectively guide the integration of AI into legal education.

As we venture through this chapter, we will explore the innovative educational models that law schools are adopting, the collaborative initiatives between tech and legal academia, and the strategies that educators are employing to stay ahead of the technological curve. We will examine case studies of institutions that are leading the way in AI-centric legal education and glean insights from the experiences of students and faculty who are at the vanguard of this educational upheaval.

The integration of AI into legal education is also redefining what it means to be a lawyer. It challenges the traditional image of a lawyer and replaces it with a new archetype: the technologically savvy legal professional, a hybrid of jurist and data scientist, whose expertise transcends legal analysis to include technological fluency. This chapter will not only chart the trajectory of this shift but will also critically analyze its implications for the legal profession, the ethical considerations it raises, and the opportunities it affords for enhancing justice and efficiency within the legal system.

Finally, this chapter will serve as a blueprint for law schools, educators, and students, outlining the steps necessary to navigate this uncharted territory. It will provide recommendations on curriculum design, teaching methodologies, and the development of competencies that align with a future where AI is an integral part of legal practice. The goal is to leave the reader with a comprehensive understanding of the intersection of AI and legal education, equipped with the knowledge and tools to embrace the impending changes and thrive within them.

In essence, this chapter is an invitation to look beyond the horizon—to anticipate and shape the nexus of law and technology, and to prepare for a revolution in legal practice that is as inevitable as it is profound.

As the legal profession stands on the threshold of this revolution, it becomes increasingly clear that future leaders in law must be as innovative and adaptable as the technologies they will employ. This new breed of lawyers will not only interpret legal texts but will also create, manipulate, and improve the algorithms that will assist in legal decision-making. They will understand the language of the law and the language of computers, fluently transitioning between complex legal reasoning and coding. This duality of skills will become the new standard of legal expertise.

The role of artificial intelligence in legal education is not just about preparing students for the future; it's about shaping that future. AI is a transformative force, and legal education is the crucible in which the future of the legal profession is being forged. Law schools have the responsibility to ensure that their students are not just passive recipients of this new knowledge but active participants in its creation and application. This requires a rethinking of legal pedagogy to foster a culture of innovation, critical thinking, and ethical engagement with technology.

To achieve this, law schools will need to partner with technologists, ethicists, and practitioners to develop curricula that are as dynamic as the field they aim to master. They must encourage interdisciplinary studies, where law students take courses in computer science, and conversely, where tech students learn about the legal implications of their work. Such cross-pollination of knowledge is vital for cultivating a generation of professionals who can bridge the divide between law and AI.

The pedagogical approach must also evolve to prioritize experiential learning, where students have the opportunity to work on real-world problems using AI tools. Simulations, clinics, and externships will allow students to apply their interdisciplinary knowledge in practical settings, ensuring they are job-ready upon graduation. Educators must employ teaching methods that promote problem-solving skills and ethical judgment, preparing students to make tough decisions at the intersection of AI and the law.

As we navigate through the changing tides of legal practice, it is crucial to maintain a focus on the ethical dimensions of AI. The power of AI to influence legal outcomes and impact lives places a heavy responsibility on the shoulders of those who wield it. Legal education must, therefore, inculcate a deep sense of moral responsibility and instill a commitment to the public good in the use of AI. This means not only teaching the principles of ethical AI but also developing a critical awareness of the societal impacts of technology.

Moreover, legal educators must anticipate the global implications of AI. The law is not confined by borders, and AI systems often operate on a global scale. International collaborations and comparative legal studies will be indispensable in preparing students to deal with transnational issues. This includes understanding different legal systems, data protection laws, and the international regulatory environment for AI.

The impact of AI on legal education extends beyond the curriculum and classroom. It challenges traditional notions of mentorship, networking, and the development of professional identity. AI tools can aid in these areas, providing personalized learning experiences and new ways of connecting with mentors and peers around the world. However, they also raise questions about the role of the teacher and the nature of the student-teacher relationship in an AI-mediated educational environment.

In conclusion, this chapter is not just an academic treatise but a call to action. It is a directive for legal educators, students, and practitioners to engage with AI not as a mere tool but as a fundamental aspect of legal practice. It urges law schools to become incubators for innovation, where the legal minds of tomorrow are empowered to use AI responsibly and creatively to advance the law and promote justice.

Through a comprehensive exploration of the challenges and opportunities at the intersection of AI and legal education, this chapter aims to lay the groundwork for a new paradigm in legal training. It is a vision of a future where the law is not only taught but evolved through the integration of artificial intelligence, ensuring that the legal profession remains at the forefront of societal progress.

In forging ahead, we recognize that this path is uncharted and the outcomes uncertain. Yet, it is within this uncertainty that the greatest potential for innovation resides. As we stand on the cusp of this new dawn, let us embrace the possibilities with courage and foresight, ready to redefine what it means to be a lawyer in the age of artificial intelligence.

Legal Academia Meets AI: The New Pillar of Legal Studies

The transformative power of artificial intelligence (AI) has reached the legal profession, heralding an era where the integration of AI into legal studies is not just innovative but essential. This section of the chapter seeks to unravel the complexities of weaving AI teachings into legal education, focusing on practical tools like Casetext, and explores how this technology is indispensable for training future legal professionals to thrive in an AI-dominant legal industry.

The necessity of infusing AI into the educational veins of law schools is beyond compelling; it is imperative. As AI reshapes the bedrock tasks of the legal profession, it also redefines the skillset required to excel. Casetext, a pioneer in legal research powered by AI, exemplifies the sort of technology that legal education must now embrace. This integration ensures that law students gain not only familiarity with AI capabilities and limitations but also the ethical foresight to wield such tools judiciously.

To assimilate AI teachings effectively, legal studies must evolve into a multidisciplinary arena, marrying the intricate knowledge of computer science, legal principles, and ethics into a unified curriculum. Incorporating platforms like Casetext within this curriculum provides a concrete understanding of how AI intersects with daily legal practice. This platform allows students to engage with advanced AI capabilities such as natural language processing and predictive analytics, revealing how these technologies can refine and expedite legal research.

Emphasizing practical experience, law schools can leverage Casetext's hands-on workshops to illustrate the application of AI in legal research. These workshops offer students an invaluable opportunity to engage directly with AI tools, navigate legal databases with increased efficiency, and comprehend how AI can streamline the search for relevant case law and statutes. The practical knowledge gained here is not merely theoretical but a skill that will serve students throughout their careers.

Casetext is not just a practical tool; it's a lens through which students can view the future of legal research. With its adoption by over 8,500 law schools and firms, Casetext has cemented its role as an educational and professional standard. Its user-centric design and exhaustive case law databases cater to a wide array of research needs, making it an exemplary model for the type of AI integration necessary in legal studies.

This platform's utilization of AI technologies like BERT and GPT-2 for natural language processing tasks provides an optimal framework for students to study AI's real-world applications. By engaging with Casetext, students learn not only the fundamentals of legal research but also the intricacies of AI technologies that drive modern legal tech solutions.

The ethical considerations of AI in law are manifold and complex. Casetext serves as a practical context to explore these issues. Courses incorporating Casetext can examine questions of algorithmic bias, privacy concerns, and data security. Through such scrutiny, students are prepared to critically assess AI technologies, fostering a responsibility to employ AI in ways that enhance the integrity and fairness of legal practice.

Moreover, the integration of AI teachings, with tools like Casetext, mandates a commitment to continuous innovation within legal education. As AI technologies evolve, so must the curriculum and the educators who deliver it. Staying current with technological advancements, legal educators can bring fresh perspectives to their courses, integrating the latest legal tech developments and fostering an atmosphere of continual learning and adaptation.

Collaboration between legal academia and the industry is crucial for an AI-informed curriculum that mirrors the realities of contemporary legal practice. Incorporating insights from practitioners and technologists into the curriculum ensures that the skills taught align closely with the competencies demanded by an AI-enhanced legal market.

In parallel with classroom education, legal educators can utilize online resources to complement formal instruction. Platforms such as Casetext offer online materials that can extend learning beyond the classroom, offering students access to an ever-expanding repository of legal knowledge and AI applications.

By integrating AI teachings in legal studies, we prepare students not as mere participants in the legal sector but as pioneers capable of leveraging AI to enhance legal services and address the nuanced ethical, professional, and societal implications of technology in the law. Such preparation is critical for an era where law and technology coalesce, ensuring that future legal professionals are proficient, ethical, and innovative in their practice.

Conclusively, integrating AI teachings, with Casetext as a primary resource, into legal studies signifies a fundamental shift towards equipping law students for the future of legal practice. As AI reshapes the legal landscape, legal education must transform accordingly, providing students not only with traditional legal acumen but also with the technological savvy and ethical grounding the contemporary legal profession demands.

The task of melding AI into legal education presents both challenges and boundless opportunities to enhance the pedagogy and practice of law. As we embrace this transformative journey, the legal profession is poised to benefit immensely, signaling the dawn of an enriched era of legal practice that is more efficient, equitable, and profoundly informed by the capabilities of AI.

The Impact of AI on Legal Research and Education

Casetext, as a leading legal research platform employing AI, stands as a testament to how the legal field is being redefined by technology. It offers a live case study for students to examine AI's influence on legal research and the broader implications for the practice of law. Students can observe firsthand how AI can parse through the complexities of legal language, analyze vast quantities of case law, and provide predictive insights that were once beyond human capacity. This exposure is critical in teaching students to become not just users but also connoisseurs of legal tech, capable of critically evaluating the tools they will inevitably use in their professional lives.

The ability of Casetext's AI to understand the natural language allows students to delve into the nuances of legal documents with unprecedented depth. By engaging with Casetext, they learn the subtleties of querying complex legal questions and receive immediate, relevant results. This experiential learning bridges the gap between theoretical legal study and the practical demands of modern legal work, equipping students with the necessary tools to excel in a tech-driven legal environment.

Forging Ethical and Technological Proficiency

Casetext's application in legal studies also serves as a springboard for discussions on ethical AI use. It offers a platform to analyze and understand the impact of AI decisions, ensuring that as students learn to harness the power of AI, they also appreciate the importance of maintaining ethical standards in its application. This duality of learning — understanding the power of AI and the importance of ethical constraints — is critical in developing a rounded legal professional in the digital age.

In the Casetext environment, students can confront and tackle the ethical dilemmas that arise in AI's application in legal research. They learn to identify potential biases in AI algorithms and understand the significance of data privacy and the safeguarding of client information. Casetext's platform can provide real-world scenarios that challenge students to consider the implications of their technological choices, ensuring they carry forward a keen sense of ethical responsibility into their professional practices.

Curriculum Evolution and Lifelong Learning

As AI continues to develop at a breakneck pace, the curriculum and the educators responsible for its delivery must also evolve. Integrating tools like Casetext into legal studies is not a one-off task but a continuous journey of learning and adaptation. It requires educators to remain at the forefront of technological advancements and incorporate these into a dynamic curriculum that responds to the changing landscape of legal practice.

Casetext's adoption by the academic community points to the need for legal education to extend beyond the classroom and into lifelong learning. By integrating such platforms, students are not confined to the knowledge within their textbooks; they have at their fingertips a constantly updated stream of legal data and AI advancements. This integration ensures that legal education is not static but a progressive journey aligned with the professional world it seeks to serve.

Shaping the Future of Legal Practice

In essence, the integration of AI teachings in legal studies, exemplified by the use of Casetext, is transformative. It is not simply about adding a new tool to the legal education arsenal; it is about reshaping the very way in which the law is taught, understood, and practiced. As we prepare a new generation of legal professionals, we are not just imparting knowledge; we are shaping the future of legal practice.

By incorporating AI technologies into the curriculum, we are pioneering a legal education that is robust, dynamic, and future-focused. The students who emerge from these AI-enriched programs will not merely be law graduates; they will be the vanguards of a new legal era. They will possess a deep understanding of AI and a strong ethical compass, enabling them to harness the power of technology to advance justice and the rule of law.

The successful integration of AI teachings into legal studies, with the practical implementation of platforms like Casetext, contributes to a broader vision: a legal profession that is not only technologically adept but also deeply committed to ethical practice. It prepares students for a world where AI is not an optional tool but an essential aspect of legal practice — a world where the most successful legal professionals are those who can navigate the interface of law and technology with confidence and integrity.

In this new dawn, where AI and legal education intersect, we stand not at the end of an era but at the beginning of a new chapter in legal history. A chapter where law schools, educators, and students together embrace the promise and potential of AI, ready to redefine the essence of being a lawyer in the age of artificial intelligence.

Technological Fluency in Law: Training for an AI-Infused Future

The legal profession stands at the brink of a revolutionary transformation, primarily steered by the relentless advancements in artificial intelligence (AI). This inflection point necessitates an unprecedented shift in the educational paradigms that groom the next generation of legal professionals. With AI's tendrils extending deeper into the fabric of legal practice, future lawyers must not only be conversant with this technology but also adept at leveraging it to propel legal outcomes and operational efficiency.

Within the hallowed halls of Michigan State University College of Law, the LegalRnD - The Center for Legal Services Innovation champions this very cause. Here, an avant-garde curriculum forges an alliance between legal doctrine and technological prowess. LegalRnD stands as a beacon of progress, teaching students to navigate the shifting sands of the legal landscape with courses that are meticulously designed to marry the rigors of legal theory with the fluidity of technological application.

The educational offerings at LegalRnD are not merely academic endeavors; they are the crucible for the transformation of legal education. The center imparts knowledge on foundational disciplines such as process improvement, project management, and data analytics, thus crafting legal innovators rather than mere practitioners. Courses imbued with technologies like the ThinkSmart Automation Platform (TAP) extend beyond conventional learning, offering students pragmatic experiences in automating legal processes. Such skills are no longer ancillary; they are quintessential to the modern lawyer's repertoire.

However, LegalRnD's commitment extends beyond the incorporation of current technologies. Through hands-on initiatives, such as the Court Kaizen project and collaborations with Elder Law of Michigan, students don't just learn about technology; they apply it in real-world scenarios that enhance access to legal services and advocate for efficiency and effectiveness in legal aid.

Furthermore, LegalRnD aspires to sculpt 'T-shaped' professionals, armed with a breadth of knowledge across disciplines, yet capable of delving deep into the complexities of law, technology, and business. By fostering such comprehensive expertise, MSU Law students are poised to be not just participants in the legal industry's evolution but its architects, constructing a future where technology and law converge seamlessly.

Turning our gaze westward, we encounter CodeX - The Stanford Center for Legal Informatics, nestled within the intellectual nexus of Stanford University. CodeX exemplifies the interdisciplinary convergence that is reshaping legal education. This collaboration between the esteemed Law School and the Department of Computer Science is a testament to the pioneering spirit that is a hallmark of Stanford's educational ethos. CodeX is the vanguard of legal technology, committed to elevating the efficiency, transparency, and accessibility of legal systems around the globe through the lens of computational law.

Here, the fusion of disciplines fosters a legal education that equips students with a diverse skill set, spanning the understanding of complex legal frameworks to the intricacies of integrating AI and machine learning into the fabric of legal practice. CodeX incubates a symbiotic relationship between legal education and technology, encouraging students to tread new ground in leveraging computational tools to refine legal processes and address challenges inherent to the legal industry.

This interdisciplinary approach is crystallized in the courses offered at CodeX, such as the class on AI for social impact, co-taught by faculties from the School of Engineering and the Graduate School of Business. Here, the future legal professional is not confined to the silos of traditional legal study but is encouraged to collaborate, innovate, and apply legal theory in concert with technological advancements to serve societal needs.

Across the border, in Canada, the Law Practice Program (LPP) at Toronto Metropolitan University (formerly Ryerson University) is redefining practical legal training. The LPP's visionary approach to embedding AI and legal tech tools within its curriculum ensures that graduates are not simply theoretically equipped with AI's potential but are also skilled in employing these tools in the legal arena. The program is pioneering, presenting an eight-month experiential learning crucible that combines rigorous online training with immersive hands-on work placements.

Graduates of the LPP emerge as multifaceted professionals who have not only mastered the theoretical implications of AI on legal frameworks but have also honed their practical skills in employing these tools through simulated environments and real-world applications. This approach is emblematic of the LPP's ethos — to produce legal professionals who are not reactive to the winds of change but are the ones setting the sails and steering the industry towards uncharted waters.

The collective narrative of LegalRnD, CodeX, and the LPP forms the tapestry of modern legal education, one that is dynamic, multidisciplinary, and thoroughly entwined with the threads of AI and technology. These institutions are not just reacting to the rise of AI; they are actively shaping how legal professionals will interact with and influence this burgeoning domain.

The pages of this chapter serve as both a narrative and a blueprint for integrating AI into the framework of legal education effectively. They present a vivid illustration of how educational models and tools like the ThinkSmart Automation Platform, computational law, and practical programs like LPP are indispensable in training jurists for an AI-infused legal industry.

At LegalRnD, the integration of technology and law is a transformative journey rather than a destination. Students engage with innovative platforms that prepare them to think like process engineers, automating and optimizing legal workflows. The ThinkSmart Automation Platform, a fixture in their curriculum, is not just a tool; it is a portal into the future of legal services. By mastering such platforms, students learn to design solutions that improve the efficiency of legal operations, paving the way for a new era of legal service delivery.

The LegalRnD curriculum is not confined to the walls of a classroom; it extends into the field, where theoretical knowledge meets practical application. In collaboration with local courts and legal aid organizations, students engage in projects that push the boundaries of traditional legal service models. These experiential learning opportunities imbue students with a pioneering spirit, enabling them to reimagine how legal services are delivered and accessed, making them a catalyst for change in the legal industry.

Similarly, at CodeX, the ethos of interdisciplinary education molds lawyers who are as adept with code as they are with case law. The synergy between legal studies and computer science fosters a new breed of legal professionals—tech-savvy, analytically proficient, and versed in the language of algorithms and legal statutes. These future lawyers are equipped to navigate and lead in a legal landscape that is increasingly complex and influenced by digital technologies.

Moreover, the curriculum at CodeX is an exemplar of the balance between theory and application. It recognizes the growing importance of AI in shaping legal frameworks and prepares students to explore this confluence with an innovative mindset. Partnerships with industry leaders and legal operations experts bring practical insights into the classroom, providing students a comprehensive view of the technological forces at play within the legal sector.

Moving towards the Law Practice Program at Toronto Metropolitan University, the emphasis on practical tech training is clear. The LPP's robust curriculum goes beyond conventional legal education by mandating the actual use of AI tools within legal contexts. This program ensures that its graduates do not just emerge with an understanding of AI's theoretical aspects but also possess a mastery over its practical applications. The LPP's graduates are ready to employ AI in various legal settings, demonstrating the effectiveness of this hands-on approach.

The Law Practice Program's success is partly attributed to its alignment with the legal community, fostering dynamic partnerships and keeping the curriculum responsive to the industry's pulse. Its candidates are proven to excel across multiple legal sectors, highlighting the practical and contemporary nature of the education they receive. The program stands as a paragon of preparing candidates for a profession that is increasingly technology-driven.

The pioneering efforts of these institutions highlight the shifting focus of legal education from traditional case law analysis to include a strong emphasis on technological proficiency and innovation. The chapter, thus, is not merely a commentary on the current state of legal education but a visionary statement on its future trajectory.

As we continue to witness the integration of AI into every facet of legal practice, it becomes evident that future lawyers must be more than just legal experts; they must also be technologists, innovators, and ethical leaders. The legal education imparted by LegalRnD, CodeX, and the LPP is setting a new standard, one that ensures future lawyers are not only prepared for the present but are also equipped to shape the future of the legal profession.

The imperative now is for legal education to be as dynamic and forward-thinking as the technology it seeks to harness. This means that law schools must continuously evolve, integrating new technological tools and legal technologies into their curricula, and fostering an environment of perpetual learning and innovation. Legal educators and institutions must anticipate and respond to the rapid advancements in AI, ensuring that their students are well-prepared to meet the challenges and opportunities that lie ahead.

As we forge ahead into this AI-dominant era, it is incumbent upon legal educators, practitioners, and students to embrace the possibilities that AI presents. The insights and frameworks provided in this chapter serve as a guiding light for this journey, illuminating the path towards a legal profession that is resilient, adaptive, and attuned to the advancements of artificial intelligence.

In the grand tapestry of legal education, the role of AI is both transformative and foundational. It is not simply an add-on to the existing curriculum but a core element that is reshaping the very nature of legal practice. The future lawyers trained under these innovative programs will emerge as the vanguards of a new legal paradigm, proficient in AI and committed to using technology to enhance justice and uphold the highest ethical standards.

In summary, this chapter does not just reflect a change in legal education; it heralds the inception of a new era in the legal profession. It is an era characterized by the seamless integration of AI into legal practice, where the next generation of lawyers stands ready to navigate the complexities of an AI-enhanced legal landscape with confidence and foresight. The evolution of legal education, as evidenced by the programs at LegalRnD, CodeX, and the LPP, is a proactive response to the digital zeitgeist—an acknowledgment that the law, in its stately progression, must now stride in lockstep with the rapid march of technology.

As we peer into the future, it is clear that AI will leave an indelible mark on the practice of law. Preparing future lawyers for this shift is not merely an academic exercise but a vital undertaking that will define the relevance and efficacy of the legal profession in the 21st century. The successful integration of AI into legal education is imperative for ensuring that the legal profession not only survives but thrives in the forthcoming landscape, marked by digital innovation and computational advancements.

The institutions highlighted in this chapter serve as pioneering examples of how legal education can adapt to meet the challenges of an AI-driven world. LegalRnD is nurturing a new generation of lawyers who are adept in legal-service delivery innovation. Through its curriculum and partnerships, it instills in students an entrepreneurial mindset, preparing them to lead the transformation of the legal industry.

Similarly, CodeX at Stanford is breaking new ground by merging legal education with computer science, creating a fertile ground for the next wave of legal technologists. Here, students learn the art of computational law and the science of algorithmic problem-solving, enabling them to craft solutions that are as legally sound as they are technologically sophisticated.

The LPP at Toronto Metropolitan University exemplifies a hands-on approach to integrating technology into legal training. Its unique blend of online instruction and real-world practice equips students with a dual expertise in law and technology. This program illustrates the importance of experiential learning in cultivating a tech-savvy legal workforce ready to embrace the tools and challenges of the modern legal environment.

This chapter, therefore, is not just a roadmap for integrating AI into legal education; it is a manifesto for change—a declaration that the legal profession will continue to evolve, innovate, and serve justice in an increasingly complex world. The inclusion of AI in legal curricula is a clarion call for educators to prepare students for careers that will be profoundly different from those of their predecessors.

As legal educators, our responsibility extends beyond imparting knowledge; we must also be catalysts for innovation, challenging students to think critically about the role of technology in law and encouraging them to be pioneers of the new legal frontier. Our students must leave our institutions not just with a degree but with a toolkit brimming with advanced technological competencies, a strong ethical compass, and a passion for lifelong learning.

The contributions of LegalRnD, CodeX, and the LPP are milestones in this journey, proving that when legal education and AI converge, the result is a robust, relevant, and dynamic legal profession. As we continue to chart this course, we are laying the foundation for a legal system that is more accessible, efficient, and equitable—attributes that are essential in a world where technology is ubiquitous.

In closing, as we stand on the threshold of a new era in legal practice, we recognize the magnitude of the task before us. It is a task that demands creativity, commitment, and collaboration. By embracing AI within our educational frameworks, we are not just preparing future lawyers for an AI-dominant industry; we are preparing them to lead it, shape it, and redefine what it means to practice law in the age of artificial intelligence. The challenge is great, but so too are the opportunities—as vast and unbounded as the potential of AI itself.

Chapter 23:

AI and the Evolution of Dispute Resolution

Embarking on the journey of Alternative Dispute Resolution (ADR) within the contemporary technological milieu represents not merely the acceptance of progress but a radical transformation towards more streamlined, accessible, and fair resolution pathways. Artificial Intelligence (AI), standing at the vanguard of this shift, heralds a new era for ADR processes, promising to refurbish traditional methods with cutting-edge, data-informed techniques. Smartsettle emerges as a paragon within this domain, exemplifying the convergence of AI and negotiation within ADR and underscoring the enhanced efficiency borne of such integration.

In the realms of mediation and arbitration, AI's infusion is not a mere acceleration of process but an elevation of analytical rigor—the kind unachievable by human endeavor alone. With its capacity to sift through extensive data sets for pertinent case precedents, applicable legislation, and dispute resolution verdicts, AI significantly augments the caliber of decision-making, guaranteeing resolutions that are both equitable and informed. Smartsettle stands out in this respect, offering an ensemble of AI-powered negotiation products that promise peaceful, cooperative outcomes in lesser time frames and are suitable for a vast range of sectors, including the terrain of labyrinthine legal disputes.

As platforms are embedded with sophisticated AI algorithms, the interactional dynamics between disputing entities, along with the mediation or arbitration milieu, are recalibrated. These platforms are engineered to cultivate more effective communication strategies, discerning mutual concessions and consensus with a proficiency previously unseen. By optimizing these initial stages, AI curtails the temporal and monetary expenditures conventionally entwined with ADR, thus elevating its allure and approachability for a diverse array of disputants. Moreover, AI's inherent capacity to evolve with each case introduction affords a progressive enhancement of dispute resolution modalities, molding methodologies to the intricacies and idiosyncrasies of every distinctive scenario.

Smartsettle's repertoire of tools serves as a beacon in this arena, fostering intelligent and efficient solutions for an array of complex negotiations, including those as intricate as mergers, acquisitions, and even peace treaties. Their platform's flexibility, allowing integration within users' proprietary websites, offers customizable solutions for estate settlements, familial resolutions, auctions, and real estate disputes, among others. It's a technology that enriches the legal space with decision-support mechanisms and consensus-building facilitation.

Nevertheless, AI's marriage with ADR presents an assortment of challenges. Ethical conundrums, transparency requisites, and the specter of algorithmic partiality demand a deliberate and circumspect methodology. It is of utmost urgency that the inception and utilization of AI in this spectrum be directed by rigorous ethical precepts, with perpetual oversight by legal experts to ensure that technology operates to enhance rather than replace the quintessentially human aspects—empathy, judgment, and a nuanced comprehension—that underlie efficacious dispute resolution.

Consequently, the quest to harness AI's full capacity within ADR is a synergistic enterprise, necessitating a collective wisdom of technologists, legal professionals, and moral philosophers. Within this voyage, the legal community stands on the cusp of a novel epoch in ADR; one wherein AI not only propels efficiency and accessibility but also nurtures a more profound, nuanced apprehension of the dispute resolution craft. Smartsettle's integration into this chapter exemplifies this potential, embodying the pinnacle of how AI can transform the very fabric of conflict resolution in the legal realm.

In conclusion, the sophisticated algorithms and methodologies encapsulated within Smartsettle's suite are pioneering a transformative shift in ADR. These AI-driven tools not only streamline the resolution of disputes but also embody the potential to transcend traditional negotiation techniques, affording a harmonious and expeditious resolution journey. As such, Smartsettle represents not just a product but a philosophy in ADR—where technology doesn't just assist but elevates the practice to an art form, ensuring that every resolution is not only reached more efficiently but also with a more profound understanding of the interests at play.

AI-Powered Enhancements in ADR

Within the dynamic domain of legal technology, the arena of Alternative Dispute Resolution (ADR) has witnessed a substantial metamorphosis, propelled by the strategic integration of Artificial Intelligence (AI). This technological tide has introduced a transformative dimension to ADR, punctuated by marked advancements in efficiency, precision, and client satisfaction. In this chapter, we probe the intricate tapestry of AI's role in ADR, unraveling the multifaceted enhancements it embeds within dispute resolution mechanisms.

The essence of ADR pivots on the expedient and effective settlement of disputes, ensuring the confidentiality and autonomy of the disputing entities. AI's prowess in data analysis surfaces as a monumental asset, processing extensive datasets with an agility that eludes human faculties, thus fortifying the pursuit of ADR objectives. The gamut of AI's influence stretches from bolstering mediation with predictive analytics to innovating the negotiation domain with automated platforms, thereby recalibrating the dynamics of dispute resolution.

At the forefront, predictive analytics stand as a seminal AI utility within ADR, endowing mediators and arbitrators with the aptitude to gauge dispute outcomes with unprecedented accuracy. AI's analytical engines dissect historical data and discern patterns, projecting resolution trajectories that enrich strategic planning and decision-making. This prescient capability not only sharpens mediator acumen in steering dialogues but also equips disputants with pivotal insights, fostering resolutions that resonate with collective interests.

Advancing beyond predictive prowess, AI-infused negotiation platforms are redefining negotiation paradigms, crafting automated negotiation instruments capable of simulating and strategizing dialogues, and conducting negotiations on the basis of established parameters and goals. The repercussions of these platforms are manifold, heralding swift settlements and mitigating the emotional and fiscal tolls tethered to elongated disputes.

The meticulous analysis of documents, an indispensable facet of ADR, witnesses a significant uplift with AI. Leveraging sophisticated natural language processing and machine learning algorithms, AI tools scrutinize voluminous documents to extract pertinent details, verify compliance with agreements, and spotlight potential discord. This technological intervention not only condenses the preparatory phase but also augments the analysis's thoroughness and precision, culminating in fairer resolutions.

Perhaps one of the most profound revolutions AI introduces to ADR is the democratization of dispute resolution access. By automating multifarious ADR facets and diminishing human intervention dependency, AI substantively curtails the costs associated with resolving disputes. Such economic alleviation is a boon for individuals and smaller establishments, for whom the fiscal weight of conventional litigation or ADR methods could be daunting.

Nonetheless, the assimilation of AI within ADR is accompanied by a suite of challenges, including algorithmic biases, the imperative of human oversight, and the ethical conundrums of legal process automation. It is imperative to cultivate AI solutions that are aligned with the core tenets of fairness, equity, and justice—principles that are the bedrock of ADR.

Consequently, legal professionals are beckoned to adopt a forward-leaning stance towards comprehending and harnessing AI. Legal practitioners, arbitrators, and mediators must acquire an essential literacy in AI's potential and limitations. Tailored training initiatives and scholarly resources focused on the intersection of AI and law are crucial in arming ADR specialists with the necessary acumen.

Collaborative synergies between technologists, legal experts, and ethical scholars are paramount to ensuring AI solutions are crafted and implemented with adherence to jurisprudential and ethical norms. Interdisciplinary dialogues are instrumental in navigating the intricacies of AI integration within ADR, guaranteeing that these innovations accentuate rather than detract from dispute resolution doctrines.

With AI continually progressing, ADR stands on the cusp of further ingenuity, from refined predictive analytics to more nuanced negotiation algorithms. The trajectory for AI to optimize dispute resolution processes is seemingly limitless. However, the legal fraternity must remain vigilant, channeling these technologies in a manner that accentuates empathy, discernment, and justice.

In synthesizing AI within ADR, we encounter a formidable opportunity to elevate the efficiency, accessibility, and efficacy of resolving disputes. AI's empowerment of ADR transcends conventional constraints, unveiling novel potentialities. This journey, however, necessitates a concerted dedication to comprehend and ethically harness AI, ensuring a symbiosis that enhances rather than substitutes the intrinsic human elements pivotal to efficacious dispute resolution.

As we venture into this nascent ADR epoch, it is the collective responsibility of the legal community to thoughtfully and strategically engage with AI technology. Doing so is crucial to ensure that the evolution of ADR remains true to our shared aspirations for justice, equity, and amicable resolution in the era of digital transformation.

Modria, a shining example of AI's integration into the ADR process, encapsulates the potential for technology to revolutionize the field. The brainchild of experts from eBay and PayPal, Modria's Online Dispute Resolution (ODR) platform was conceived to transcend traditional boundaries of dispute resolution, making it more accessible, user-friendly, and efficient. By offering ready-made flows for common disputes, such as those pertaining to debt or landlord/tenant conflicts, Modria has pioneered a path for the swift and amicable settlement of disputes online.

The practicality of Modria’s platform lies in its seamless connection with existing court systems, providing a fluid transition between informal online negotiation and formal litigation. The integration showcases how AI can enhance case efficiency, not in isolation, but as a part of a larger judicial ecosystem. This aspect of Modria’s design reflects a profound understanding that the future of legal dispute resolution is not one that replaces traditional systems but one that complements and interweaves with them.

Modria’s platform is equipped with features that are emblematic of the next generation of ADR tools. Persistent chat functions, customizable templates, and real-time dashboards allow for a level of engagement and monitoring that were previously challenging to achieve. Such features exemplify the nuanced ways that AI can enhance the dispute resolution process, offering clarity, consistency, and convenience to all parties involved.

Moreover, Modria’s ability to support mediation and evaluation by neutral parties represents the human-AI collaboration that is essential in contemporary ADR practices. While AI can handle the processing and organization of information, human judgment remains at the core of dispute resolution. The technology's capacity to facilitate this collaboration is a testament to its design, which respects the nuances of human interaction and the complexity of legal disputes.

The fact that Modria has adeptly managed over a million cases globally speaks to its scalability and efficacy. It demonstrates that AI-powered ODR is not a distant future concept but a present-day reality with a proven track record. This success is a beacon for other ADR professionals and institutions, highlighting that the integration of AI in dispute resolution is not only possible but also advantageous.

As we project into the future, we envision a landscape where AI, like Modria, becomes an integral part of ADR infrastructure. The legal community's role in this transition is multifaceted. It involves fostering an environment where innovation is encouraged and adopting a mindset of continuous learning to stay abreast of technological advancements. As AI capabilities expand, so too must our strategies for incorporating them into legal practices.

Training and education in AI must become staples in the legal profession, ensuring that all stakeholders have a comprehensive understanding of these technologies' capabilities and ethical implications. The development of AI in the legal sector must be conscientious, with a keen eye towards inclusivity, ensuring that the benefits of technology are accessible to all segments of society.

Furthermore, it is not just about integrating AI but also about evolving with it. As AI tools like Modria become more sophisticated, they will present new possibilities and challenges. The onus is on the legal sector to remain adaptable, balancing the scales between technological efficiency and the indispensable human elements of ADR.

In conclusion, Modria's incorporation into the sphere of ADR illustrates a broader narrative of innovation and progress. It exemplifies the potential for AI to not only support but also significantly enhance the dispute resolution process. The legal community, now equipped with such advanced tools, stands at the threshold of a new era. It is a time that calls for strategic engagement with technology, ensuring that as we harness the power of AI like Modria, we continue to uphold the values of justice and fairness that are the cornerstone of ADR.

By embracing this technological shift, we can anticipate a future where ADR is more efficient, equitable, and accessible. AI, when thoughtfully implemented, has the potential to augment the very fabric of dispute resolution. It promises a future where the resolution of conflicts, big and small, is no longer tethered solely to the traditional, often cumbersome mechanisms of the past but is propelled into the future with innovation, agility, and an unwavering commitment to justice.

The Swift New Age of ADR: AI's Accelerative Impact

In the vanguard of legal innovation, artificial intelligence (AI) is profoundly transforming Alternative Dispute Resolution (ADR)—a sphere traditionally encompassed by mediation, arbitration, and other methods outside the courtroom. This transformation heralds not just an enhancement of efficiency but also a revolutionary overhaul in the methodologies used to resolve disputes.

At the heart of AI's invaluable contribution to ADR is its unmatched capability to sift through and analyze extensive datasets at velocities unreachable by human effort alone. Traditional dispute resolution techniques are effective yet often marred by their slow pace. AI steps into this realm as a beacon of expeditiousness, ready to tackle and possibly alleviate these longstanding criticisms.

Foremost in the arsenal of AI's contributions to ADR is the advent of predictive analytics. Through the mining of historical data, AI systems can forecast dispute outcomes with remarkable accuracy. This predictive prowess is not only instrumental in setting grounded expectations early in a dispute but also guides the negotiation process with insights drawn from a well of data.

Further advancing ADR is AI's ability to refine communication between disputants. Leveraging advancements in natural language processing, AI platforms can interpret the subtlest of nuances and emotions behind textual communication. This clarity in dialogue diminishes the potential for misunderstandings, expediting negotiations and ushering in timely resolutions.

Arbitration reaps AI's benefits, as well. Assisting arbitrators in sifting through precedents and statutes, AI slashes the time traditionally spent on legal research. Additionally, by comparing case facts against historical counterparts, AI supports the crafting of decisions that are both fair and equitable, potentially instilling greater consistency within arbitration outcomes.

However, threading AI into the fabric of ADR does not come without its hurdles. There are pertinent concerns, including algorithmic bias, data privacy, and the pressing need for transparency in AI-mediated processes. It is crucial, as we venture further, to confront these issues to safeguard the integrity and fairness of AI-aided dispute resolution.

Confronting these issues begins with crafting AI systems that are robust, accurate, and capable of elucidating their rationale in terms easily grasped by humans. The domain of Explainable AI (XAI) is crucial here, focusing on demystifying AI decision-making—a critical need in the legal field where the reasoning behind verdicts can be as vital as the outcomes.

The successful integration of AI into ADR calls for a synergistic effort. Legal practitioners, technologists, and ethicists must unite to forge AI tools that are technically adept while aligned with the moral tenets of legal practice. Collaborative, interdisciplinary teams are essential to ensure that these tools are instilled with a deep understanding of legal principles and ethical codes.

Education and skill development are pivotal in this technological embrace. Legal professionals must be armed with the knowledge to harness AI effectively, which entails understanding the workings, capabilities, and limits of AI. Tailored training initiatives and workshops are key to bridging this knowledge gap, ensuring the smooth assimilation of AI into ADR practices.

As we gaze into the future, AI's potential to metamorphose ADR is boundless. With ongoing advancements, we can expect even more refined AI applications that will further enhance the processes of dispute resolution. From intelligent negotiation assistants to sophisticated analytics forecasting resolutions with increasing precision, the horizon is replete with promising developments for ADR.

Moreover, as AI embeds itself more deeply within ADR, it paves the way for more democratic dispute resolution services. Diminishing the costs and duration associated with ADR, AI enables a broader swath of society to access these services, thereby contributing to more widespread justice.

In summation, AI's role in expediting dispute resolutions within ADR is transformative and manifold. By tapping into AI's potential, legal professionals anticipate a future of more efficient, transparent, and just dispute resolution. Yet, realizing this vision mandates a collective effort to confront AI's challenges, uphold ethical standards, and continue investing in education. As we tread this path, the promise of AI-augmented ADR shines as a beacon of innovation within the legal profession, heralding an era of both efficacy and inclusivity.

AI's assimilation into ADR is an inevitable tide, not a mere possibility. By approaching AI with a conscientious stance towards its challenges and promises, the legal community stands at the cusp of unveiling a landscape of dispute resolution that is unprecedented in its capability to streamline and democratize the pursuit of justice. This foray into AI-enhanced ADR is not just about harnessing new technology—it's about reinvigorating our commitment to the core tenets of justice in an age marked by rapid technological advancement.

The World Intellectual Property Organization (WIPO) recognizes the pivotal role of AI in dispute resolution and has accordingly tailored its Alternative Dispute Resolution (ADR) services to address the nuanced complexities of AI-related disputes. As AI technologies permeate various sectors, from healthcare to entertainment, the resultant intellectual property (IP) disputes are inevitably growing both in frequency and complexity. WIPO's ADR framework is specifically designed to provide a structured and efficient alternative to the often protracted and public court litigations.

In the context of AI, WIPO's ADR services shine in addressing disputes over AI-generated content, collaboration contracts, and copyright issues, among other areas. These disputes encompass a broad spectrum, including data privacy, intellectual property infringement, licensing agreements, and technology transfer. Given the pace at which AI is evolving and the diverse applications it finds itself in, such disputes require resolution mechanisms that are as dynamic and innovative as the technology itself.

The potential parties to these disputes are as varied as the applications of AI, extending from individual academics, artists, and developers to large entities like government agencies, media organizations, and start-ups. WIPO's ADR services cater to this diverse clientele, offering resolutions that are not only expedient but also cognizant of the specialized nature of each dispute.

Administering these disputes under the WIPO Mediation, Arbitration, and Expert Determination Rules ensures a process that is well-suited for the intricacies of IP and technology disputes. These rules are crafted with a keen understanding of the need for confidentiality, especially when it comes to sensitive information like trade secrets.

WIPO's role is multifaceted—it not only provides a procedural framework but also offers guidance on ADR clauses, appropriate fees, and administrative services like video conferencing and document management. Its commitment to neutrality and independence is evident in the way it administers disputes, ensuring fairness and impartiality.

Moreover, the effectiveness of ADR proceedings is heavily dependent on the quality of the mediators, arbitrators, or experts presiding over them. WIPO maintains an open-ended list of experts with specialized knowledge in IP and AI, affording parties the luxury of choosing professionals who bring the necessary technical expertise to the table. This flexibility in the appointment of experts is crucial for the nuanced resolution of AI-related disputes.

WIPO extends beyond the administration of disputes by providing procedural guidance and training. The organization offers recommended contract clauses and submission agreements to help parties integrate ADR processes into their contracts proactively. Furthermore, WIPO's training and workshops in mediation and arbitration are instrumental in equipping stakeholders with the necessary skills to navigate AI-related disputes effectively.

WIPO's collaborative approach, particularly through initiatives like WIPO's Intellectual Property and Frontier Technologies program, aims to develop ADR procedures that not only expedite dispute resolution but also preserve the underlying business relationships. In an ever-evolving industry such as AI, WIPO's willingness to cooperate with stakeholders and its continuous enhancement of services are testaments to its commitment to adapting to the changing landscape.

The proactive approach of WIPO in tailoring its ADR services for AI-related disputes is a clear indication that the legal landscape is advancing in tandem with technological progress. As we move forward, the alliance between AI and legal professionals, guided by institutions like WIPO, promises a future where dispute resolution is not only more efficient but also more aligned with the rapid pace of innovation. The legal community's journey with AI at its side is one of both challenge and opportunity, with WIPO's ADR services serving as a cornerstone in the evolving edifice of dispute resolution.

As AI continues to influence every facet of ADR, the legal profession stands before a transformative force that can redefine access to justice and the efficiency of legal processes. The confluence of AI and WIPO's ADR services symbolizes a paradigm shift—a transition from traditional methods to a new era characterized by swiftness, precision, and accessibility. This is the beacon guiding the legal community into a future that promises not only to evolve with the technologies of the day but also to uphold the enduring values of fairness, impartiality, and justice for all.

Chapter 24:

The AI Imperative: Transforming Legal Practices for Future Excellence

The landscape of legal services is undergoing a monumental shift with the advent of artificial intelligence (AI), marking a transition from traditional processes to a future where legal outcomes are enhanced through the power of AI. At this transformative juncture, legal professionals face the imperative to not only adopt AI but to integrate it into the fabric of their practice. This necessitates the creation of a robust business case for AI, one that aligns technological advancement with strategic business objectives, optimizing operational efficiency, enhancing client satisfaction, and, ultimately, improving case outcomes.

This intricate journey commences with the vision to implement AI, bolstered by a concrete plan that convinces stakeholders of its inherent value. Legal practitioners must engage in comprehensive due diligence, evaluating both the tangible and intangible benefits AI brings. A critical analysis is necessary to determine how AI can refine workflows, such as automating the sifting of documents or using predictive analytics for litigation tactics, thus reallocating valuable attorney hours to complex, high-value tasks.

AI's capability to analyze patterns and predict outcomes from extensive datasets presents unparalleled insights, enabling strategic decision-making and enhancing the firm's competitive stance. The thorough evaluation of these benefits will anchor the business case, showcasing both the immediate operational efficiencies and the strategic long-term benefits of AI adoption.

Developing an AI adoption roadmap requires a strategy that addresses technological, cultural, and operational facets. This entails a unified effort across the firm to secure buy-in at all levels by highlighting AI's direct impact on service delivery and case outcomes. Firms must invest in training and culture development to prepare their team for AI adoption, fostering an environment that embraces innovation. Moreover, the roadmap must cover data governance and security protocols to maintain compliance and protect client confidentiality. Throughout this journey, the business case for AI shifts from a hypothetical proposition to a tangible strategic asset, laying the groundwork for a future where AI and legal acumen combine to set new benchmarks in legal service excellence.

Within this framework, Altman Weil's expertise becomes invaluable. As a pioneer in management consulting services dedicated to legal organizations, Altman Weil stands as an indispensable guide for law firms traversing the complexities of AI adoption. With a heritage spanning over fifty years, Altman Weil commands a unique understanding of the legal industry's challenges and opportunities. They provide bespoke consultancy in areas crucial for the integration of AI into legal practices, such as strategy, governance, financial management, and service innovation.

Altman Weil's adeptness in steering law firms towards sustainable growth, operational efficiency, and increased profitability renders it an authoritative voice in the legal consultancy sphere. The firm's depth of knowledge in legal management and lawyer coaching allows it to provide custom solutions that include the development of competitive compensation models, executive search services, financial stewardship, crisis handling, and succession planning. By leveraging Altman Weil's expertise, law firms can ensure that their journey towards AI integration promotes heightened efficiency, service quality, and client satisfaction — the cornerstones of a thriving legal practice in the competitive sector.

In the course of defining a clear direction for the firm, Altman Weil plays a crucial role in setting specific priorities to reach the firm's objectives. Their extensive experience with law firms facilitates the resolution of a multitude of management and business challenges, positioning firms for success in the digital age. Altman Weil's consultancy transcends typical advisory roles, delivering innovative strategies that shape not only the future of individual firms but also the future of legal service delivery as a whole.

For law departments, Altman Weil ensures that operations are streamlined for peak performance, achieving internal harmony within departments and aligning their function with the broader corporate strategy. Their expertise in aligning legal department performance with the strategic objectives of executive teams and business units is particularly valuable in the context of AI adoption, where the alignment of technology with business strategy is paramount.

Altman Weil's special services, which showcase their experience and innovative expertise, are tailored to the unique challenges faced by law firms and corporate and government law departments. Their advisory encompasses every aspect of legal operations, from strategic planning to the implementation of complex technologies like AI. By tapping into this rich vein of expertise, law firms can craft a compelling business case for AI, underscoring the potential for transformative change and competitive advantage in legal services.

As legal practices navigate the intricate path of adopting AI, with Altman

Weil's guidance, they can anticipate not just incremental improvements but significant advancements in service quality, client satisfaction, and operational efficiency. Law firms equipped with a robust business case for AI — one that incorporates the deep industry insights and strategic acumen of Altman Weil — are poised not only to adapt but also to lead in the burgeoning legal-tech landscape.

In sum, the quest to integrate AI into legal practices is a strategic imperative that demands expert navigation. Altman Weil, with its extensive experience in legal consultancy and a deep comprehension of the legal market's nuances, emerges as the quintessential ally for law firms on this transformative journey. Their expertise in implementing service strategies that meet the distinct needs of legal teams solidifies their role as the architects of success in an AI-influenced legal sector.

Altman Weil's proficiency in guiding legal entities through the AI evolution ensures that the transition from traditional legal practice to a technology-enhanced future is not only seamless but also strategically advantageous. The business case for AI in legal practices, enriched by Altman Weil's consultancy, becomes more than a narrative of adoption; it is a blueprint for innovation, a testament to the firm's enduring commitment to legal excellence and client service in the age of artificial intelligence.

Articulating the Value Proposition of AI to Legal Stakeholders

In the dynamic landscape of legal services, the integration of Artificial Intelligence (AI) marks a pivotal transition, signifying not merely a shift in operational mechanics but also a fundamental transformation in the professional ethos and strategic outlook. For stakeholders entrenched in traditional paradigms—be they partners, associates, paralegals, or most critically, clients—the proposition of AI is one fraught with uncertainty. To navigate this terrain and to convincingly advocate for the value AI imparts requires a nuanced, multi-pronged strategy that is underpinned by empirical evidence, strategic foresight, and unequivocal communication of the potential returns on investment. In this context, the strategic consulting expertise of the Boston Consulting Group (BCG) becomes an invaluable ally.

The essence of BCG's engagement lies in its venerated capacity to demystify AI for the legal profession. AI, despite its encroachment into various sectors, retains a sheen of mystique, at times prompting skepticism within legal circles. BCG, through meticulous data-driven analyses, dissects the AI mystique, elucidating its functionality, delineating the specific legal quandaries it addresses, and elucidating its application in daily legal tasks. This educational approach is instrumental in diminishing resistance, paving the way for a broader acceptance and integration of AI technologies in legal workflows.

Understanding each stakeholder group's distinct concerns and aspirations is pivotal. Attorneys and paralegals may gravitate towards AI for the promise it holds in expediting research and due diligence, thereby liberating them from the tedium of repetitive tasks. Conversely, clients may be more captivated by the prospect of expedited services, enriched outcomes, and heightened transparency in billing. BCG's tailored narrative addresses these disparate points of engagement, showcasing AI not as a distant technological behemoth but as a pragmatic tool poised to enhance their professional sphere.

The potency of BCG's argument for AI adoption is bolstered by their repository of case studies, chronicling legal entities that have triumphantly woven AI into their fabric. These narratives serve a dual purpose: they validate AI's applicability in the legal domain, and they enumerate the palpable benefits—operational efficiency, cost-effectiveness, and strategic foresight—that AI confers. Such success stories resonate deeply, countering skepticism and fortifying stakeholder confidence.

Economic considerations are inextricable from the discourse on AI integration. Stakeholders are rightly concerned with the financial implications—both the initial investment and the ancillary costs associated with adapting to new technological ecosystems. Herein, BCG's acumen is manifest. They adeptly elucidate the investment in terms of return on investment (ROI), contrasting it against the status quo to illuminate the financial gains that accrue from AI adoption.

A forward-looking vision is indispensable in BCG's narrative. The legal profession, no stranger to evolution, is now at the cusp of a transformation driven by unprecedented technological advancement. BCG's strategic insights underscore that adopting AI is not a mere reaction to current challenges but a proactive measure to secure relevance and competitiveness in the impending future. This foresight is particularly compelling for stakeholders in sectors besieged by intense competition.

The stakeholder engagement does not conclude with the strategic blueprint; it extends into the planning and execution phase. BCG champions a collaborative approach, soliciting input from stakeholders at every echelon, which not only cultivates a sense of ownership but also ensures that the integration of AI is attuned to the actual needs and preferences of its users. This collaborative strategy, informed by BCG's experience, maximizes the utility and acceptance of AI.

The proposition of AI necessitates a comprehensive framework for training and support—an endeavor that BCG identifies as a cornerstone of successful integration. The prospect of mastering new AI-infused processes can be daunting, and BCG's commitment to providing extensive training and robust support mechanisms assuages apprehensions and facilitates a smoother transition.

Further, BCG accentuates the ethical imperative and professional responsibility towards judicious AI integration. The onus to embrace AI, as framed by BCG, transcends operational efficiency; it embodies the legal profession's commitment to uphold the highest standards of service and integrity. This ethical dimension of AI integration, when communicated effectively, reinforces its necessity and virtue.

Confronting the implications of AI on employment and the prevailing work paradigms is an obligation that BCG does not eschew. The pervasive anxiety that AI may supplant human legal professionals necessitates a forthright dialogue. BCG elucidates that AI is conceived as a complement to human intellect, destined to augment rather than substitute. Moreover, they elucidate the emergent roles and prospects birthed by AI, thereby reframing the discourse around AI as one of opportunity and advancement.

The cultivation of a compelling narrative—one that resonates with the ambitions, allays the fears, and illuminates a trajectory towards a more streamlined, efficacious, and competitive practice—is central to BCG's methodology. By tailoring this narrative to the aspirations and trepidations of each stakeholder group, BCG transforms skepticism into a palpable enthusiasm for the untapped potential of AI.

Engagement does not conclude with mere advocacy; BCG instills a culture of continuous feedback, integrating regular check-ins, surveys, and open forums to discuss AI-driven initiatives. This inclusive feedback loop ensures that AI integration remains attuned to the evolving needs of stakeholders, preemptively addresses potential concerns, and uncovers further opportunities for AI to add value.

BCG also understands the power of demonstrating AI's impact on client satisfaction—a compelling argument in a profession where client expectations are constantly escalating. AI's capability to deliver services that are more tailored, precise, and expedient can position a legal practice as a pioneer, distinctly advantageous in a highly competitive market.

In the delicate balance of innovation and confidentiality intrinsic to legal services, BCG does not underestimate the importance of data security and privacy. The sanctity of client information is paramount, and stakeholders must be assured that AI systems not only protect this sensitive information but also comply scrupulously with existing data protection legislations.

In the broadest sense, BCG's mission in championing AI within legal services is an exercise in creating an ecosystem that fosters education, engagement, and robust confidence in the tangible benefits of AI. It is about constructing and communicating a vision—a vista of the future where AI is not merely a tool for enhancing efficiency but is integral to fostering a culture of innovation, dynamism, and client-centric service delivery.

The profundity of BCG's approach lies in its capacity to align AI's capabilities with the strategic imperatives of legal practices. BCG aids in envisioning a future where legal practitioners are not ensnared in the minutiae of process-driven tasks but are empowered to exercise their cognitive skills more fully, benefiting from AI's precision and analytical prowess.

To summarize, BCG's role in convincing stakeholders of AI's value is not limited to showcasing immediate benefits. It extends to crafting a strategic narrative that aligns with the long-term goals of legal practices. This narrative is rooted in a clear understanding of the current landscape and an astute anticipation of future trends. By interweaving the perspectives and insights of various stakeholders into this narrative, BCG ensures that the integration of AI into legal practices is not just a fleeting trend but a well-considered strategic decision aimed at long-term success and sustainability.

In the realm of strategic consulting, particularly within the legal sector, BCG stands as a beacon of innovation and strategic transformation. Their advocacy for AI is predicated on a deep understanding of the technology's potential and the nuanced dynamics of the legal industry. The firm's methodical approach to stakeholder engagement—marked by an emphasis on educational initiatives, demonstrable case studies, and an inclusive feedback system—renders AI a less daunting and more inviting prospect.

Through its involvement, BCG does not merely dispel the enigma of AI but positions it as an essential facet of modern legal practice. The consulting firm's message is unequivocal: AI in legal services is not just about adapting to change; it's about embracing and leading it. By guiding stakeholders through the complex terrain of AI integration with strategic acumen, BCG plays a pivotal role in transforming legal services into a domain where technology and human expertise converge to create unparalleled value for clients and practitioners alike.

In charting this course, BCG elucidates that the value of AI extends beyond operational efficiencies—it encompasses the realization of new service paradigms, the cultivation of innovative problem-solving approaches, and the reinforcement of the legal profession's commitment to excellence. It is a journey that BCG navigates with the assurance of experience, the precision of data-driven insight, and the clarity of strategic vision, propelling legal practices toward a future where AI is not merely an adjunct but a fundamental pillar of legal expertise.

Thus, convincing stakeholders of AI's value, with BCG's strategic input, is not a matter of persuasion alone but one of profound transformation—reshaping the contours of legal practice into a more agile, insightful, and client-responsive profession.

Operationalizing AI: Strategic Planning and Implementation

The integration of Artificial Intelligence (AI) into law firms signifies a pivotal shift in the legal industry, moving from traditional methodologies to innovative, technology-driven practices. This transformation requires a comprehensive strategy, encompassing both the theoretical understanding of AI's potential and the practical application within the legal domain. Law firms embarking on this journey must navigate through various stages of AI adoption, each necessitating meticulous planning and execution.

Strategic Planning for AI Adoption

The journey begins with strategic planning, which lays the groundwork for successful AI integration. This phase involves a thorough analysis of the firm's current processes, identifying areas where AI can add value. The objective is to understand how AI can enhance efficiency, reduce costs, and improve the quality of legal services. Law firms must consider various AI technologies, such as machine learning, natural language processing, and predictive analytics, evaluating their potential impact on different legal tasks, from document analysis to case outcome predictions.

In this strategic phase, Thomson Reuters Legal emerges as an indispensable resource. With its comprehensive suite of legal technology solutions and in-depth market insights, Thomson Reuters Legal aids law firms in aligning their AI initiatives with industry trends and best practices. The firm's access to cutting-edge legal AI technologies, combined with analytical tools, provides a solid foundation for law firms to develop a nuanced understanding of AI's capabilities and potential applications in their specific legal contexts.

Operational Execution and Implementation

After defining the strategic framework, law firms must focus on the operational execution of AI initiatives. This involves selecting appropriate AI technologies, integrating them into existing workflows, and managing the change process within the organization. Operational execution is where the theoretical planning translates into tangible actions, with law firms implementing AI tools for tasks like legal research, contract analysis, and client advisories.

Successful implementation requires a structured approach, ensuring that AI tools are seamlessly integrated into the legal practice without disrupting existing operations. Training programs and workshops become crucial at this stage, as they help legal professionals acclimate to new technologies and understand how AI can augment their work. Thomson Reuters Legal supports this phase by offering training resources, technical support, and consultancy services, ensuring that law firms can effectively leverage AI tools in their daily operations.

Navigating Challenges and Embracing Future Opportunities

As law firms progress in their AI adoption journey, they encounter various challenges, from technical hurdles to ethical considerations. However, these challenges also present opportunities for growth, learning, and future-proofing the practice.

Addressing Ethical and Governance Issues

One of the most significant challenges in AI integration is managing the ethical and governance aspects of technology use in legal practice. AI systems, while powerful, can raise concerns related to bias, transparency, and accountability. Law firms must establish robust governance frameworks to ensure that AI applications adhere to ethical standards and regulatory requirements. This includes developing policies on data privacy, AI transparency, and accountability mechanisms.

Thomson Reuters Legal plays a crucial role in guiding law firms through these ethical and governance complexities. By providing resources on legal AI ethics, data security, and compliance standards, Thomson Reuters Legal helps firms develop and implement governance structures that ensure responsible AI use. Additionally, the firm's insights into regulatory trends and best practices in legal AI governance aid law firms in staying ahead of the curve, ensuring that their AI systems are both effective and ethically sound.

Cultural Transformation and Workforce Enablement

Another challenge lies in effecting a cultural transformation within the law firm, fostering an environment where AI is embraced as a tool for enhancement rather than a threat to traditional practices. This cultural shift requires educating and enabling the workforce, ensuring that legal professionals are equipped with the skills and knowledge to leverage AI effectively. The transition involves breaking down resistance to change, and promoting a mindset that views AI as a collaborator in the legal process.

Thomson Reuters Legal contributes to this cultural transformation through educational initiatives and training programs designed to enhance AI literacy among legal professionals. These programs demystify AI, illustrating its practical benefits and how it can be integrated into various legal tasks. Furthermore, Thomson Reuters Legal's commitment to supporting law firms through this cultural shift, with ongoing support and training, ensures that the integration of AI becomes a part of the firm's DNA, fostering a forward-thinking, technology-savvy workforce.

Future Trends and Continuous Learning

Looking ahead, law firms must stay attuned to the evolving landscape of AI technologies and their implications for legal practice. The rapid advancement in AI presents both challenges and opportunities, requiring law firms to remain agile and proactive in adopting new technologies. Continuous learning and adaptation are essential, with firms needing to monitor emerging AI trends, assess their potential impact, and explore new applications that can further enhance legal services.

Thomson Reuters Legal, with its pulse on the latest developments in legal technology, provides law firms with the insights and tools needed to navigate these future trends. Through research reports, industry analyses, and technology forecasting, Thomson Reuters Legal helps law firms anticipate and prepare for changes in the AI landscape, ensuring that they remain competitive and innovative. This foresight enables law firms to strategically plan for the integration of next-generation AI technologies, ensuring they are well-positioned to leverage advancements in AI for legal analytics, automated decision-making, and enhanced client services.

Sustaining AI Integration and Leveraging Competitive Advantage

The long-term success of AI integration in law firms hinges on their ability to sustain and evolve their AI initiatives. This requires not only maintaining the existing AI systems but also continuously evaluating and enhancing them to meet the changing needs of the legal market. Law firms must establish mechanisms for regular review and improvement of their AI tools, ensuring they remain effective and aligned with the firm's strategic objectives.

Thomson Reuters Legal aids in this ongoing process by providing access to the latest legal technology updates and innovations. With its comprehensive analytics and reporting tools, law firms can measure the performance and impact of their AI solutions, identifying areas for refinement and expansion. This continuous evaluation and enhancement process ensures that law firms can maintain a competitive advantage, using AI as a lever for growth, client satisfaction, and market leadership.

Moreover, the collaboration between law firms and Thomson Reuters Legal facilitates the exchange of knowledge and best practices, enabling law firms to learn from the broader legal tech community. This collaborative ecosystem fosters innovation and co-creation, leading to the development of cutting-edge AI solutions that address the unique challenges of the legal sector.

Embracing AI as a Catalyst for Transformation

In conclusion, the journey of AI integration in law firms is a comprehensive and dynamic process, encompassing strategic planning, operational execution, ethical governance, cultural transformation, and continuous learning. By partnering with Thomson Reuters Legal, law firms can navigate the complexities of AI adoption with confidence, leveraging the firm's resources, insights, and technology solutions to maximize the benefits of AI.

As law firms continue to evolve with AI, they not only enhance their operational efficiencies and service quality but also position themselves as leaders in the innovative landscape of legal services. The successful integration of AI in law practices represents a strategic move towards a future where technology and law converge, creating a legal ecosystem that is more efficient, transparent, and client-focused.

This extensive journey of AI adoption, supported by the strategic partnership with Thomson Reuters Legal, equips law firms with the tools and insights necessary to leverage AI as a transformative force in the legal industry. With a commitment to continuous improvement and innovation, law firms can harness the power of AI to redefine their practices, offering unparalleled service to their clients and setting new standards of excellence in the legal profession.

Chapter 25:

Advancing Legal Practices Through Ethical AI

At the intersection of artificial intelligence (AI) and the legal profession lies a terrain ripe with ethical considerations and an urgent need for the careful application of AI technologies. Within this evolving landscape, Fox Rothschild has emerged as a leader, actively shaping the discourse and implementation of AI within the industry. Their dedication to addressing the multifaceted challenges of AI in legal practice is manifest in their insightful articles, such as "Ethical and Thoughtful Use of AI in the Legal Industry" and the firm's significant step in establishing a formal AI Practice, accompanied by the appointment of its first Chief Artificial Intelligence & Information Security Officer. This appointment reflects a broader trend where law firms are increasingly recognizing the importance of dedicated roles to navigate the complex interplay between AI, legal practice, and information security.

The transformative potential of generative AI in legal practices is evident through its capacity to streamline operations and bolster productivity. By automating routine and repetitive tasks, legal professionals are afforded the liberty to redirect their focus towards more nuanced and strategic endeavors. Nevertheless, the rapid progression of AI technologies calls for a balanced and judicious approach to their deployment within the legal sphere, aimed at curtailing risks and upholding ethical standards. The narrative that AI technologies could potentially supplant lawyers is being reframed; it's not the technology that poses a threat but rather the lawyers who fail to harness the power of AI that risk obsolescence. It is the adept use of AI that promises to accelerate legal professionals ahead of the curve, transforming the practice and business of law.

A critical distinction must be drawn between the publicly available AI chatbots and the specialized, legally oriented AI tools. While the former, such as ChatGPT and Microsoft Bing, have gained widespread popularity, they may not always align with the stringent requirements of legal practice concerning data confidentiality and security. On the other hand, tools like Casetext's Co-Counsel and Henchman's contract drafting tool are designed with the specific needs of the legal profession in mind, ensuring reliability and security in their application.

Fox Rothschild has recognized these nuances and, in response, has formalized its nationwide Artificial Intelligence Practice. Under the adept leadership of Mark McCreary, a partner with the firm since 1998 and now its Chief Artificial Intelligence & Information Security Officer, the practice group has expanded to over 25 members. This initiative illustrates the firm's commitment to making sure that knowledgeable attorneys, who are abreast of the fast-evolving AI landscape, are at the helm of AI-related decision-making. McCreary's leadership is poised to help both the firm and its clients navigate the intricate legal issues spawned by the proliferation of AI. The firm's AI Practice concentrates on fostering responsible, ethical, and legal AI usage, encompassing regulatory compliance, intellectual property protection, and the cultivation of ethical AI principles.

In light of the pivotal role AI plays in legal operations, Fox Rothschild has been proactive in helping its clients harness AI technology responsibly. The firm's AI team collaborates with businesses to formulate policies that align with applicable laws and regulations, mitigate risks, and propel innovation. This guidance extends to grappling with the intricate data privacy, security, and transparency issues that AI can engender, further encompassing aspects such as regulatory compliance, contractual agreements, intellectual property protection, dispute resolution, litigation, and ethical AI.

Mark McCreary, with his prior role as Fox Rothschild's first Chief Privacy Officer and his certifications from the International Association of Privacy Professionals, stands as a testament to the firm's leadership in the privacy and AI realms. His expertise is sought after in ensuring compliance with data security and privacy laws, rules, and regulations, and in assisting businesses to fortify their defenses against data breaches and to respond effectively when they occur.

In synthesizing Fox Rothschild's approach and the broader legal industry trends, several guidelines for the ethical and effective integration of AI in legal practices emerge:

1. **Policy Adoption**: It is imperative for law firms to institute policies that uphold ethical obligations and establish parameters for AI usage. Fox Rothschild's endeavor to navigate the complex legal implications of AI is characterized by its development and implementation of policies that ensure compliance and foster an ethical framework for AI application.

2. **Training and Education**: To ensure AI is used responsibly and its ethical implications are fully grasped, lawyers must receive thorough training. Such educational initiatives are a staple at Fox Rothschild,

emphasizing the firm's belief in equipping its professionals with the knowledge required to navigate the AI landscape effectively.

3. **Augmentation, not Replacement**: The role of AI should be to enhance the capabilities of legal practitioners, not to supplant them. Fox Rothschild's approach underscores the importance of AI as a tool to support, not replace, the expertise and judgment of its legal professionals.

4. **Accuracy and Compliance**: A core tenet of Fox Rothschild's AI policy is ensuring that all AI-generated documents undergo meticulous review for accuracy and compliance with legal standards. This process safeguards the integrity of legal work and upholds the reputation of the firm for maintaining rigorous quality control.

5. **Transparency with Clients**: Transparency is central to client relationships at Fox Rothschild. Clients are informed about the use of AI in the creation of their legal documents. This openness is essential to building trust and facilitating a comprehensive understanding of the technological tools at work in their legal services.

6. **Continuous Learning**: The legal field, especially in the context of AI, is ever-evolving. Fox Rothschild prioritizes staying abreast of AI advancements and industry best practices to ensure their application of AI remains ethical, responsible, and cutting-edge.

Through these principles, Fox Rothschild has established itself as a vanguard of the ethical application of AI in the legal profession. The firm's initiatives echo a broader commitment to navigate the complex intersection of law and technology with foresight and integrity. This is vital in an era where the capabilities of AI are rapidly expanding, and the implications of its application are increasingly significant.

The integration of AI in legal practices is not without its challenges, as illustrated by a New York lawyer's overreliance on ChatGPT for case law, leading to the inclusion of fictitious citations in a legal brief. This incident underscores the critical need for a cautious and educated application of AI in legal research to prevent such ethical breaches. In response to these kinds of challenges, Fox Rothschild's formalized Artificial Intelligence Practice provides an exemplary model for the industry, showcasing a firm that is not only prepared for the future of legal practice but is actively shaping it.

The AI Practice at Fox Rothschild, with over 25 experienced professionals, offers a comprehensive suite of services designed to ensure that AI is utilized in a manner that is not only innovative but also adheres to ethical and legal principles. This proactive stance is crucial in addressing the growing concern within the legal profession regarding the responsible use of generative AI, as evidenced by U.S. District Judge Brantley Starr's order. His directive, while controversial, signals a heightened awareness and scrutiny of AI in legal proceedings and the imperative for law firms to operate within carefully considered ethical frameworks.

Fox Rothschild's commitment to ethical AI is further exemplified in the leadership of Mark McCreary. His extensive experience and thought leadership in AI and privacy space serve as the backbone of the firm's strategy in navigating AI-related legal challenges. McCreary's role as Chief Artificial Intelligence & Information Security Officer underscores the firm's forward-thinking ethos, emphasizing the importance of specialized knowledge and strategic oversight in the utilization of AI technologies.

The dynamic nature of AI in the legal sector demands that firms like Fox Rothschild remain vigilant and adaptable. The firm's initiatives serve as a blueprint for the legal industry, advocating for responsible innovation while maintaining a steadfast commitment to ethical standards. By doing so, Fox Rothschild not only safeguards its clients' interests but also contributes to the sustainable advancement of legal practice in the age of AI.

The future of AI in legal practice is undeniably bright, with its potential to redefine efficiency, unlock new capabilities, and enrich the quality of legal services. However, this future hinges on the responsible stewardship of AI, as demonstrated by Fox Rothschild's pioneering efforts. As AI continues to evolve, it will undoubtedly present new ethical dilemmas and regulatory challenges. Law firms that follow in the footsteps of Fox Rothschild, embracing a culture of continuous learning, ethical application, and transparent communication, will not only thrive but will also ensure that the legal profession continues to serve its fundamental role in society with integrity and excellence.

Fox Rothschild's strategic approach to AI in legal practice serves as a testament to the potential of technology to enhance the profession. By prioritizing ethical considerations and responsible application, the firm positions itself at the leading edge of legal innovation. This dedication ensures that AI is not a force for displacement within the industry but a transformative tool that augments the expertise and effectiveness of legal practitioners. The forward-looking initiatives and robust AI framework established by Fox Rothschild set a benchmark for the legal industry, ensuring that AI is leveraged to its full potential while adhering to the highest ethical and legal standards.

As the legal profession continues to adapt to the advancements in AI, the example set by Fox Rothschild provides valuable insights into the ethical incorporation of these technologies into legal practices. By maintaining a vigilant stance on AI's evolving capabilities and potential pitfalls, the firm ensures that its application of AI technologies remains ethical, legal, and at the forefront of the industry's best practices. In an era characterized by rapid technological change, Fox Rothschild's approach to AI in legal practice exemplifies a commitment to excellence, innovation, and unwavering ethical standards that will undoubtedly shape the future of the legal profession.

Ensuring Responsible AI Applications in Legal Practices

The AI Now Institute stands at the forefront of a movement toward responsible and ethical artificial intelligence, with its mission deeply rooted in the betterment of AI's societal impacts. Since its inception in 2017, it has been instrumental in promoting the thoughtful and conscientious use of AI technologies. Their mission, firmly grounded in enhancing AI's societal impacts, places a particular emphasis on the need for AI applications to be transparent, fair, and accountable. In a profession that upholds justice and equity as its cornerstones, the interweaving of AI introduces not only efficiencies but also novel ethical complexities.

AI Now's extensive research and scholarship illuminate the potential risks and biases present within AI systems employed in the legal field. Such work is critical in educating policymakers, legal practitioners, and the general public on the intricacies of AI, advocating for a greater understanding and revealing of the processes behind AI decision-making. By advocating for the demystification of the 'black box' of AI technologies, AI Now encourages practices that support, rather than undermine, the foundational principles of the legal system.

The institute's contributions extend beyond mere advocacy. They provide concrete guidance and best practices for the ethical deployment of AI in legal settings. Their research delves into the delicate interplay between technological innovation and ethical considerations. For legal practitioners, AI Now's guidelines serve as a beacon to navigate the use of AI tools, such as those used for legal research, case predictions, or document analysis. The aim is not merely to utilize AI for its capabilities but to enhance service quality without compromising ethical standards. AI Now's stance on transparency advocates for AI systems that assist in decision-making to be understandable and their reasoning accessible, aligning with the legal system's commitment to justice and individual rights.

The influence of AI Now extends to policy and regulation, with the institute's findings often contributing to shaping the discussions around legal frameworks for AI. Their advocacy has been recognized by significant policy bodies, including the US Federal Trade Commission, demonstrating the depth and impact of their work. AI Now's contributions are directed towards establishing regulations that protect against bias and preserve fairness and due process in the use of AI within legal contexts. The institute's insights ensure that the governance of AI in legal applications respects the right to a fair trial and the integrity of legal proceedings.

In the arena of educational initiatives and stakeholder engagement, AI Now's influence is equally pronounced. Through specialized workshops and educational programs, they enhance legal professionals' understanding of AI's ethical dimensions. Recognizing the need for an interdisciplinary approach, AI Now fosters collaboration between technologists, legal experts, and policymakers. This cooperative effort is crucial for legal professionals to not only understand the responsible use of AI but also to develop the acumen to critically evaluate its role in the legal system. AI Now's vision for ethical AI in law resonates at multiple levels, from individual practice to broader institutional policies, thanks to its active engagement with various stakeholders.

As the legal profession increasingly incorporates AI into practice, the AI Now Institute's advocacy for responsible AI utilization becomes ever more important. They call for a continuous learning and adaptation approach to ensure that as AI technologies evolve, so does the profession's understanding and application of these tools. The institute's work serves as a reminder that adopting AI in law is not simply a technical upgrade but a significant ethical commitment. The journey of AI in the legal sector is emblematic of the enduring need for vigilance and dedication to the values of justice and fairness.

Furthermore, the development of AI in law is a profound ethical undertaking, signifying more than just a technological evolution. It is a journey that reiterates the importance of upholding justice and fairness, pillars of the legal profession, in the face of advancing technology. The legal community's commitment to these values must be steadfast as it harnesses AI to enhance the quality and efficiency of legal services. By emphasizing ethical AI use, the legal profession maintains its fundamental values even as it embraces innovation.

Partnerships between AI researchers, developers, and legal professionals are critical for advancing the ethical use of AI in legal settings. Such collaborations can foster the development of AI tools designed specifically for legal applications, embedding ethical considerations from their inception. A partnership approach encourages ongoing improvement of AI technologies, ensuring they serve the dynamic needs of legal practice effectively and ethically.

Regulatory frameworks play an essential role in guiding responsible AI use, and legal professionals are key players in shaping policies that promote ethical AI development and use. This engagement with regulatory authorities is crucial to ensure that AI applications in the legal domain comply with ethical standards, professional obligations, and societal expectations.

In summary, promoting responsible AI use in legal settings requires a collective effort dedicated to principles of transparency, accountability, fairness, and privacy. These principles form the bedrock of ethical AI application in law, ensuring that as the legal profession ventures further into this technological frontier, it remains anchored to its core values.

The AI Now Institute has become a beacon for this movement, fostering an environment where legal professionals can not only deploy AI technologies but also imbue them with the values of the legal tradition. Through ongoing research, they provide a critical examination of how AI technologies are developed and applied, highlighting areas where the law must evolve alongside these advancements to ensure that rights and liberties are protected in the digital age.

Their work does not stop at theoretical research; it extends to practical engagement with the global legal community. By pushing for robust educational programs, AI Now ensures that legal professionals are equipped with the knowledge and tools needed to understand and navigate the complexities of AI. The institute's advocacy for interdisciplinary collaboration has resulted in a shared language and understanding between technologists and legal experts, bridging gaps and fostering a cohesive approach to AI governance.

The policy implications of AI Now's work cannot be understated. Their engagement with policymakers and regulatory bodies shapes the very frameworks within which AI operates in legal settings. The institute's insights have contributed to policy changes that prioritize ethical considerations and the safeguarding of rights within the realm of artificial intelligence. This policy influence is crucial in an era where AI can have profound implications for justice and fairness.

Educating legal professionals and engaging with a broader coalition of stakeholders positions AI Now as a leader in shaping the ethical use of AI. By equipping legal practitioners with the knowledge to question, critique, and ultimately harness AI responsibly, AI Now ensures that the legal industry remains at the forefront of ethical AI application. The institute's commitment to public engagement has fostered a more informed dialogue about AI and its role in society, creating a more receptive environment for the responsible development and use of these technologies.

The future directions that AI Now advocates for are rooted in the understanding that the journey with AI is one of continuous evolution. The institute emphasizes the importance of staying ahead of the curve, not only in technology but also in ethical considerations. As AI systems become more complex, the legal profession's understanding and governance mechanisms must also advance. AI Now's role in this is crucial, as they continue to provide the research, advocacy, and education needed to inform these developments.

In conclusion, the work of the AI Now Institute encapsulates a critical truth: AI's integration into law is not only about the power of technology to transform practices but also about ensuring that this power is harnessed ethically. The legal sector's engagement with AI reflects a commitment to using these new tools in ways that honor the legal profession's enduring principles. Through the institute's efforts, the legal community is reminded of its duty to not only embrace technological advancements but to do so with a conscientious approach that upholds justice, fairness, and integrity.

The continuous collaboration between legal professionals and AI experts, guided by the research and advocacy of institutions like AI Now, is essential for ensuring that AI tools serve the cause of justice. This collaboration aids in the creation of AI applications that are not only technologically advanced but also ethically sound and legally robust. Legal professionals are thus empowered to use AI as a means of enhancing the delivery of legal services while also protecting the core values of their profession.

In light of the above, it becomes clear that the legal profession's relationship with AI is not merely about adopting new technologies. It is about shaping these technologies in a way that is congruent with the ethos of the legal field. The AI Now Institute has been a catalyst in this process, ensuring that as the capabilities of AI expand, so does our understanding of its ethical dimensions.

The institute's emphasis on transparency and accountability in AI applications helps demystify these technologies, making them more accessible and understandable to legal professionals and their clients alike. This transparency is essential for building trust in AI systems and for ensuring that they are used in a manner that complements, rather than complicates, the pursuit of justice.

Moreover, AI Now's work in addressing the issue of bias in AI is of paramount importance. By highlighting the ways in which AI can inadvertently perpetuate existing societal biases, the institute plays a critical role in advocating for the development of AI tools that are fair and equitable. This focus on combating bias ensures that AI serves to enhance the fairness of legal processes, rather than detracting from it.

The AI Now Institute also places a significant emphasis on data privacy and security, recognizing that the integration of AI into legal services raises new concerns about the protection of sensitive information. Legal professionals deal with confidential client data daily, and the institute's guidelines help ensure that AI tools are developed with the strongest data protection measures in mind.

The importance of ongoing education and stakeholder engagement in the responsible use of AI in legal settings cannot be overstated. AI Now's commitment to these efforts ensures that the legal community is not only informed about the latest developments in AI but also actively involved in shaping how these technologies are used. This commitment to education and dialogue ensures that the integration of AI into legal practice is done thoughtfully, with a comprehensive understanding of both its potential and its limitations.

Engagement with technology providers and regulatory bodies is another area where the AI Now Institute has been particularly effective. By fostering dialogue between these groups and the legal profession, the institute helps to ensure that the development and deployment of AI in legal settings are in line with ethical principles and legal standards. These collaborative efforts are essential in creating a shared vision for the future of AI in law, where innovation is balanced with the need to protect individual rights and uphold the rule of law.

The role of AI Now in developing ethical standards and best practices for AI in legal applications is a testament to the need for industry-specific guidelines. These standards consider the unique challenges and ethical dilemmas that arise in legal practice, such as ensuring client confidentiality, maintaining privilege, and adhering to the adversarial nature of legal proceedings. By providing a framework for ethical AI use, AI Now aids legal professionals in navigating the complex landscape of modern legal technology.

AI Now's influence is also evident in the increasing attention given to the post-deployment monitoring and evaluation of AI tools. Regular assessments of the impact of AI on legal practice, client outcomes, and the broader justice system are vital for ensuring that AI strategies are aligned with ethical and professional standards. These evaluations serve as a feedback mechanism, allowing for the refinement of AI tools to better meet the needs of the legal profession and the clients it serves.

The institute's advocacy for such rigorous evaluation and adaptation of AI tools in law underscores the dynamic nature of technology and the necessity for the legal profession to remain flexible and responsive. As new AI applications emerge, they must be reviewed not only for their technical efficacy but also for their alignment with ethical standards and legal norms.

The AI Now Institute's role in the legal sector thus extends far beyond research and policy. It is about fostering a culture of responsibility and ethics in the use of AI. The institute's work encourages legal professionals to take a proactive stance, not just reacting to technological advancements but actively shaping them to serve the greater good.

In recognizing the multifaceted nature of promoting responsible AI use in legal settings, the legal profession must appreciate the importance of a multidisciplinary approach. Lawyers, judges, and legal scholars must engage with AI developers, ethicists, and data scientists to ensure that AI tools are not only technically competent but also ethically designed and socially beneficial. The AI Now Institute's commitment to fostering such interdisciplinary engagement is critical to achieving this goal.

The institute's influence is expected to continue growing as AI becomes an increasingly ubiquitous part of the legal landscape. The ethical challenges posed by AI will only become more complex, and the guidance provided by AI Now will be essential for navigating these challenges. Their ongoing work will help ensure that as AI technologies evolve, they do so in a way that respects the core values of the legal profession and the rights of the individuals it serves.

In conclusion, the AI Now Institute's contributions to the legal field are invaluable. They not only inform and guide the profession's approach to AI but also shape the very fabric of legal ethics in the age of artificial intelligence. As legal professionals, there is an obligation to ensure that the use of AI in law not only improves efficiency and service delivery but also adheres to the highest ethical standards. The institute's research and advocacy are pivotal in this regard, helping to create a future where AI is used responsibly, ethically, and in service of the fundamental principles of justice and fairness.

The AI Now Institute underscores the importance of a proactive and informed legal profession that does not shy away from technology but embraces it with a clear understanding of its potential and pitfalls. By upholding these principles and continuing to engage with the work of the AI Now Institute, the legal community can ensure that AI serves as a partner in the quest for justice, enhancing the legal profession's ability to serve society while remaining steadfast in its commitment to ethical practice. This collaboration between human expertise and artificial intelligence, guided by the principles and values that define the legal profession, promises not only a more efficient and effective legal system but one that continues to uphold the rights and dignity of all individuals.

Developing Ethical Guidelines for AI in Legal Systems

The intersection of Artificial Intelligence (AI) with legal practice marks an epochal shift, heralding untold potential for streamlining the delivery of legal services and enhancing the procedural efficiencies of the sector. However, with this rapid integration comes a compelling imperative to establish a set of comprehensive ethical standards. These standards must serve as stalwart guardians of the profession's foundational principles: justice, equity, and professionalism in an age of relentless technological change.

At this juncture, the indispensable roles of Data & Society and The Future of Life Institute in forging the discourse on ethical AI in the legal sphere must be acknowledged. Data & Society, through its incisive empirical research, dissects the multifaceted societal implications of data-centric technologies and AI. It champions the cause of crafting equitable technological futures, ones that honor human dignity and ensure that AI applications serve the cause of justice rather than exacerbate existing societal fault lines.

Simultaneously, The Future of Life Institute (FLI) embarks on its mission with a strategic foresight that underscores AI's safe and beneficial deployment. It stands as a bulwark against the risks associated with transformative technologies, navigating the potential perils while maximizing AI's profound capabilities for enhancing life. Together, these institutions weave a comprehensive tapestry of insights that lay the groundwork for the ethical scaffolding necessary for the judicious integration of AI into legal practice.

Delving deeper, Data & Society's influence on legal AI has been nothing short of transformative. As a think tank at the forefront of examining AI's societal consequences, it casts a critical eye on the often opaque interplay between technology and social structures. Its mission, firmly grounded in meticulous research and proactive engagement, seeks to unravel the intricate tapestry of values and decisions that underlie the design and deployment of AI systems. In doing so, the institute underscores the imperative of guiding technological innovation with a steady hand, ensuring that our collective future is shaped with a conscientious regard for ethical principles.

The institute's findings on the nuanced intrusion of biases within AI systems have been revelatory, particularly for the legal profession, which stands as a sentinel of justice. Its scholarly endeavors serve as a repository of wisdom, guiding legal practitioners in sifting the ethical grains from the chaff of mere efficiency. By promulgating a balanced approach that tempers the enthusiasm for innovation with an unwavering commitment to social equity, Data & Society's research becomes an ethical compass for legal AI applications.

It is this alignment of Data & Society's empirical research with the legal field's ethical imperatives that offers a blueprint for AI technologies that resonate with the core principles of legal practice. The institute's independence from commercial influence ensures its contributions to the discourse on legal AI ethics are devoid of the biases that often underpin technologically driven agendas, proffering a vision firmly rooted in the pursuit of communal well-being and justice.

In a parallel vein, The Future of Life Institute's work on AI safety brings to the forefront the importance of vigilance and proactive measures in the deployment of AI within the legal sector. FLI's commitment to steering transformative technologies like AI toward beneficial ends while mitigating inherent risks is of paramount relevance to the legal field. The institute's advocacy for a rigorous framework for AI safety—which emphasizes reliability, security, and control—provides an ethical template for the development and application of legal AI tools.

FLI champions the proactive measures to preclude AI's potential for misuse and adverse consequences. This emphasis on safety and precaution aligns with the legal sector's duty to anticipate and forestall situations where AI might inadvertently transgress rights, favor certain demographics, or otherwise act counter to the public's interest. FLI's work in this area is a touchstone for those at the helm of developing and implementing AI in legal systems, consistently reminding them to consider the broader societal implications of their technological choices.

Upon comparative analysis, the approaches of Data & Society and The Future of Life Institute to AI ethics, though distinct in their trajectories, are remarkably synergistic. Data & Society's granular research into the societal intricacies of AI development complements the strategic, broad-stroke initiatives of FLI. In the domain of law, this dual focus translates to AI systems that are socially responsible and operationally secure, ensuring adherence to the rule of law and fostering a foundational public trust.

This synergy offers a holistic framework for the development of ethical AI standards in the legal field. The amalgamation of Data & Society's deep-dive analyses and FLI's strategic vision suggests a model for legal AI that withstands the dual tests of social conscience and technical rigor, advocating for AI systems that are not only efficacious but are also subjected to the scrutiny of public transparency and ethical accountability.

Indeed, the real-world implications of the principles espoused by Data & Society and The Future of Life Institute are palpable in a number of illustrative case studies. Consider, for instance, the realm of AI in predictive policing and bail-setting—a space where Data & Society's advocacy for equity and inclusivity has proven to be instrumental. By integrating rigorous checks against historical biases, AI-assisted legal decisions are beginning to transcend the confines of past inequities, aiming for a jurisprudence that is fair and just. Similarly, FLI's principles of AI safety and precaution are mirrored in the burgeoning field of document review and legal analytics, where advanced AI systems are being designed with an overarching emphasis on safeguarding client confidentiality and upholding the integrity of legal processes.

These case studies are more than academic musings; they represent the tangible confluence of research and practice. As legal professionals engage with advanced AI technologies, they draw upon the foundations laid by Data & Society and FLI, navigating the ethical complexities and operational challenges that accompany AI adoption. The legal sector's commitment to these ethical benchmarks illustrates a resolve to uphold societal values and the sanctity of professional integrity.

Looking to the future, the influence of Data & Society and The Future of Life Institute on AI ethics within the legal profession is poised to continue shaping the ethical landscape. As AI technologies become more sophisticated and deeply integrated into legal practice, the interplay between empirical research and strategic foresight will become increasingly crucial in guiding the development and refinement of ethical standards.

This will necessitate a legal profession that is not only technologically adept but also ethically astute. Data & Society and FLI will likely continue to play pivotal roles in advocacy and education, ensuring that legal professionals are prepared to face the nuanced challenges of AI with discernment and responsibility. Moreover, as AI tools become more prevalent in legal settings, these institutions' emphasis on public engagement and inclusivity will become critical in ensuring that AI systems are developed and deployed transparently and reflect a broad range of societal values and perspectives.

The potential for AI to contribute positively to the legal profession cannot be overstated. It promises enhanced access to justice, greater efficiency in legal processes, and the augmentation of human legal expertise. However, the full realization of this potential hinges on the meticulous crafting and application of ethical standards that are as rigorous as they are flexible. The frameworks and philosophies provided by Data & Society and FLI will undoubtedly play a critical role in shaping these standards, advocating for an approach to AI that is transparent, accountable, non-discriminatory, and aligned with the public interest.

As the legal profession continues to evolve in response to AI's increasing role, the standards established through the collaborative efforts of thought leaders and organizations like Data & Society and FLI will serve as both anchor and compass. They will ensure that the development and deployment of AI in the legal domain are rooted in ethical considerations, guiding its application towards the betterment of the legal system and the welfare of society as a whole. The journey ahead will require continuous vigilance, adaptability, and a commitment to the principles that define the legal profession and the broader society.

The establishment of ethical AI standards in law is thus an iterative, ongoing process that necessitates constant dialogue, reassessment, and refinement. Through the concerted efforts of institutions like Data & Society and The Future of Life Institute, along with the engagement of the legal community with their research, this endeavor is set to progress robustly. As AI reshapes the legal landscape, all stakeholders must ensure that this transformative technology is harnessed with integrity, foresight, and an unwavering commitment to the ethical principles that underpin the pursuit of justice and equity.

In conclusion, crafting standards for ethical AI in law is a foundational commitment to justice, equity, and professionalism, as well as a practical engagement with the realities of technological innovation. It requires legal practitioners to integrate AI into their practices thoughtfully, transparently, and in a manner that serves the greater societal good. By adhering to the ethical standards shaped by the research and advocacy of Data & Society and FLI, the legal profession can navigate the complexities of AI with confidence, ensuring that its transformative potential is realized for the benefit of all.

This rich tapestry of ethics and technology, woven through the collective efforts of the legal community, heralds a new era for legal services—one where the efficiency and efficacy promised by AI are matched by an equally strong commitment to ethical practice. This is the guiding light that will ensure AI's role in law not only enhances the practice but also does so with a steadfast dedication to ethical integrity and the welfare of society. It is a collective journey that calls for engagement, diligence, and a commitment to the highest standards of ethics and professional conduct, forging a future where AI is unequivocally a force for good in the legal sphere.

Chapter 26:

AI's Role in Democratizing Justice: Bridging the Legal Divide

The dawn of Artificial Intelligence (AI) in the justice system signals a transformative era where the potential for broadening access to justice has never been more tangible. The current justice landscape, with its formidable complexities, presents an array of barriers that obstruct the equitable distribution of legal services. Yet, in this challenging setting, AI emerges as a beacon of hope, promising to level the playing field for those previously marginalized by economic, social, and geographical disparities.

As we unpack the concept of access to justice, it's clear that the obstacles it encompasses—financial burdens, convoluted legal procedures, and a lack of legal literacy—are deeply entrenched. The intersection of AI within this realm brings forth a vision where such barriers can be not just mitigated but dismantled.

Historically, the pursuit of justice has mirrored the progression of society, evolving as a testament to the values and technological milestones of each era. Access to justice, once a privilege of the few, has seen a gradual democratization, particularly with the advent of digital technology. The legal system's initial tryst with technology—through computers and the internet—laid the groundwork for what AI is poised to further revolutionize. AI's entry into the legal sector isn't merely an advancement; it's a redefinition of how legal services are conceived and delivered.

The promise of AI extends beyond mere automation to fundamentally democratizing legal services. By leveraging advancements such as machine learning and natural language processing, AI is set to overhaul the availability and affordability of legal aid. It is a harbinger of a future where legal support is not an exclusive service but a universal right. The capability of AI to sift through and analyze legal data with precision heralds a new chapter where legal advice becomes an accessible resource for the broader public. In doing so, AI has the potential to bridge the gap between legal services and those who need them most.

The role of AI in overcoming the traditional barriers to justice is multifold. Economically, it introduces efficiencies that drastically reduce costs, making legal services more accessible to individuals of all financial backgrounds. AI's ability to optimize resource allocation in legal practices means that the benefits of legal advice can be extended to a wider audience at a fraction of the traditional cost. Social barriers, too, are addressed as AI reaches out to marginalized communities, providing legal tools and services specifically designed to meet their unique requirements. Geographically, AI conquers the limitations of distance, delivering legal services to remote and underserved regions through sophisticated digital platforms, thus transcending the physical confines of traditional law offices.

AI's impact is palpable within the domain of legal assistance, as evidenced by its integration into legal research and document analysis. AI-powered tools are redefining these processes, not only by enhancing efficiency but by augmenting the accuracy of legal work. The development of AI chatbots and virtual assistants, capable of dispensing preliminary legal advice, marks a significant stride in the provision of legal services. These AI systems, empowered with the ability to understand and process natural language, make personalized legal information more accessible, thereby demystifying the complexities of the law. Moreover, the advent of self-help legal tools, driven by AI, enables individuals to address and resolve straightforward legal issues on their own. This empowerment is a crucial step in expanding the reach and impact of legal assistance.

The integration of AI in public legal education marks another critical stride towards achieving universal access to justice. Through interactive AI platforms, the public is afforded a means to grasp their legal rights and the intricacies of legal processes in a manner that is both engaging and accessible. This educational role of AI is indispensable; it lays the foundation for a society where individuals are equipped to make informed legal decisions and engage with the legal system with confidence and understanding.

In breaking down accessibility barriers, AI introduces platforms that simplify the legal process. Initiatives employing AI to simplify legal jargon and procedures are not only enhancing public understanding but are also revolutionizing the manner in which legal knowledge is imparted and perceived. These platforms, designed with user-friendliness in mind, play a pivotal role in fostering an inclusive environment where justice is genuinely accessible to everyone.

However, the path to integrating AI into legal practices is fraught with challenges that must be carefully navigated. Foremost among these is the safeguarding of data privacy, a critical consideration when handling the sensitive information inherent in legal matters. The reliability and accuracy of AI's predictions and the counsel it provides are also areas that demand meticulous scrutiny and ongoing calibration to ensure justice is upheld. Beyond these technical challenges lie broader ethical concerns—the moral implications of automated legal advice must be reconciled with the fundamentally human context of justice.

Looking ahead, the future of AI in the legal domain is poised for greater sophistication and deeper integration. The responsibility to embrace and integrate AI into legal practices falls squarely on the shoulders of legal professionals. They must remain vigilant and proactive in ensuring AI is utilized as a tool to enhance rather than replace the human elements of the legal profession—empathy, ethical judgment, and the nuanced understanding of human affairs.

In conclusion, the integration of AI into the legal system holds substantial promise for closing the justice gap, offering new pathways to enhance the efficiency and reach of legal services. This potential is not without its caveats; a measured and conscientious approach is necessary to ensure that AI serves as a complement to, rather than a substitute for, the nuanced judgment and empathetic engagement that lie at the heart of the legal profession.

AI's journey within the realm of law is at a nascent yet rapidly advancing stage. Its integration has begun to recalibrate expectations and redefine potential within the justice system. By leveraging AI's capabilities, legal services can be rendered more cost-effective and time-efficient, enabling a redistribution of legal professionals' focus towards more impactful and complex cases that require a deeper level of human insight and interaction.

The burgeoning relationship between AI and the legal profession presents an opportunity to reexamine and reinforce the ethical frameworks that guide legal practice. As AI applications become more prevalent, ensuring that these systems are governed by principles that uphold justice, equity, and confidentiality becomes paramount. Legal practitioners must be at the forefront of this discourse, advocating for and shaping the implementation of AI in ways that reflect the core values of the profession.

In an ideal future, AI will act as an equalizer in the legal field, addressing not just the symptoms but the root causes of the justice divide. It could potentially transform legal systems into more transparent, responsive, and user-centered institutions. Such a transformation would not only benefit those seeking legal assistance but also invigorate the practice of law itself.

The responsibility of legal professionals in adopting AI extends to a commitment to continuous learning and adaptation. As AI evolves, so too must the knowledge and skills of those who wield it. This entails understanding not only the capabilities of AI but also its limitations and fostering a symbiotic relationship where technology is wielded as a tool for human betterment.

AI's full impact on access to justice will ultimately be measured by the extent to which it enables individuals to exercise their rights and navigate the legal system with confidence. Initiatives that simplify legal procedures, provide affordable legal services, and promote legal literacy are fundamental to this endeavor. In parallel, there must be a concerted effort to monitor and mitigate any unintended consequences of AI deployment, such as biases in algorithms or the erosion of personal interactions in the legal process.

The way forward requires a collaborative effort among technologists, legal professionals, policymakers, and society at large. Together, these stakeholders must engage in an ongoing dialogue to ensure that AI's development is aligned with the goal of enhancing access to justice. This includes not only addressing the technical and ethical challenges but also actively working to ensure that the benefits of AI in the legal domain are widely and equitably distributed.

As we reflect on the role of AI in the justice system, it is essential to maintain a balance between enthusiasm for the opportunities it presents and caution for the responsibilities it entails. The advent of AI in law is not an endpoint but the beginning of an ongoing process of improvement and refinement. The ultimate goal is not to replace the human element of the legal profession but to empower it, ensuring that justice is more than an ideal—it is a tangible, accessible reality for all.

In sum, AI represents a pivotal shift in the landscape of legal services, one that carries the potential to transform access to justice into a more inclusive and equitable reality. Yet, the path to achieving this vision is complex and requires a nuanced approach that respects the delicate balance between technological innovation and the timeless human aspects of the legal profession. With careful consideration and responsible integration, AI can and should be harnessed as a powerful ally in the ongoing quest for a more just society.

Transforming Legal Access: The Rise of AI-Driven Platforms like DoNotPay

In the realm of legal services, artificial intelligence (AI) has emerged as a transformative force, particularly through platforms like DoNotPay, which exemplify the democratization of legal assistance. Traditionally, accessing legal services was fraught with challenges such as prohibitive costs, intricate legal jargon, and the daunting complexity of legal procedures. However, the advent of AI-driven solutions like DoNotPay has started to dismantle these barriers, offering a glimmer of hope for universal legal access. DoNotPay, originally conceived to help individuals contest parking tickets, has evolved into a comprehensive legal assistant, addressing a wide spectrum of legal issues from consumer rights disputes to small claims court cases. This evolution signifies a paradigm shift toward making legal assistance not just a privilege for the affluent or well-informed but a fundamental right accessible to all.

The genesis of DoNotPay reflects a broader mission to democratize legal services, encapsulating the belief that the complexities of the law should not impede justice. Starting with the simple yet impactful task of contesting parking tickets, DoNotPay has expanded its scope to encompass various legal domains, illustrating the potential of AI to streamline and simplify legal processes. This journey from a singular service to a multifaceted platform highlights the scalability and adaptability of AI in addressing diverse legal needs. DoNotPay's mission resonates with the principle that legal services should be comprehensible, affordable, and accessible, aligning with the broader objective of ensuring that legal support is not an exclusive commodity but a universal service.

AI's role in legal technology, particularly through innovations like

DoNotPay, is revolutionary. The integration of AI in legal services transcends mere automation; it involves the intelligent analysis of legal documents, the generation of legal correspondence, and the provision of legal advice through sophisticated algorithms. DoNotPay leverages these AI capabilities to decode the complexities of legal systems, making them more navigable for the layperson. This AI-driven approach not only enhances the efficiency of legal services but also makes them more inclusive, bridging the gap between legal expertise and the common individual. The result is a more streamlined, less intimidating process that encourages people to assert their legal rights and navigate the legal system with greater confidence and understanding.

The accessibility and user-friendliness of legal services are at the core of DoNotPay's approach. The platform addresses the common perception of legal processes as inaccessible and perplexing, especially to those without a legal background. Through its intuitive interface and AI-driven interactions, DoNotPay simplifies the legal landscape for users, guiding them through the necessary steps to resolve their legal issues. This simplification extends to the automation of document preparation and submission, reducing the potential for human error and the time-consuming nature of traditional legal processes. By making legal services more approachable and manageable, DoNotPay empowers individuals to take legal action without the fear of navigating complex legal frameworks.

DoNotPay's Broad Spectrum of Services and Its Industry Impact

DoNotPay's suite of services illustrates the broad potential of AI in addressing various legal challenges. Beyond its initial offering of contesting parking tickets, the platform now facilitates a range of legal services, including but not limited to disputing bank fees, securing refunds for unsatisfactory services, canceling unwanted subscriptions, and navigating small claims court proceedings. Each service reflects DoNotPay's commitment to simplifying legal processes, embodying the principle that individuals should have the power to resolve their legal matters without undue stress or financial burden. The platform's ability to distill complex legal actions into manageable, user-driven processes marks a significant step toward democratizing legal services, offering a practical and user-friendly alternative to traditional legal avenues.

The impact of DoNotPay on the legal industry and its consumers has been both profound and multifaceted. For the industry, DoNotPay represents a disruptive force, challenging the status quo and prompting a reevaluation of how legal services are delivered and consumed. For consumers, it has been a liberating tool, providing a means to address legal issues without the traditional barriers of cost, complexity, and accessibility. The platform's success in democratizing legal services goes beyond individual victories in legal disputes; it signifies a shift towards a more equitable legal landscape where justice is more readily available to all. The effectiveness of DoNotPay's AI-driven approach, coupled with its growing popularity, underscores the potential of technology to foster a more just and accessible legal system.

However, the journey of integrating AI into legal services, as exemplified by DoNotPay, is not devoid of challenges and limitations. Concerns regarding data privacy, algorithmic bias, and the ethical implications of AI in legal practice are critical issues that need addressing. The development of comprehensive and robust regulatory frameworks is essential to ensure that the application of AI in legal services is conducted ethically and responsibly. These challenges necessitate a collaborative approach, involving legal practitioners, technologists, and policymakers to navigate the complex interplay between technology and law, ensuring that the benefits of AI-driven legal services are realized without compromising ethical standards or the integrity of the legal system.

In addressing these challenges, the role of continuous innovation and ethical oversight cannot be overstated. As DoNotPay and similar platforms evolve, they must remain vigilant in ensuring that their AI algorithms are transparent, unbiased, and respectful of privacy and legal ethics. This will involve not only technological advancements but also a commitment to ethical practices, regular audits, and adaptations based on legal and societal feedback. The objective is to harmonize the efficiency and accessibility brought by AI with the fundamental principles of justice and equity.

The potential future developments in AI and the democratization of legal services are vast and promising. As AI technology becomes more sophisticated, its integration into legal services is expected to deepen, offering even more innovative solutions to legal issues and further reducing barriers to legal access. This future vision encompasses AI systems capable of handling more complex legal reasoning and decision-making, potentially transforming the landscape of legal advice, representation, and dispute resolution.

In conclusion, DoNotPay's journey from a parking ticket contesting bot to a comprehensive legal service platform epitomizes the transformative power of AI in democratizing legal services. By making legal assistance more accessible, understandable, and affordable, DoNotPay has not only changed how individuals approach legal issues but has also started to shift the traditional dynamics of the legal industry. While challenges remain, particularly in ensuring the ethical and fair use of AI in legal services, the path forward is marked by the potential for greater justice accessibility and empowerment of individuals in legal matters. As the legal profession and technology continue to intersect and evolve, the promise of AI in democratizing legal services offers a hopeful outlook for a legal system that is more inclusive, efficient, and just for all, ensuring that the right to legal assistance is not a luxury but a standard provision accessible to everyone.

Breaking Educational Barriers: How AI Platforms Like Upsolve Empower Users

In the quest for justice and legal empowerment, technology, particularly Artificial Intelligence (AI), has emerged as a pivotal ally. It is reshaping the terrain of public legal education and accessibility, forging new pathways to knowledge and justice that were once inconceivable. This discussion delves into the transformative role AI plays, with a specific focus on how platforms like Upsolve utilize AI to democratize legal assistance for the many, not just the few.

Public legal education is not simply about imparting knowledge; it's about enabling action. Upsolve embodies this philosophy, employing AI to demystify the complexities of bankruptcy—a realm fraught with legal intricacies and societal stigma. Through its AI-driven assistance, Upsolve offers more than just a navigational tool for legal procedures; it presents a comprehensive resource replete with articles, guides, and interactive tools. These resources are carefully crafted to illuminate the nuances of bankruptcy law, enabling individuals to grasp their legal circumstances and make informed decisions.

Upsolve's AI-driven platform operates at the intersection of technology and compassion, extending a lifeline to those drowning in debt. Its user-friendly interface simplifies the convoluted process of filing for bankruptcy, which typically poses a significant financial and intellectual barrier. By streamlining this procedure, Upsolve's AI not only makes the act of filing more accessible but also imbues users with the confidence to self-advocate within the legal system. The platform's genius lies in its ability to translate legal complexities into actionable knowledge, thus equipping individuals with the autonomy to independently navigate the legal framework.

The provision of such a platform is revolutionary, considering the prohibitive costs often associated with legal services. Upsolve leverages the efficiency of AI to provide services at no cost, thereby addressing a critical gap in the justice system. By integrating algorithmic efficiency with legal expertise, the platform not only informs individuals of their rights and options but also provides the means to act on them. This amalgamation of education and actionable guidance signifies a new era where technology and legal proficiency unite to redefine public access to legal systems.

As we examine the specific role of AI in public legal education, Upsolve emerges as a poignant example of the potential for technology to dissolve the barriers of complexity and cost that often accompany legal issues like bankruptcy. The platform transforms a traditionally intimidating process into one that is approachable and manageable. It's not merely about operational efficiency; it's about creating an inclusive environment where legal education is synonymous with empowerment.

The ripple effects of such innovation are vast. AI in platforms like Upsolve does not only assist in the logistical aspects of filing documents or navigating procedural hurdles. More fundamentally, it changes how individuals perceive and interact with the law. The essence of this technological empowerment is the fostering of legal self-efficacy—the belief in one's ability to understand and influence one's legal standing. Upsolve, through its AI-driven tools, provides a scaffold upon which individuals can build a more profound understanding and a more robust approach to their legal challenges.

In summary, the role of AI in public legal education, as exemplified by Upsolve, is one of profound transformation. The platform exemplifies how AI can transcend traditional barriers to justice, highlighting a future where legal processes are not only more efficient but also more equitable. The case of Upsolve affirms the power of AI to revolutionize access to legal knowledge and services, fundamentally changing the landscape of public legal education and accessibility.

The elegance of AI systems like those employed by Upsolve lies in their capacity to distill vast, complicated legal texts into clear, navigable pathways for the layperson. By harnessing advanced computational linguistics and machine learning, AI interprets and simplifies legal language, rendering it into digestible content for users. This is of immeasurable value, as the legalese that often obfuscates understanding is stripped away, making the law not a gatekeeper but a guidepost. In doing so, AI serves as a democratizing force, stripping away the layers of intimidation that can distance justice from those it serves.

The transformative nature of AI is particularly evident when it comes to breaking down socioeconomic barriers. Traditional legal aid, while invaluable, often has capacity limits, unable to meet the sheer scale of need. AI stands in this gap, scalable and unimpeded by the physical constraints that limit human-centric services. With platforms like Upsolve, access to legal guidance is not a commodity rationed to the few but a widespread service available to the many. Herein lies the essence of technology as an equalizer – it does not tire, it does not discriminate, and it does not charge by the hour.

Moreover, Upsolve and similar platforms exemplify how AI facilitates a more inclusive and personalized approach to legal education. By analyzing user interactions and feedback, AI can tailor content and recommendations to fit individual needs, reflecting the unique circumstances and concerns of each user. It is a nurturing of legal understanding that is bespoke, not generic – a nurturing that is critical for effective empowerment.

In essence, AI brings forth a new paradigm in which legal support is not a scarce resource but a ubiquitous tool, as readily available as any basic service. It enables individuals to transcend the barriers of ignorance and helplessness, fostering a sense of agency that is pivotal for meaningful participation in the legal process. Through platforms like Upsolve, AI manifests not only as a source of information but also as an enabler of justice, aligning with the ideal that knowledge of one's legal rights and remedies should be a universal truth, not a privileged secret.

At its core, AI in public legal education is about connectivity. It connects knowledge to need, information to inquiry, and empowerment to the everyday individual. The intersection of AI and law, as exhibited by platforms like Upsolve, is more than a mere convenience; it is a profound expansion of legal consciousness. It is about creating a well-informed populace that can advocate for themselves and others, who can approach the legal system not as outsiders but as informed participants.

This chapter has explored the breadth and depth of AI's role in public legal education and accessibility, as highlighted by the pioneering work of platforms like Upsolve. It has emphasized the transformative potential of AI in making legal knowledge and services available to all, regardless of socioeconomic status, and in bridging the gap between legal systems and the public they are designed to serve.

AI's place in the domain of law signals a shift towards a future where the labyrinthine complexities of legal processes become navigable for everyone. It heralds a time when empowerment through knowledge becomes the cornerstone of a just society. The work of Upsolve is not just a testament to the power of technology but also a call to action for continued innovation and commitment to public legal education and accessibility.

As we stand on the cusp of this new era, we are reminded that the potential of AI in law is not fully realized. The journey ahead requires vigilance to ensure that AI remains a force for good, an advocate for the disenfranchised, and a beacon of hope for a fairer, more inclusive justice system. As legal professionals, technologists, and educators, our collective endeavor must be to harness AI not only to enhance our practices but, more importantly, to bridge the chasms in public legal education and accessibility. It is a road filled with promise, and with the responsible application of AI, the legal landscape can indeed be reshaped into one where justice is accessible to all.

In closing, the role of AI in public legal education and accessibility is a tapestry woven with the threads of innovation, empowerment, and inclusivity. Upsolve stands as a beacon within this tapestry, showcasing how AI can be harnessed to serve the greater good. As we forge ahead, we must carry the mantle of responsibility to ensure that AI continues to be a key player in the quest for an informed and accessible legal system for all.

Technology and Tenant Rights: The JustFix.nyc Model

JustFix.nyc stands as a groundbreaking organization at the nexus of artificial intelligence (AI) and public legal education, with a dedicated focus on advancing housing justice. This organization equips tenants with sophisticated AI-powered tools, significantly streamlining the process of understanding and asserting housing rights. The AI functionalities embedded within JustFix.nyc's system excel at examining comprehensive datasets related to housing, facilitating tenants in meticulously documenting their living conditions, identifying recurring instances of landlord neglect or harassment, and adeptly navigating the intricate landscape of housing laws. This technological empowerment plays a critical role in leveling the legal playing field, particularly for tenants who may not possess the resources to fully comprehend their legal circumstances.

Exploring further the utility of JustFix.nyc's AI, it becomes evident that it plays a vital role in public legal education. The AI technology is intricately designed to decode the complex legal terminology found in rent histories, building code violations, and other housing-related documents. This clarification process is crucial for tenants, who might otherwise find the legal language and proceedings daunting or completely foreign. The AI-driven tools that JustFix.nyc offers provide a structured and analytical method for gathering evidence in housing disputes, which can lead to more equitable and just resolutions.

Beyond providing individual legal assistance, JustFix.nyc is catalyzing a broader cultural shift toward transparency and accountability in the housing market. By granting tenants a clearer understanding of their rights and the legal procedures available to them, JustFix.nyc not only empowers individuals but also fosters a collective movement toward equitable housing justice. This shift is crucial for ensuring that landlords and property managers adhere to fair housing practices and respect tenants' rights. The organization's innovative use of AI exemplifies the potential of technology to serve as a force for social good, assisting individuals in documenting their housing situations and gaining a comprehensive understanding of their legal rights.

The underlying philosophy of JustFix.nyc's approach is the belief that technology, especially AI, should be utilized to serve the public interest, particularly in critical areas such as housing. The work of JustFix.nyc stands as a testament to the power of AI in bridging legal knowledge and access gaps, contributing to a more just and equitable society. The system not only aids tenants in understanding their legal situations but also plays a significant role in the broader context of housing justice, aiming to ensure the universal upholding of fair housing standards.

This initiative is indicative of a wider trend where AI is increasingly applied to social challenges, providing innovative solutions to longstanding issues in public legal education and access to justice. JustFix.nyc's model exemplifies how AI can be utilized to empower disadvantaged communities, enabling them to assert their rights and navigate the legal system more effectively. The organization's efforts offer a beacon of hope, signaling a transition towards a more just and equitable society where technology is integral to legal empowerment and housing stability.

JustFix.nyc represents not merely a technological platform but a movement towards democratizing legal access and achieving justice in housing. Its use of AI to interpret and manage legal information translates into tangible, actionable support for tenants confronting legal challenges. By simplifying the legal process and making it more accessible, JustFix.nyc is contributing to the transformation of the housing justice landscape, making it more inclusive and equitable. The organization's dedication to using AI for social good serves as a potent reminder of technology's potential to drive positive change, particularly in critical sectors like housing and legal access.

Considering the broader impact of JustFix.nyc's work, it is evident that the organization is addressing immediate needs while also fostering a fundamental shift in the provision and consumption of legal services in the housing sector. The integration of AI into public legal education and housing advocacy represents a progressive approach that is set to redefine the dynamics of tenant-landlord relations, ensuring a more balanced administration of justice. Through its innovative application of technology, JustFix.nyc is establishing a model for how AI can be leveraged to address complex societal issues, paving the way for a future where technology and justice are intertwined, creating a more equitable society.

People Clerk: Simplifying Small Claims with AI

In an era where access to justice is often hampered by the complexities of legal procedures, technological advancements have presented a beacon of hope, notably through platforms like People Clerk. This innovative platform stands at the intersection of artificial intelligence (AI) and legal services, extending a lifeline to those embarking on the often daunting journey through the small claims process. The intricacies of legal systems can be unnerving for the uninitiated. Navigating the labyrinthine passages of small claims court, with its filing requirements and procedural nuances, is a formidable challenge that can deter many from seeking the justice they deserve. People Clerk emerges as a formidable ally in this landscape, democratizing the legal process by equipping individuals with a powerful AI-driven toolkit designed to simplify and streamline their legal journey.

At its core, People Clerk harnesses the transformative power of AI to dissect and demystify the small claims process. The platform is engineered to intuitively guide users through each step, from initial document preparation to the intricate dance of filing and managing evidence. Its AI algorithms are meticulously programmed to personalize the experience, tailoring guidance and documentation to the specific circumstances of each case. In essence, People Clerk functions as a personal legal assistant, one that is available around the clock and grounded in the logic of machine learning and natural language processing. The sophisticated AI not only crafts the necessary legal documents but also evolves with each interaction, continuously refining its assistance based on a growing repository of legal data and user feedback.

The democratizing influence of People Clerk cannot be overstated. For many, legal representation is a luxury beyond reach, leaving them to face the Goliath of the legal system alone. This disparity has given rise to what is often referred to as the 'justice gap,' a void where the unrepresented struggle to find their footing. People Clerk fills this void by granting users unfettered access to a repository of legal knowledge and the tools to wield it effectively. Its AI components work in concert to create a user-friendly environment that transcends the traditional barriers of legal language and procedure. This not only empowers users but also imbues them with the confidence to engage with the legal system on their own terms.

The AI algorithms are the silent sentinels behind People Clerk's efficiency. They perform a plethora of functions, from automating the drafting of court documents to offering predictions on court dates and deadlines. The technology also encompasses a dynamic evidence organizer, which uses pattern recognition and data sorting to help users assemble their cases with cogency and precision. This approach not only saves time but also optimizes the chances of success by ensuring that every relevant piece of information is presented in a clear and orderly fashion.

People Clerk's impact on access to justice is measured not just in its technological prowess but in its human-centric design. The platform's interface is a study of empathy and user experience, recognizing that behind every case is a person seeking fairness and resolution. It strips away the intimidation factor of legal jargon and processes, presenting users with clear, step-by-step guidance that is both approachable and actionable. This guidance is not static; it is responsive, adapting to the evolving needs of the user and the dynamic landscape of small claims courts.

In leveraging AI for legal aid, People Clerk confronts a crucial ethical consideration: the imperative to ensure that technology serves the interests of justice without bias or error. To this end, the platform is built on a foundation of algorithmic transparency and continual refinement, ensuring that its AI systems operate with the highest standards of accuracy and fairness. This ethical commitment is embedded in the platform's DNA, with safeguards designed to detect and correct any potential biases in the decision-making processes.

The social implications of People Clerk's intervention in small claims courts are profound. By lowering the barriers to legal action, it has ignited a quiet revolution in the pursuit of justice. Citizens who might once have foregone their rights due to the cost or complexity of legal action are now stepping forward, armed with the knowledge and tools provided by the platform. This has far-reaching effects on the collective consciousness surrounding legal rights and the pathways to justice. It engenders a culture of empowerment, where the legal system is seen not as an insurmountable fortress but as an accessible avenue for redress and accountability.

Moreover, the People Clerk's role in the ecosystem of AI legal aid is indicative of a larger trend towards the integration of technology in law. The platform exemplifies how AI can be harnessed to not only automate tasks but also to bridge the human-technology divide, fostering a symbiotic relationship where each enhances the capabilities of the other. It stands as a case study of how AI can be deployed responsibly and innovatively to serve the greater good within the legal sector. This intertwining of human insight with machine efficiency paints a promising picture for the future of legal tech, where the nuanced complexities of the law are made comprehensible and navigable for all.

The integration of AI into the realm of small claims court via People Clerk also serves as a critical innovation in judicial efficiency. Courts, often burdened with high volumes of cases and limited resources, benefit from the streamlined and well-organized submissions that the People Clerk facilitates. The platform's capacity to prepare and file court documents in accordance with the precise requirements of specific jurisdictions alleviates the administrative load on court staff, reducing the likelihood of procedural delays due to clerical errors or incomplete filings. This, in turn, contributes to swifter resolutions and a more fluid legal process, benefitting not only the individuals involved but the justice system as a whole.

People Clerk's contribution extends beyond individual cases; it has a cumulative effect on the legal landscape. As more people use the platform, the data collected serves as a rich resource for understanding trends and identifying systemic issues within the small claims process. This data, when analyzed, can reveal insights into common legal challenges faced by individuals, potentially informing policy decisions and prompting reforms that make the justice system more equitable and effective.

The educational aspect of People Clerk is another facet of its contribution to access to justice. Through the platform, users gain a deeper understanding of legal principles and procedures, which empowers them to be more proactive and informed in their legal endeavors. The ripple effects of this education are significant; it leads to a more legally literate populace who are better equipped to assert their rights and navigate various legal scenarios beyond the small claims court.

Reflecting on the broader societal impact, the role of platforms like People Clerk in fostering legal empowerment cannot be underestimated. In an age where inequality often manifests in the form of legal disenfranchisement, providing tools that simplify the legal process is an act of social justice in itself. It represents a leveling of the playing field, where technology acts as an equalizer, mitigating the advantage that wealth and legal expertise typically confer in legal disputes.

The utility of People Clerk, however, is not without its challenges. The very nature of AI, with its reliance on data and patterns, requires a continuous and vigilant approach to ensure inclusivity and adaptability. The platform must be conscious of the diverse backgrounds and unique circumstances of its users, ensuring that the guidance it provides is not only legally sound but also contextually relevant. This necessitates an ongoing dialogue with users, legal professionals, and the courts, making sure that People Clerk remains a living system, evolving in step with the needs of its community.

In conclusion, the emergence of the People Clerk as a pivotal tool in the small claims court arena signals a transformative shift in how individuals engage with the legal system. Its AI-driven assistance embodies the promise of technology as a force for good in the quest for justice. By simplifying the small claims process, providing accessible legal education, and ensuring a fair and unbiased application of its algorithms, People Clerk stands as a paragon of innovation in legal tech. It not only closes the justice gap for those without legal representation but also paves the way for a future where access to justice is a universal reality, not a privilege. In this capacity, People Clerk does not just offer a service—it champions a cause, serving as a vanguard for a more inclusive and equitable legal landscape for all.

Chapter 27:

The Digital Transformation of Legal Notifications: AI-Powered Services of Process

The service of process is a cornerstone of the legal system, a critical conduit through which the machinery of justice ensures that individuals are informed about legal actions against them. The advent of Artificial Intelligence (AI) and legal technology is set to redefine this foundational element, promising a transformation characterized by enhanced precision, elevated efficiency, and an unwavering adherence to the tenets of due process. This discussion aims to unpack the revolutionary impact of AI and legal tech on the service of process, highlighting the myriad ways in which these technologies are reshaping this vital legal function.

At its core, the service of the process involves a series of meticulously coordinated steps designed to deliver legal documents in a manner that is both verifiable and consistent with jurisdictional mandates. The process ensures that all parties are duly notified and can prepare for, or respond to a legal proceeding. Traditionally, this has necessitated a high degree of manual coordination, with process servers navigating an often complex web of regulations, geographical barriers, and potential for human error. These challenges have not only impeded the speed at which legal notifications can be delivered but have also introduced vulnerabilities that could potentially undermine the integrity of legal outcomes.

Enter AI and legal tech, heralds of an era marked by digital transformation and the automation of processes that have long been dependent on human intervention. At the forefront of this transformation is the deployment of sophisticated algorithms capable of parsing through extensive datasets to determine the most effective strategies for reaching recipients. By analyzing historical data, AI can identify patterns and predict the optimal times and locations for service, thereby increasing the probability that the intended individual receives the notice. This predictive power is further enhanced by machine learning, which enables systems to adapt and improve their accuracy over time, learning from each interaction to refine future approaches.

One of the most striking enhancements offered by AI in the realm of service of process is in route optimization for process servers. By considering real-time traffic data, historical success rates, and even the personal habits of recipients, AI-driven systems can chart the most efficient paths for servers, ensuring that documents are delivered promptly while minimizing operational costs. This not only saves valuable time but also translates into economic benefits for law firms and their clients, as resources can be allocated more judiciously.

The role of AI in authenticating identities during the process of service is another critical advancement. By integrating with public and proprietary databases, AI can assist in verifying the identities of individuals, thereby reducing the likelihood of disputes regarding the validity of the service. This capability is particularly pertinent in an age where identity theft and privacy concerns are prevalent, ensuring that legal notifications are served to the correct individuals and safeguarding against fraudulent activities.

Automation extends to the realm of documentation, where AI systems can meticulously generate and manage records of service, ensuring that every step of the process is captured and stored in compliance with legal standards. This not only facilitates a seamless audit trail but also frees legal professionals from the burden of manual record-keeping, allowing them to focus their expertise on more complex and strategic legal tasks.

The implications of AI in the service of process are profound, touching upon the very essence of justice and fairness within the legal framework. By significantly reducing the instances in which individuals are unaware of legal proceedings against them, AI enhances the accessibility of the legal system, reinforcing the principle that justice should be within reach of all, irrespective of their circumstances. Legal professionals who embrace these technological innovations are at the vanguard of a movement toward a more agile, responsive, and just legal structure.

This evolution of the service of process through AI and legal tech is not just a narrative of operational enhancement; it is a testament to the legal field's adaptive capacity and its commitment to meeting the intricacies of the modern world. The intersection of technology and law is generating new paradigms for how legal services are delivered and experienced, with AI-driven service of process exemplifying a shift towards a more enlightened, equitable, and efficient pursuit of legal redress.

As we consider the broader impact of AI on the legal landscape, the service of process emerges as a pivotal area of transformation. The strides made in this domain reflect a wider trend in the legal profession towards innovation and technological integration. With the continued advancement of AI and legal tech, the service of process will likely see further enhancements, including the integration of natural language processing for automated communication with recipients, the use of blockchain for immutable service records, and perhaps even the advent of autonomous drones for physical document delivery. Each of these potential developments not only represents a leap forward in the mechanics of legal notification but also underscores a deeper ethos of continuous improvement within the legal sector.

In sum, the integration of AI and legal tech into the service of process is more than a mere procedural update; it is a significant stride toward a future where legal processes are not only more efficient and less error-prone but also more aligned with the evolving expectations of society. It signals a future where the law is not only practiced with greater precision and care but also with an eye towards harnessing the power of technology to uphold and advance the principles of justice.

As we embrace this future, it is incumbent upon legal practitioners, technologists, and policymakers to forge a collaborative path that honors both the letter and spirit of the law. The synergy between AI and the service of process is emblematic of the potential for technology to enhance the administration of justice, ensuring that legal proceedings begin on a foundation of timely and reliable notification.

The benefits of integrating AI into the process of service extend to the recipients of legal notifications. With greater precision in service, individuals are afforded adequate time and information to prepare their response, an essential aspect of due process. The certainty of delivery reduces ambiguity and the anxiety that can accompany legal disputes, fostering a more transparent and predictable judicial experience. Furthermore, the use of AI to facilitate understanding of the legal documents served—by providing summaries or highlighting critical information—can demystify legal jargon, making the legal system more accessible to laypersons.

For law firms, the transition to AI-enabled service of process solutions represents a strategic investment in the future of legal practice. Firms that adopt these technologies signal to their clientele a commitment to innovation and efficiency. This not only strengthens client relationships but also positions these firms as leaders in a legal market that is increasingly driven by technological proficiency. Moreover, the data harvested through AI-enhanced processes can yield insights into legal trends and client needs, informing more personalized and proactive legal services.

However, this transition is not without its challenges. The implementation of AI in legal processes must be navigated with a conscientious understanding of the ethical implications. Concerns around privacy, data security, and the potential for algorithmic bias must be rigorously addressed to maintain trust in the legal system. The design and deployment of AI tools in the service of the process require a careful balance—leveraging the capabilities of AI while ensuring that technology serves the interests of justice and not the other way around.

This balance is particularly crucial given the diversity of legal environments across jurisdictions. The global nature of many legal practices necessitates AI systems that are adaptable to different legal standards and cultural contexts. This adaptability must be built into the DNA of legal tech solutions, ensuring that they are not only robust and scalable but also sensitive to the nuances of local legal landscapes.

In contemplating the future of AI and legal tech in the service of process, one can envision a landscape where these technologies act as facilitators of a more connected and coherent legal ecosystem. As process serving becomes more integrated with digital court filing systems and online dispute resolution platforms, the role of AI will likely expand to enable seamless handoffs between different stages of the legal process. The potential for AI to interface with emerging technologies, such as virtual reality, could further revolutionize the experience of receiving and understanding legal notifications.

Ultimately, the measure of success for AI and legal tech in the service of process will be the degree to which they enhance the human elements of the legal profession. By assuming the repetitive and time-intensive tasks, AI frees legal professionals to apply their critical thinking and empathetic skills where they are most needed—in advising clients, negotiating settlements, and advocating in courtrooms. It is this augmentation of human capability that lies at the heart of AI's promise in the legal realm.

As we forge ahead, the evolution of the service of process through AI and legal tech is not an endpoint but a journey—one that will continue to unfold as technologies advance and the legal profession adapts. It is a journey that holds great promise for the elevation of legal services, one that requires a thoughtful approach to the integration of technology with the time-honored principles of law. For those at the intersection of AI, legal tech, and the service of process, the path ahead is one of exciting possibilities, filled with the potential to redefine the pursuit of justice in the digital age.

From Manual to Machine: AI's Overhaul of Legal Notification Systems

The service of process, integral to the legal system, marks a party's entry into legal proceedings. Traditionally rooted in manual and physical delivery methods, this area is primed for an overhaul. As the 21st century progresses, AI's potential to transform the service of process is not just promising but groundbreaking. AI technologies can fundamentally alter how legal notifications are managed, served, and acknowledged, offering legal practitioners a chance to streamline operations and reduce human error margins significantly.

AI-driven systems can automate party identification and verification for service, ensuring legal documents are dispatched accurately and quickly. AI's ability to process data rapidly, cross-reference information, and predict optimal service times revolutionizes the service of process into a precision-driven operation. Furthermore, integrating AI brings accountability and transparency previously unseen, with digital logs, AI-powered tracking, and real-time updates providing a clear record of each step, enhancing the legal process's integrity.

AI's role is crucial in navigating the complex regulations of service across different jurisdictions. With legal requirements constantly evolving, AI systems can adapt and ensure compliance through updated databases and algorithmic adjustments, reducing procedural error risks that could invalidate service. Critics argue that AI integration might sacrifice the personal touch and overlook nuances a human could catch. However, AI doesn't replace but rather augments the human element, handling routine tasks to free up legal professionals for more strategic case aspects.

AI in service of process also offers significant cost benefits by automating tracking, delivery, and acknowledgment phases, making legal services more accessible to a wider population segment. Adapting to AI-driven service necessitates a mindset and operational shift, with legal professionals needing to understand AI systems and their legal applications. The learning curve is manageable, and the benefits far outweigh the initial investment.

Machine learning ensures that AI systems continually improve in accuracy, efficiency, and reliability. This evolution in service of process will become more refined and tailored to the legal community's specific needs. Privacy and security of sensitive legal documents are critical, with AI systems for legal applications built on stringent security measures, including encryption and secure access protocols, maintaining service integrity.

The future of service of process aligns with AI enhancements, moving towards fully integrated AI systems. Legal professionals embracing this evolution will lead to a more efficient, accurate, and equitable legal process. Collaborating across technology developers, legal experts, and regulatory bodies is vital to navigate AI's ethical implications in legal processes, creating an innovation-supportive framework while upholding justice and fairness principles.

The practical, necessary evolution in service of process with AI enhancements addresses current limitations and future legal system needs. AI empowers legal practitioners to provide more reliable, efficient, and equitable services, heralding a new law practice era. As this transformation unfolds, recognizing the ongoing integration journey of AI into legal practices is crucial. Continued innovation, education, and regulation will be paramount in unlocking AI's full potential in revolutionizing the service of process, promising a revolutionary future for adaptable legal professionals.

ServeManager's Existing Technologies:

ServeManager, a cloud-based software platform for Process, is pivotal in this transformation. It assists law firms in managing cases, documents, and billing and can be integrated with AI tools to automate tasks, saving time for critical business processes. ServeManager includes integration capabilities with AI tools like Google Gemini (Bard AI), which automates tasks; Browse AI, offering over 20 million ready automation and live chat support; Leap AI, which integrates crucial business parts; and Nekton AI, linking ServeManager with numerous online apps and tools like Formstack, QuickBooks Online, Google Calendar, Cognito Forms, and Google Sheets.

ServeManager enhances service process efficiency by providing a centralized platform for managing all aspects of legal service delivery. Its automation capabilities, powered by AI, streamline case management, document handling, and billing processes. The platform's ability to integrate with advanced AI tools like Google Gemini, Browse AI, Leap AI, and Nekton AI means that law firms can automate repetitive tasks, freeing up legal professionals to focus on more complex, value-added activities.

Furthermore, ServeManager's potential for integrating predictive analytics and route optimization represents the future of legal service delivery. AI's role in enabling dynamic decision-making and improved management of service attempts is increasingly critical. Predictive analytics can forecast the best times and methods for serving documents, while route optimization can streamline the process of physical document delivery, reducing costs and improving efficiency.

In conclusion, ServeManager stands at the forefront of revolutionizing the process of service with AI enhancements. Its existing technologies and potential for future integration with AI-driven predictive analytics and route optimization tools offer a glimpse into a more efficient, accurate, and streamlined legal service delivery future. This evolution, powered by AI and enabled through platforms like ServeManager, signifies a monumental shift in how legal services are rendered, promising a more accessible, transparent, and equitable legal system.

Redefining Service of Process Standards with AI in Legal Tech

In the sphere of legal services, the precision and timeliness of notifications are pivotal. This fact has never been more pronounced than in the current landscape, where Artificial Intelligence (AI) stands as a transformative force. At the forefront of this revolution is ABC Legal, a company whose nearly half-century of expertise has been reinvigorated by AI's capabilities, setting a new benchmark for accuracy and compliance within the legal notification process.

The essential function of AI within ABC Legal's operational matrix is multifaceted, primarily enhancing the accuracy of legal notifications. This enhancement is not merely incremental; it is a radical elevation in the reliability of the service of process. By integrating AI, ABC Legal orchestrates an array of complex tasks with exceptional precision. This includes the classification of legal documents, identification of jurisdictional requirements, and ensuring that each step of the notification process adheres to the stringent regulatory frameworks that govern legal proceedings. The implications are profound; AI does not only perform tasks but does so with an error margin that narrows almost to the point of nonexistence. This ironclad accuracy is crucial in legal proceedings, where the stakes are high, and the tolerance for error is practically nil.

AI's role within ABC Legal's suite of services emerges as a cornerstone of its operational strategy. The company's AI algorithms are designed to streamline service strategies, tailoring them to the unique demands of each jurisdiction. This ensures that every service or process is executed in strict compliance with local, state, and federal regulations. However, the capabilities of AI extend beyond compliance. In scenarios fraught with complexities—be it a multi-jurisdictional case or serving individuals who are difficult to locate—AI empowers ABC Legal with predictive analytics and pattern recognition, enabling the formulation of strategies that can navigate and manage these complexities with surprising agility.

One of the most salient benefits of AI integration in ABC Legal's services is automated tracking. The traditional process of serving is a labyrinthine undertaking that requires meticulous tracking and documentation. AI revolutionizes this aspect by automating the tracking process, ensuring real-time updates, and maintaining a comprehensive log of every action taken. The complex scenario management that AI affords allows ABC Legal to foresee potential pitfalls and proactively address them, thereby preserving the integrity and uninterrupted flow of the service process. This predictive approach mitigates the risks of delays and non-compliance, which are not uncommon in the nuanced realm of legal services.

Moreover, ABC Legal's utilization of data through AI does more than enhance service accuracy—it informs it. The rich data insights gathered from historical performance, combined with real-time analytics, grant ABC Legal an unparalleled ability to fine-tune its processes. These insights lead to informed decision-making, which is critical in an environment where every second count and every action can have far-reaching legal implications. By leveraging data, ABC Legal not only elevates its service accuracy but also fortifies its compliance posture, ensuring that its clients are safeguarded against the risks associated with non-compliance.

The technological prowess of ABC Legal is embodied in its mobile app, a marvel of AI integration that provides process servers with the tools to deliver proof of service with incontrovertible evidence. This includes time-stamped photos and precise GPS coordinates for every service executed. Such granular evidence is the gold standard in the legal profession, as it can decisively refute claims of improper service. This level of detail, enabled by AI, underpins the company's commitment to delivering an undisputed, reliable service that resonates with integrity and precision.

The ethos of innovation that permeates ABC Legal is championed by its leadership. CEO Tim Dinehart and CTO Brandon Fuller embody the company's commitment to setting new performance and quality standards, spearheading initiatives that weave advanced technical solutions into the fabric of legal services. Their vision is clear: to create a legal service process that is not just efficient but is also reflective of the highest standards of quality and reliability.

In the broader context of the legal profession, AI's introduction by entities like ABC Legal signals a paradigm shift. The landscape is being redefined by technologies such as Natural Language Processing (NLP), which automates the interpretation and synthesis of large volumes of text, and Machine Learning (ML), which enables legal professionals to engage with their work in ways that were previously untenable. These technologies are not merely auxiliary tools; they are becoming intrinsic to the legal profession's functionality.

The use of ML has evolved from its nascent applications in billing to becoming a powerful analytical tool that processes vast amounts of court data to inform strategic decision-making. The legal sector, historically reticent in embracing change, now finds itself at an inflection point, where the adoption of AI, ML, and related technologies is not an option but a necessity. The pandemic's impact has amplified this necessity, bringing to light the critical role that technology plays in maintaining the continuity of legal processes. As legal entities grapple with large backlogs of cases, the integration of AI has proven to be a lifeline, lessening the burden on court staff and enabling greater access to legal support for those in need.

ABC Legal's approach to embracing emerging technologies, particularly during times of unprecedented challenges, underscores the importance of ease-of-use in technology adoption. Their platform's design philosophy pivots around user-friendliness, ensuring that the transition to digital solutions does not impose additional burdens on their clients. This emphasis on simplicity and accessibility is paramount, as the perception of complexity can be a significant deterrent to technology adoption in the legal sector. ABC Legal's solutions, therefore, not only provide operational advantages but also serve to accelerate the industry's broader acceptance of technological tools.

The company's focus on ease-of-use extends to a suite of focused solutions, such as deep skip tracing and electronic filing, which have gained even more prominence during times of limited physical interaction. By ensuring that these tools are intuitive and integrate seamlessly into existing workflows, ABC Legal facilitates a smoother digital shift for their clients, allowing them to circumvent the backlogs and inefficiencies that have plagued others in the industry.

However, the impact of ABC Legal's technology adoption extends beyond operational productivity. By leveraging tools like digital document management, they enhance customer service and satisfaction, fostering better experiences for all stakeholders involved in the legal process. Processes that once were opaque and unwieldy are now transparent and expedient, thanks to the electronic service of process documentation, e-signing, and e-filing. Such advancements are not merely changing the mechanics of legal work; they are reshaping the public perception of the justice system itself.

No longer is the judiciary viewed solely as a venue for legal redress; it is increasingly perceived as a service provider that facilitates easy access to justice. This transformation has been driven in part by ABC Legal's commitment to technology and client-centered service, a commitment that has made the legal system more approachable and user-friendly.

The embrace of digital solutions during the pandemic has demonstrated that with the right tools, many legal professionals can maintain or even enhance their effectiveness when working remotely. Modern solutions, including video conferencing, virtual notaries, and digital documentation, have catalyzed an improvement in the operational efficiency of firms and simultaneously elevated customer satisfaction and retention.

As the legal profession looks to the future, the trajectory is clear. Legal professionals must adapt to technological advancements or risk obsolescence. Consumer expectations are evolving; they no longer tolerate the lethargic pace once synonymous with the legal profession. They demand the kind of responsiveness and efficiency that tools like NLP, ML, AI, and automation have introduced. These technologies have established a new normal, shaping consumer expectations and setting a standard that legal professionals are now expected to meet.

In anticipation of the future, ABC Legal is poised to continue pioneering innovative changes. The integration of AI and other emerging technologies presents opportunities not only to improve productivity but also to enhance security, efficiency, and marketability. As legal services become more interconnected with technological advancements, the focus will inevitably shift to harnessing these tools to drive not just compliance and accuracy, but also to deliver superior client experiences.

ABC Legal's evolution and continued investment in AI highlight the firm's understanding that the future of law is inextricably linked with legal tech. The convergence of technology and legal services is creating an environment where adaptability and innovation are not just valued but are imperative for success. ABC Legal stands as a testament to the potential of AI to redefine the industry, emphasizing that the true measure of technology's value lies in its capacity to transform not just how services are delivered, but also how they are experienced by those who rely on the legal system the most.

This dedication to innovation and client satisfaction encapsulates the spirit with which ABC Legal approaches its work. As legal professionals and firms navigate the complexities of the modern legal landscape, they would do well to take a page from ABC Legal's playbook, where AI-driven accuracy and efficiency are not just goals but realities that are reshaping the legal notifications space. With nearly 50 years of expertise under its belt and a forward-thinking embrace of technology, ABC Legal is not just participating in the evolution of legal services; it is actively steering it towards a future where legal processes are synonymous with precision, efficiency, and unwavering reliability.

Disrupting Legal Delivery: The Intersection of Blockchain and AI Technologies

The intersection of blockchain technology and artificial intelligence (AI) is revolutionizing the way legal services, particularly the service of process, are managed and executed. Companies like CaseMail are at the forefront of this revolution, leveraging the robust capabilities of blockchain to enhance the security, transparency, and efficiency of legal communications and document handling. With CaseMail's groundbreaking innovation, the United States Postal Service (USPS) has certified the first blockchain-generated ePostage labels, which are critical to ensuring a verifiable chain of custody for both digital and physical legal assets. This is not a minor upgrade to existing systems but a significant leap towards an immutable and trustable framework for legal proceedings.

CaseMail's integration of blockchain takes advantage of the technology's inherent security features to offer proof of document authenticity, ownership, and the linking of wet signatures to electronic ones. Their documents, once entered into the blockchain, are immutable, meaning they cannot be altered or challenged, which is a foundational requirement for the legal industry's rigorous standards. This is further complemented by CaseMail's Certified.Email service, which extends the power of blockchain to email communications, allowing for seamless authentication of accounts across all email providers and the certification of an entire email conversation, including all attachments.

But it's not just about security and transparency; it's also about the effectiveness and reach of the service of process. With the advent of blockchain technology, services like these can expand globally, breaking down geographical barriers and accelerating the speed of legal processes. This has significant implications for law firms, government agencies, and global commerce, where time is often of the essence and where the verifiable and expedient service of process is paramount.

Now, let's delve deeper into the innovative work being done by Holland & Knight. This law firm has become synonymous with the incorporation of AI, blockchain, and non-fungible tokens (NFTs) into legal practices. They've been recognized for their prowess in Blockchain & Cryptocurrencies and have been trailblazers in the use of NFTs for legal notices. The method approved by the New York Supreme Court of serving a temporary restraining order through an NFT represents a novel, groundbreaking approach to serving legal documents to anonymous defendants involved in virtual asset disputes.

Holland & Knight's expertise doesn't stop at blockchain. The firm's team has harnessed AI to provide comprehensive support for clients in diverse industries, anticipating future regulations and ensuring compliance with current policies. They're not just practitioners but influencers, actively shaping federal AI policy and legislation. This proactive stance means they're well-placed to tackle legal issues related to privacy, intellectual property, consumer protection, and antitrust as they intersect with AI technology.

Their foresight is particularly evident in their approach to the metaverse and related litigation, which includes a focus on labor and employment law, brand protection, and privacy. Holland & Knight understands that as our digital and physical realities become increasingly intertwined, the legal implications become more complex and nuanced. By navigating these challenges, they're setting precedents that will shape the legal landscape for years to come.

The practical applications of these technologies are already being realized, as evidenced by the service of process via NFT executed by Holland & Knight. In what can be described as a legal-tech milestone, the firm became the first to airdrop an NFT containing legal notice to a defendant's blockchain address, a method sanctioned by the New York Supreme Court. This strategy not only demonstrates innovative thinking but offers a template for future legal actions, particularly in addressing the service of process to elusive parties in the digital age.

This case, involving a high-profile cryptocurrency exchange hack, illustrates the agility of legal tech in responding to new challenges. With the defendants' identities obscured by the anonymous nature of blockchain transactions, traditional service methods were impractical. Holland & Knight's solution — serving the defendants via a blockchain token — was a stroke of legal and technical ingenuity, ensuring that the service of process remained robust in the face of novel adversities.

As blockchain and AI continue to evolve, we can expect to see further integrations like these, where legal practices not only adapt to new technologies but also harness them to create more efficient, secure, and equitable systems. CaseMail and Holland & Knight exemplify the potential for legal services in the digital age, showcasing how innovation can be both a response to emerging challenges and a catalyst for systemic change.

What we are witnessing with CaseMail and Holland & Knight's forays into blockchain and AI is not merely a set of technological advancements but a redefinition of the capabilities of legal firms. As we march towards a future where digital and physical assets are indistinguishable in their legal significance, the innovations these organizations are pioneering will serve as blueprints for an entire industry's transformation. The legal profession is often seen as one bound by tradition and precedent, but as these developments show, it is also capable of leading the charge in the adoption of cutting-edge technologies.

The broader legal community can draw inspiration from these examples, exploring how blockchain and AI can be used not just for streamlining existing processes but for creating entirely new paradigms for the service of process and other legal functions. As we look ahead, the possibilities for how these technologies can be deployed are vast and varied. Firms that can harness these tools effectively will undoubtedly gain a significant competitive advantage, offering their clients a level of assurance and efficiency that was previously unattainable.

The potential for blockchain and AI in the legal industry also extends to the realm of ethics and compliance. With these tools, the accuracy of records, the authentication of signatures and documents, and the traceability of legal actions become far more reliable. This is not just a matter of convenience; it is a fundamental shift in how trust and veracity are established in legal proceedings. By creating an immutable ledger of actions and communications, blockchain technology ensures that each step in the legal process is recorded and verifiable, which could dramatically reduce the opportunity for fraud and malfeasance.

In addition to enhancing the integrity of legal processes, the use of blockchain and AI can greatly improve accessibility and inclusivity. With these technologies, legal services can be rendered more efficiently and at a lower cost, potentially expanding access to justice for individuals and organizations that may have previously found legal costs prohibitive. Moreover, by automating certain aspects of legal work, attorneys can focus their expertise on more complex and nuanced legal matters, where human judgment and creativity are indispensable.

Furthermore, the use of blockchain and AI is not confined to the borders of any single country; these are inherently global technologies. As such, they offer a way to harmonize legal processes across jurisdictions, facilitating international transactions and cooperation. With the right frameworks in place, blockchain could allow for a standardized global process for service of process, document authentication, and other legal functions, which would be a significant boon for international trade, cross-border litigation, and regulatory compliance.

In conclusion, the integration of blockchain and AI technologies by pioneering companies like CaseMail and forward-thinking law firms like Holland & Knight represents a significant development in the legal industry. These technologies promise not only to enhance the security and efficiency of legal processes but also to foster innovation, inclusivity, and global collaboration. As the legal profession continues to navigate the complexities of the digital age, embracing such innovations will be crucial to meeting the evolving needs and expectations of clients and society at large. The digital service of process is just the beginning; the future will undoubtedly unveil even more transformative applications of these remarkable technologies in the legal sphere.

Legal Notifications in the Digital Age: Blockchain and AI as Catalysts for Change

The integration of blockchain and artificial intelligence (AI) in the legal tech industry represents a transformative convergence of technologies, delivering strategic advantages that are reshaping the practice of law. In this section, we will explore how blockchain serves as a backbone for a trustworthy management system for legal notifications, the advanced capabilities of AI in ensuring accuracy and task automation, and the synergistic effect these technologies have on expanding legal service accessibility. We will conclude by looking ahead to the future of Service of Process and how these technologies may continue to revolutionize this critical aspect of the legal system.

Blockchain's Function in Establishing a Trustworthy Management System for Legal Notifications

At the core of the legal tech revolution is blockchain technology. Known for its robust security and immutability, blockchain is an ideal tool for creating an infallible management system for legal notifications. Legal notifications, a cornerstone of the judicial process, require a high level of veracity and traceability to ensure that all parties are informed and that due process is followed. Blockchain, with its decentralized ledger, offers an unprecedented level of transparency and security.

When a legal notification is issued, it is recorded as a block on the blockchain with a unique, tamper-evident digital signature. This block is then linked to the preceding entry, creating a chain that is, by design, resistant to modification. This ensures that once a notification has been recorded, it cannot be altered or deleted without detection, instilling confidence in the authenticity of the information.

Moreover, the decentralized nature of blockchain means that no single entity has control over the entire system, reducing the risk of manipulation and increasing trust among users. The parties involved can access the blockchain to verify the issuance and receipt of legal notifications, ensuring all procedural steps are visible and verifiable.

AI's Advanced Capabilities in Ensuring Accuracy and Task Automation

The contribution of AI to the legal tech space is equally significant. Through its advanced data processing capabilities, AI enhances accuracy in the legal field, an area where precision is paramount. Machine learning algorithms, a subset of AI, can analyze vast quantities of legal documents to identify relevant information, predict outcomes, and even draft basic legal documents. This not only increases efficiency but also reduces the possibility of human error, leading to more reliable and consistent results.

Task automation is another arena where AI shines in legal tech. Routine tasks such as document review, legal research, and case prediction can be automated using AI. By handling these repetitive and time-consuming tasks,

AI frees legal professionals to focus on more complex and strategic aspects of their work. Furthermore, AI's ability to learn and improve over time means that its accuracy and efficiency will continue to grow, offering even greater value in the future.

The Combined Effect of Blockchain and AI in Expanding Legal Service Accessibility

The intersection of blockchain and AI is where their true potential is realized, particularly in expanding the accessibility of legal services. Blockchain ensures that legal documents and notifications are secure and verifiable, while AI automates and streamlines processes. This combination is especially beneficial for remote and underrepresented groups who may lack traditional access to legal resources.

Through blockchain and AI, legal services can be made available online, eliminating the need for physical presence and reducing costs. This democratizes the availability of legal assistance, making it feasible for individuals in remote areas or those with limited financial resources to receive the same level of service as those in more privileged circumstances.

Moreover, AI-driven chatbots and automated online platforms can provide preliminary legal advice and assistance around the clock, removing barriers of time and accessibility. Blockchain can be utilized to record these interactions, providing a secure and transparent record of the legal aid provided.

Conclusion: AI, Legal Tech, and Blockchain and the Future for Service of Process

As we look to the future, the implications of AI and blockchain in the realm of Service of Process are profound. Blockchain's capability to provide a secure and transparent system for managing legal notifications ensures that the Service of Process can be tracked and verified with a level of certainty previously unattainable. This will significantly reduce instances of fraud and errors, leading to a more efficient and just legal system.

AI's role in task automation will continue to evolve, potentially reaching a point where the Service of Process can be managed almost entirely autonomously. With continuous advancements in AI, we can expect to see smarter systems capable of navigating complex legal frameworks and regulations with ease.

As legal tech continues to advance, the combination of blockchain and AI holds the promise of making the law more equitable and accessible. It is a future where legal processes are not only more efficient and transparent but also more inclusive, extending the protection of the law to those who were previously beyond its reach. The strategic advantages of these technologies in legal tech are not just incremental improvements but are leading to a fundamental rethinking of how legal services are delivered and how justice is administered.

In conclusion, the fusion of blockchain and AI is not merely an enhancement to current legal tech—it's a redefinition of the landscape. It represents a future where the legal profession is not just about the law but also about technology's role in ensuring justice is both served and accessible. As these technologies become more integrated into legal workflows, we can anticipate a paradigm shift in how legal services are provided, characterized by increased speed, efficiency, and transparency.

The legal profession stands on the cusp of a new era, where blockchain's immutable record-keeping capabilities ensure the integrity of legal proceedings, and AI's cognitive powers enhance the analytical and decision-making tasks traditionally performed by humans. This synergy could potentially automate the entire lifecycle of legal processes, from the initiation of cases to the execution of legal instruments and enforcement of judgments.

In the realm of Service of Process, this means the implementation of sophisticated digital systems capable of issuing, recording, and verifying legal notices with minimal human intervention. Smart contracts—self-executing contracts with the terms of the agreement between buyer and seller directly written into lines of code—may automate many aspects of the legal notification and acknowledgment processes. When integrated with blockchain technology, smart contracts can offer an unassailable record that a service has been executed according to the stipulated legal requirements.

Moreover, as the algorithms underpinning AI become more adept at understanding natural language, we can envisage AI systems that not only serve notices but can also assess the recipient's understanding and acknowledgment, ensuring that the spirit of the Service of Process—informing individuals of legal actions against them in a manner that they comprehend—is upheld.

However, with these advancements come new challenges and considerations. The need for stringent privacy controls, ethical standards, and regulatory compliance will be paramount. Ensuring that the use of blockchain and AI aligns with the principles of justice and the rule of law will require ongoing vigilance and adaptability from legal professionals and technologists alike.

The future of the Service of Process and legal notifications through blockchain and AI technologies may also lead to a more globalized legal system. Cross-border legal processes could become more streamlined, with blockchain providing a universally recognized platform for legal transactions and AI offering translation and cultural context for international Service of Process.

Ultimately, the integration of blockchain and AI signifies a departure from traditional legal service models, prompting a reevaluation of the skills and roles of legal professionals. Lawyers and legal practitioners will need to become adept at working alongside these technologies, harnessing their capabilities to provide better, more accessible legal services.

As the adoption of blockchain and AI within the legal industry matures, we can expect a legal system that is not only more efficient and secure but also more just and equitable. The strategic advantages of these technologies lie in their potential to remove barriers, reduce biases, and provide a level of service hitherto impossible. While the full realization of this potential is still unfolding, the foundation being laid today is a testament to the transformative power of blockchain and AI in legal tech. With continued innovation and thoughtful implementation, the future of Service of Process is set to be revolutionized, ensuring that justice is both seen and served in the digital age.

Conclusion: AI as the New Standard of Legal Workflow Efficiency

As this book draws to its final chapter, it's evident that the journey into the world of Legal AI and Legal Tech is far from an academic exercise—it is a clarion call for transformation within the legal profession. Throughout the pages of "Legal AI," we've ventured deep into the heart of AI's impact on the practice of law, unearthing its burgeoning influence and its significant stride from auxiliary tool to a linchpin of modern legal practice. It is a journey that has not merely been about recounting potential, but about tracing the undeniable fingerprints of AI and Legal Tech across the fabric of legal work.

This exploration has been both broad and intricate, charting the trajectory of AI from its infancy to a formidable force reshaping the pillars of legal operations. The integration of AI has not been presented as a fanciful addition but as a crucial advancement for law firms seeking to thrive in an era marked by unrelenting change. The drive towards efficiency, precision, and enhanced client service is not just an ideal; it is the new benchmark of excellence, a threshold that continuously rises as more firms adopt and refine their use of AI and Legal Tech.

Firms that choose to embrace this technology are positioning themselves on the frontier of innovation, primed to unlock newfound levels of operational agility and service delivery. The dichotomy is stark: adopt and evolve, or hesitate and risk obsolescence. Firms clinging to the status quo may soon discover they are unable to compete with contemporaries who harness AI to optimize everything from case analysis to client engagement. This integration is not a mere advantage—it is rapidly becoming the baseline expectation.

Navigating the Path Forward: Challenges and Solutions in AI Adoption

In our discourse, it is not enough to laud the merits of AI without confronting the substantial challenges its integration poses. These challenges—ethical considerations, data privacy, the need for perpetual learning and adaptation—are not minor footnotes. They are substantial barriers that must be approached with sagacity and integrity.

The adoption of AI in law is not without its ethical quagmires. Fairness in algorithmic decision-making, the safeguarding of client data, and ensuring the equitable application of AI tools necessitate a framework that prioritizes ethical considerations alongside technological advancement. The cultivation of AI within the legal profession demands a steadfast adherence to principles that have long defined legal excellence—justice, confidentiality, and a commitment to the client's best interests.

Moreover, the incorporation of AI is not merely a technical adjustment but a cultural shift within firms. It calls for legal professionals who are not just technologically proficient but also philosophically aligned with the evolving nature of their practice. Legal education, both formal and continuing, must pivot to accommodate this shift, ensuring that the next generation of lawyers is as comfortable with the nuances of AI as they are with the intricacies of legal precedent.

The Imperative of Continuous Learning and Adaptation

The discourse around AI in the legal profession often orbits around the idea of transformation. Yet, transformation is not a single event—it is an ongoing process. As such, the journey towards fully embracing AI is punctuated by continuous learning, a sustained commitment to innovation, and an adaptable mindset. Legal professionals must become lifelong learners, ever-responsive to the evolving capabilities of AI technologies.

Understanding the capabilities and, equally important, the limitations of current AI technologies is paramount. Firms must navigate these waters with a strategic outlook, discerning the most opportune areas for AI's application. The ability to automate mundane tasks, while simultaneously employing sophisticated algorithms for data analysis, holds the promise of freeing legal professionals to engage more deeply with the complexities of legal strategy and client advocacy.

The Collaboration between Technology and Tradition

The future of legal practice is unlikely to be one where AI operates in isolation. Instead, it will be characterized by a collaborative synergy between technology and the traditional practice of law. This partnership promises to elevate the quality of legal services, not only through heightened operational efficiency but through a redefined client experience that melds the best of technology with the irreplaceable human touch of the lawyer-client relationship.

Yet, resistance to change is a well-known characteristic of the legal profession. To overcome this inertia, firms must not only demonstrate the efficiencies and cost savings brought about by AI but also illuminate how these technologies can enhance the quality and delivery of legal services. It is a message that needs to resonate not just with managing partners and IT departments but with every stakeholder in the legal process—from paralegals to clients.

The Global Perspective on AI in Law

The global legal community is not monolithic in its approach to AI. The pace of adoption and the level of innovation vary widely across different countries and legal systems. Nonetheless, there is a growing consensus on AI's potential to transform legal practices on a worldwide scale. International collaboration and the exchange of knowledge will be crucial in leveraging AI's full potential across diverse legal landscapes. Understanding and navigating the variations in regulatory and ethical guidelines for AI in different jurisdictions is essential for globally-oriented legal practices and reinforces the importance of a nuanced approach to AI adoption.

Crafting Comprehensive AI Strategies for Law Firms

To thrive in this new terrain, law firms and legal departments must develop robust AI strategies that encompass skill-building, fostering a culture of innovation, and staying abreast of technological advancements. A well-considered AI strategy extends beyond mere technology acquisition—it involves cultivating a mindset within the firm that embraces continuous improvement and the reevaluation of traditional workflows. The financial and strategic advantages of AI in legal services are clear, evidenced by improved efficiency, cost reduction, and elevated client satisfaction, which in turn drive the firm's market position and profitability.

Case studies and success stories of AI integration in legal practices serve as illustrative roadmaps for those embarking on this journey. They highlight not only AI's capacity to enhance various legal functions but also the strategic underpinnings that ensure the effective deployment of these technologies. By examining these narratives, legal professionals can extract valuable insights and best practices that can be adapted to their unique contexts.

The Hybrid Future: AI and Human Lawyers

As we reflect on the transformative journey of AI in the legal sector, it is apparent that we stand on the brink of a new epoch where AI does not supplant lawyers but rather complements and augments their expertise. The prospect of a hybrid model—where AI tools and human lawyers operate in concert, each amplifying the other's strengths—represents the pinnacle of this evolution. It is a vision of the future where AI's analytical prowess is matched with the lawyer's nuanced judgment and empathetic client interaction.

Data as the Keystone of AI in Law

The lifeblood of any AI system is data. Thus, for AI to be effectively integrated into legal practices, a strong emphasis must be placed on data management. Implementing sound strategies for the collection, storage, and processing of data is not just a technical necessity but a cornerstone for maintaining client trust and meeting the ethical obligations inherent to legal practice. Firms must navigate the delicate balance between utilizing data to fuel AI systems and protecting the sanctity of client confidentiality.

Addressing the Ethical Landscape of AI in Legal Practices

The ethical landscape surrounding AI in law is rich and complex. The integration of AI necessitates grappling with questions of bias, job displacement, and the overarching responsibility for AI-generated outcomes. Addressing these issues calls for a concerted effort to develop AI technologies that are not only effective but also adhere to the highest standards of transparency, fairness, and accountability.

Enhancing Client Relationships Through AI

AI has the potential to revolutionize client relationships by enabling more personalized and efficient services. However, it is crucial to balance technological interactions with the personal, human connections that are the hallmark of legal service. AI should be viewed as a tool to enhance these interactions, not replace them, ensuring that the essence of the lawyer-client relationship remains intact.

Forging Symbiotic Relationships with AI Developers

To shape the future of AI in law, a close collaboration between legal professionals and AI developers is essential. Lawyers' intricate understanding of legal processes and requirements can inform the development of AI tools that are finely attuned to the specific demands of legal work. This synergy between legal expertise and technological innovation can drive the creation of tools that are both cutting-edge and deeply integrated into legal practice.

Adapting to Rapid Technological Advancements

The legal profession must be agile and responsive to the rapid advancements in AI technology. Proactivity is key, as is a willingness to explore and experiment with emerging AI applications. This proactive stance entails not only adopting existing technologies but also forecasting and preparing for future innovations that may redefine the practice of law.

AI and the Democratization of Legal Services

One of the most profound opportunities presented by AI integration is the democratization of legal services. By making legal knowledge more accessible and routine tasks more efficient, AI can lower barriers to legal services, making them more affordable and timely for a broader segment of the population. This aligns with the broader societal goal of ensuring equitable access to justice.

The Path Ahead: A Call to Action for Legal Professionals

As this book's narrative concludes, we recognize that the integration of AI in legal practices is a complex and ongoing journey. It is one that holds the promise of redefining the legal profession in significant ways. Legal professionals have a responsibility to lead this transformation, ensuring that as we embrace the future, the foundational values of justice, ethics, and human compassion remain at the forefront of our profession.

This is not a journey to embark on with trepidation but with purpose and anticipation. The integration of AI into legal practices offers a means to achieve greater efficiency, more accurate outcomes, and improved access to legal services for all. The road ahead is one of discovery and opportunity, and it beckons us to join in shaping a future where AI and the legal profession evolve together, harnessing the best of both technology and humanity.

The ongoing integration of AI in legal practices demands that we not only acknowledge the transformative potential of these technologies but also take active steps to integrate them into the fabric of legal work. It is imperative for legal professionals to maintain the highest standards of their profession while navigating the new terrain shaped by AI. This will require a sustained commitment to ethics, ongoing education, and a strategic approach to innovation.

The Strategic Imperative: Building AI Competency in Law

For law firms and legal departments worldwide, the strategic imperative is clear: to build competency in AI and Legal Tech that aligns with their specific needs and goals. This competency goes beyond mere familiarity with AI products; it encompasses a deep understanding of how AI can be strategically applied to enhance decision-making, optimize resource allocation, and deliver legal services more effectively.

As AI continues to mature, its applications within the legal field will only become more sophisticated, demanding an equally sophisticated response from legal practitioners. It's no longer sufficient to be merely proficient in traditional legal skills. The modern legal professional must also be adept at working alongside AI, harnessing its analytical capabilities to deliver services that are at once innovative, efficient, and client-centric.

Embracing the New Frontier: AI as a Catalyst for Legal Innovation

The integration of AI into legal practices is not an end but a starting point—a catalyst for broader innovation within the field. Law firms that embrace AI are finding themselves at the leading edge of a new frontier, where the boundaries of legal service delivery are being redefined. In this new landscape, the most successful firms will be those that view AI as a partner in the creative process, a tool that can spark new ways of thinking about and practicing law.

The Holistic Approach: AI as Part of a Larger Ecosystem

In order to truly leverage AI, it is critical for law firms to consider it as part of a larger ecosystem. AI does not operate in a vacuum; it interacts with a firm's culture, workflows, and client relationships. The challenge is to integrate AI into this ecosystem in a way that is seamless and enhances each component. Firms must take a holistic approach, ensuring that AI adoption is matched by changes in organizational culture, training, and processes that collectively move the firm toward a more innovative and responsive way of working.

Legal AI and the Public Good: Access to Justice and Ethical Practice

AI's potential extends beyond the profit motive and touches upon the public good, particularly in terms of access to justice. The democratization of legal services through AI can help to bridge the gap in legal representation and make the law more accessible to underserved communities. This noble aim reinforces the need for ethical practice in AI development and deployment, ensuring that the benefits of AI in law reach all strata of society and contribute to a more just and equitable legal system.

The Inclusive Future: A Profession Shaped by Diversity and AI

Looking forward, the legal profession will increasingly be shaped by a combination of diversity and AI. The inclusion of diverse perspectives in AI development will lead to more robust, fair, and effective legal technologies. In turn, AI can support the profession's commitment to diversity by minimizing biases in hiring, case management, and legal decision-making.

Conclusion: A Blueprint for the Future

As "Legal AI" concludes, it's clear that the integration of AI into legal practices represents a pivotal moment in the evolution of the profession. For legal practitioners, the challenge is to embrace this change, recognizing that AI can serve as a powerful ally in the pursuit of excellence. The future of legal practice, underpinned by the judicious use of AI, is one where efficiency, effectiveness, and equitable access to justice are not merely ideals but practical realities.

The journey ahead for legal AI is one of potential and promise. It calls on every legal professional to take part in shaping a future that is reflective of our shared values and aspirations—a future where the integration of AI into legal practices continues to evolve, fostering a profession that is as dynamic as the technology that shapes it.

In this spirit, we conclude not with an endpoint but with an invitation to continue the journey. Together, we can harness the transformative power of AI to craft a legal landscape that is innovative, efficient, and, above all, just. Let us carry forward the blueprint laid out in these pages and build a future where technology and human insight combine to create a legal profession that is ready for the challenges and opportunities of the 21st century and beyond.

Appendix A: Acknowledgments

In the pursuit of exploring the interplay between artificial intelligence and the legal profession, many individuals and organizations have played pivotal roles. Their contributions, whether through insights, encouragement, or critique, have been invaluable in shaping this comprehensive narrative. This section is dedicated to expressing gratitude to those who have made this exploration not only possible but also meaningful.

First and foremost, acknowledgment is due to the academic experts who have graciously shared their profound knowledge of AI, machine learning, and their potential applications within legal practices. Their expertise has been instrumental in framing the discussions around AI's transformative capabilities and limitations in the legal domain.

Gratitude is also extended to the numerous legal professionals – lawyers, paralegals, and attorneys – who have provided real-world perspectives on integrating AI into legal workflows. Their firsthand experiences have offered invaluable insights into the practical challenges and benefits of AI in legal settings, making the advice and strategies outlined in this book grounded in practical reality.

The legal tech community deserves special mention for their enthusiastic engagement with this project. From emerging startups to established companies, their innovative AI solutions and willingness to share knowledge have illuminated the path towards technology-driven legal practices. Their contributions have highlighted the dynamic nature of legal tech and the ever-evolving possibilities it presents.

To my peers, both in academia and the legal profession, whose conversations and debates have continually pushed the boundaries of this work, I extend my warm thanks. Your critical feedback and thought-provoking questions have been essential in refining the arguments and viewpoints presented.

Special appreciation goes out to the legal clients who have shared their experiences and expectations regarding AI-enhanced legal services. Your perspectives have underscored the importance of client-centric approaches in the evolution of legal practices, emphasizing the need for AI to enhance, rather than replace, the human elements of legal service.

The team behind the scenes – including editors, researchers, and support staff – have my heartfelt gratitude. Your dedication, attention to detail, and tirelessness have been crucial in bringing this book to life, ensuring that the narrative remained coherent, engaging, and accessible.

Additionally, I must acknowledge the global AI ethics community for their critical work in highlighting the ethical considerations of AI in law. Their efforts have ensured that discussions around AI implementations remain grounded in principles of fairness, accountability, and transparency.

Finally, I extend my sincere appreciation to you, the reader. Your interest in the junction of AI and law reflects a shared vision of a future where technology amplifies the impact of legal services. It is my hope that this book, informed by the collective wisdom and contributions of many, serves as a valuable guide in your exploration of AI in legal practices.

In a heartfelt personal acknowledgment, I must express my deepest gratitude to my family. To my wife, Olesia, whose steadfast belief in the concept of this book has been a beacon of support; to my son, Ocean SirWolf, whose youthful vigor and enthusiasm have often shed new light on my work; and to my daughter, Alice, whose patience and understanding seem to know no bounds. The many long nights, days, and weekends that you have all sacrificed to make this book a reality have been the silent foundation of its completion. Your love, your sacrifice, and your unwavering faith in this endeavor have not only been my support system but also my inspiration. For this, and for so much more, I am eternally grateful.

In closing, while this book seeks to illuminate the path towards AI integration in legal practices, it is the continued engagement, innovation, and ethical consideration of the entire legal community that will ultimately shape this future. To all who are contributing to this exciting journey, your work is sincerely appreciated and deeply respected.

May we all continue to strive for a legal profession that leverages AI to improve access, efficiency, and justice in our societies. Thank you once again for your invaluable contributions and for joining me in this exploration of AI's potential within the legal realm.

About the Author

Vas stands at the forefront of technological innovation and SaaS marketing, a beacon whose career began at the venerable halls of Yahoo. His journey through the tech landscape is marked by strategic foresight and transformative roles at startup ventures such as Apiphany, acquired by Microsoft, and API Fortress, absorbed by Sauce Labs. Vas is not only a participant in this ever-evolving field but a pioneer; he founded LegalWeb.AI, an avant-garde Customer Engagement AI platform, and StartUP MQL, a cutting-edge SaaS Demand Generation and Legal Marketing Agency. As Vice President of Digital Marketing for a leading Legal Tech company that is redefining Service of Process, Vas has harnessed the revolutionary potential of AI to enhance legal domains and streamline operational efficiencies. His profound contributions have catalyzed over $500 million in pipeline opportunities, securing $100 million in Annual Recurring Revenue (ARR). Vas's career is a testament to his unmatched strategic acumen and his relentless pursuit of innovation. An ardent advocate for Legal AI, he tirelessly promotes its transformative impact on business operations, continually pushing the boundaries of what is possible in technological advancement and operational efficacy.

Printed in Great Britain
by Amazon

47419621R10245